The Future of Life and the Future of our Civilization

JOHANN WOLFGANG GOETHE
UNIVERSITÄT
FRANKFURT AM MAIN

The Future of Life and the Future of our Civilization

Edited by

Vladimir Burdyuzha
Russian Academy of Sciences, Moscow, Russia

Springer

A C.I.P. Catalogue record for this book is available from the Library of Congress.

ISBN-10 1-4020-4967-6 (HB)
ISBN-13 978-1-4020-4967-5 (HB)
ISBN-10 1-4020-4968-4 (e-book)
ISBN-13 978-1-4020-4968-2 (e-book)

Published by Springer,
P.O. Box 17, 3300 AA Dordrecht, The Netherlands.

www.springer.com

Printed on acid-free paper

Dedicated to the memory of Carl Sagan (USA) and Joseph Shklovskii (Russia)

Contents

Part V What is our Future

Preface

Our second Symposium "The Future of Life and the Future of our Civilization" was held in May of 2005 year in Germany (Frankfurt am Main). The first one "The Future of the Universe and the Future of our Civilization" was held in July of 1999 year in Hungary (Budapest and Debrecen). The late professor George Marx from Eotvos University was Chairman of Local Organizing Committee (LOC) in Budapest. In Debrecen professor Denes Berenyi from Institute of Nuclear Research was Chairman of LOC.

After Symposium in Hungary a thought was created to hold a new one to discuss the Future of Life. Practically 6 years came after Budapest's meeting. At first the thought was to hold such Symposium in two stages. The Symposium might be begun in India (Rerikh's places in Himalaya) and then might be continued in Greece since Greece is a motherland of our Civilization. But India and Greece were rejected owing to financial problems. It is necessary to remember the late French professor Michel Bounias from University of Avignon who took in participation on all stages of organization of these Symposiums. Professors Jiannis Seiradakis also as and Athina Geronikaki of Thessaloniki University helped me very much before transfer of Symposium of Greece in Germany.

In January of 2004 year was a critical moment since Symposium in Thessaloniki might be wrecked (any money were absent). In this moment professor Claudius Gros of Saarland University proposed to transfer our Symposium in Germany. We quickly wrote the proposal to Volkswagen Foundation and in November of 2004 year a positive reply was already. Thus Claudius Gros saved this situation and our Symposium was held in Frankfurt/Main in May (2-6) of 2005 year where Claudius was Chairman of Local Organizing Committee.

The Scientific Org. Committee (SOC) had 13 members: Mohammed Al-Malki (S.Arabia); Michel Bounias (France); Vladimir Burdyuzha (Russia); Claudius Gros (Germany); Rajan Gupta (USA); Mohammad Reza Hafeznia (Iran); Hans Haubold (Austria); Vadim Kvitash (USA/Ukraine); Elia Leibowitz (Israel); Nils-Axel Morner (Sweden); Bidare Subbarayappa (India); Helena Tlaskalova (Czech Rep.) and Maciej Zalewski (Poland).

Advisory Committee had 11 members: Georgio Bertorelle (Italy); John Cairns (USA); Anatoliy Dmitrievsky (Russia); Lev Fishelson (Israel);

brey de Grey (UK); Guillermo Lemarchand (Argentina); Emil Skamene (Canada); Marcelo Sorondo (Vatican); Jiannis Seiradakis (Greece); Hoi-Lai Yu (Taiwan) and Raymond Zeltz (France).

The Symposium in Frankfurt/Main "The Future of Life and the Future of our Civilization" (modeling and predictions) was dedicated to the memory of outstanding Scientists: CARL SAGAN (USA) and JOSEPH SHKLOVSKII (RUSSIA).

This Symposium was an interdisciplinary one. Questions of Life safety were discussed in detail. As the professor George Marx in Budapest as and the professor Claudius Gros in Frankfurt have shown self sacrifice practically in all questions of holding of the Symposiums. In Frankfurt more than 50 scientists of different countries have participated. Among participants a Nobel Prizeman -Christian de Duve (Belgium) was also.

After long discussions we have prepared a Declaration "Protection of Life and our Civilization" which is opened for a signature in Internet - **http://archive.future25.org/Symposium05/declaration.html**

In preparation of the Declaration Ruediger Vaas (Germany), Bidare Subbarayappa (India), Lev Fishelson (Israel), Aubrey de Grey (UK) and Rajan Gupta (USA) have taken in part very actively.

The main thought of the last Symposium can be formulated as: conservation of the natural biota in the sufficient volume is the key problem of Life preservation and stability of our Civilization. Besides, an ecological sustainable development in our anthropogenic systems is impossible in principle after excess of a carrying ecological capacity. Unfortunately we have already overstepped the first ecological limit. Other words a new history of our Civilization must be built that would agree with laws of the biosphere.

A superior science and technology and an inferior moral are a combination that dynamically unstable. Unfortunately this is our modern world. The analysis of Dr. G. Lemarchand has shown that we would need to wait between 30-500 years in order to generate a violent event at which the whole human population will disappear.

In conclusion the quotation of Dr. J. Kleczek is appropriate: "The Earth is our cosmic home moving around Sun like a lonely spaceship with 6 billion human beings on board. It is a blue fragile Beauty with its own energy sources which are limited and soon will be depleted"

Oleg Ivaschenko (Ukraine) and Alexei Alakoz have helped me very much for the preparation of the Proceedings. Besides, I would like to thank Professor C. Gros and Frau M. Kolokotsa for excellent organization of our Symposium in Goethe University of Frankfurt/Main in May of 2005 year.

Vladimir Burdyuzha, Chairman of SOC, 20 March of 2006 year.

DECLARATION
PROTECTION OF LIFE AND OUR CIVILIZATION

Preamble

The world-wide community of scientists forms a unique society that aims to discover universal truths. Academic Andrei Sakharov once noted that "The formula $E = mc^2$ holds true on all continents". Sadly, some important discoveries of this community continue to be used for destructive purposes. The three particularly alarming issues that we wish to raise are:

(I) The application of science for the production of armaments and other military technology that have not succeeded in preventing human conflicts and wars;

(II) The fast pace of application of technology leading to degradation of the environment and emerging ecological imbalances that continue to erode the symbiotic relationship between man and nature; and

(III) The continued poor state of value placed on life and on the basic human right to live with dignity, freedom and harmony.

Today we are witness to many bloody conflicts and other destructive activities (trans-national criminals, drug-dealers, terrorism, etc). Even the basic human right - the right to life - is often not guaranteed. It is important that scientists all over the world collaborate to protect and improve our lives and our civilizations, based on the common moral principles as defined in the UN declaration of the rights of man. Many of these intertwined issues were deliberated upon by a group of scientists, technologists and environmentalists at an International Symposium "The Future of Life and the Future of our Civilization" held at the Goethe University in Frankfurt/Main (Germany) during 2-6 May 2005 and they unanimously adopted the following Declaration.

Declaration

Being deeply concerned over the poor state of national and international affairs, global changes, ecological catastrophes, continued investment in ever more powerful weapons, arms races and international conflicts that are undermining the future of mankind; and recognizing the paramount importance of the active involvement of scientists, technologists and other intellectuals in addressing these problems and finding viable solutions to them:

We, the participants in the Symposium on "The Future of Life and the Future of our Civilization", unanimously recommend at an "INTERNATIONAL CENTER FOR STUDY OF THE FUTURE" be established under the auspices of the UN or one of its Agencies.

This Center should draw upon the expertise of leading scientists, technologists, economists, sociologists, and others from different parts of the world to suggest effective measures for the protection of life and our civilization.

Such a Center should be international and independent of political and religious influences, and through careful and unbiased analysis create consensus that advises governments and other policy making bodies or individuals.

The Center should track, analyze and provide long term prognosis of the spectrum of events that may affect the safety and well being of life on the Earth. These should include catastrophes due to military applications in space or to events of ecological, technological, atmospheric or hydrospheric nature.

The Center should seek to bring together preeminent specialists from throughout the world, and bring to bear all available methodologies to provide solid, scientifically-based predictions and proposed solutions.

THE OBJECTIVES OF THE PROPOSED CENTER SHOULD BE:

a. to collect and analyze information about any events with the potential to threaten the safety of life on Earth;
b. to provide scientific and analytical support to the many branches of the UN;
c. to reduce and mitigate losses due to ecological and socio-economic disasters;
d. to reduce and eventually to stop the production and distribution of all weapons. In particular international controls and restrictions on the weapons of mass destruction should be adhered to, starting first by the actions of scientists.

e. to strengthen efforts towards promotion of peace and harmony among all different peoples of the world.

f. to protect basic human rights and to evolve an international order built upon a solid foundation of openness, dialogue, transparency and good-neighbourly relations.

We express our hope that international laws will be enacted and acknowledged by all nations. Violation of the safety of our planet should be considered a grave crime against humanity.

We propose that all countries reduce their military investments by 0.1 % every year and the money thus saved should be used to provide health care and education to all people. In addition, individuals, foundations, and non-governmental organizations should be encouraged and empowered to continue to work with the poor and the needy.

We consider that a scientific approach to analyzing, anticipating and resolving issues related to the long term health and safety of our planet and its citizens must constitute the very foundation of the center.

We have a choice: to stress the environment beyond its capacity for regeneration and to live with conflict, war, poverty, hunger and disease, as we are doing today, or to embrace a new dawn and create freedoms and opportunities for all. It is time to take a stand!

In full concurrence and support thereof, we affix our signature below.

Affiliation	Name and Family Name	Country	Signature

This variant of Declaration was prepared by Rajan Gupta. It is practically the same as and in Internet and it is open for a signature.The Internet address is: **http://archive.future25.org/Symposium05/declaration.html**

I

LIFE AS A SPACE PHENOMENON

The Spread of Life Throughout the Cosmos

Chandra Wickramasinghe

Cardiff Center for Astrobiology, Cardiff University, 2 North Road, Cardiff, CF10 3DY, UK

There is growing evidence for the widespread distribution of microbial material in the Universe. A minuscule ($<10^{-21}$) survival rate of freeze-dried bacteria in space ensures the continual re-cycling of cosmic microbial life in the galaxy. Recycling and amplification occurs within comets during the early phases of the formation of a new planetary system. Evidence that terrestrial life may have come from comets has accumulated over the past decade. The implications of this point of view, which was developed in conjunction with Fred Hoyle since the 1970's, are now becoming amenable to direct empirical test by studies of cometary material collected in the stratosphere.

1. Introduction

The standard theory for the origin of life begins with a primordial atmosphere on the Earth in which the synthesis of the chemical building blocks of life (e.g. amino acids) from inorganic gases occurs through the action of solar ultraviolet radiation and electric discharges. Organic molecules so formed are then supposed to rain down into the primitive oceans producing a dilute soup. In such an exceedingly dilute solution reaction rates would be minimal and biochemistry hard put to produce all the complex chemical transformations needed for an origin of life. It is therefore proposed that evaporation of water from shallow lakes and ponds, and at the margins of the sea led to sufficient concentrations of organics for prebiotic chemistry to proceed, and after millions of years a self-replicating living cell is postulated to arise. This was essentially the model proposed by Oparin and Haldane [1], which was at least partly vindicated by the work of S.L. Miller [2] over 50 years ago. The formation of the relatively simple chemical

V. Burdyuzha (ed.), The Future of Life and the Future of Our Civilization, 3–21.

building blocks of life was demonstrated, but the further steps leading to life remained elusive.

The highly specific and exceedingly intricate complexity of living organisms at a molecular level is self-evident, and it is clear that no significant progress has been made in 50 years towards understanding how the information gap between a non-living mixture of organic molecules and life could be achieved. For instance, Margulis [3] has stated that to proceed "....from a bacterium to people is less of a step than to go from a mixture of amino acids to that bacterium". Moreover we have no definite knowledge by which we can assert that this step ever happened on the Earth – namely that a *de novo* origin was localized to a tiny speck of 'cosmic dust' that is the Earth. The super astronomical information gap between non-life and life provides, in the view of the author, the main justification for considering theories of origin that involve the universe as the whole. The final information content of life could have been arrived at by a cumulative addition of truly minuscule probabilities in self-replicable partial stages that are repeatedly and serially impressed on a vast cosmic system. This process must be envisaged to take place over cosmological timescales, and to involve the resources of all the stars in all the galaxies in the entire universe.

2. Rationale for Panspermia Theories

There is no logic that demands an origin of life on the Earth. The fact that life is found on the Earth does not mean that life necessarily started here. The Earth is not disconnected from the wider Universe, or sealed away from cosmic contaminants. Even today cometary organic molecules arrive here plentifully, at an average rate of tens of tones per day. Thus a chain of connection:

Earth → Comets → Presolar nebula → Interstellar clouds → stars → Galaxy → Universe can be envisaged.

Living material contains about twenty different types of atoms, the most important being carbon, nitrogen, oxygen and phosphorus. The ultimate source of origin of these chemical elements is stellar nucleosynthesis – the process by which the primordial element H is converted first to He and then to C, N, O and heavier elements in the deep interiors of stars. Thus at the level of the constituent atoms we are indisputably creatures derived from the cosmos.

From the 1970's onwards, astronomers discovered a host of organic molecules in interstellar clouds, and since 1986 similar molecules were also found in comets. These discoveries prompted Fred Hoyle and the present author [4-5] to re-examine the ancient theory of panspermia, which insists that life is a cosmic phenomenon, and that life on Earth is derived from a vast cosmic system.

Louis Pasteur's classic experiments [6-7] in the 1850's and 1860's, in which he showed that microorganisms are always derived from preexisting microorganisms, provided perhaps the most important experimental basis for panspermia. Indeed this was a conclusion that was reached quite early in 1874 by the German physicist Hermann Von Helmholtz [8]:

"It appears to me to be fully correct scientific procedure, if all our attempts fail to cause the production of organisms from non-living matter, to raise the question whether life has ever arisen, whether it is not just as old as matter itself, and whether seeds have not been carried from one planet to another and have developed everywhere where they have fallen on fertile soil...." (these ideas were also discussed in [9].)

The next noteworthy proponent of panspermia at the dawn of the 20th century was the Swedish Chemist Svante Arrhenius [10]. In 1908 Arrhenius noted that microorganisms possess unearthly properties, properties that cannot be explained by natural selection against a terrestrial environment. The example for which Arrhenius himself was responsible for taking seeds down to temperatures close to zero Kelvin, and of then demonstrating their viability when reheated with sufficient care. Arrhenius conceived of microorganisms travelling individually through the galaxy from star system to star system. He noticed that organisms with critical dimensions of 1 micron or less are related in their sizes to the typical radiation wavelengths from dwarf (sun-like) stars in such a way that radiation (light) pressure can have the effect of dispersing these particles throughout the galaxy.

3. The Survival Problem

Space-travelling individual bacteria would be susceptible to deactivation and damage from the ultraviolet light of stars, and this was already known in the first decades of the twentieth century. Becquerel [11] criticized panspermia on the basis of possible ultraviolet damage of space-travelling microbes, and similar criticisms were repeated ever since, even in modern times [12]. These arguments are either flawed or highly insecure, however. Even under normal laboratory conditions microorganisms are not readily killed by ultraviolet, they are mostly deactivated due to the dimerization of pyrimidine bases. No genetic information is lost in the process and in many instances the damage can be repaired by the operation of specialized enzymic systems. Dimerization of bases distorts the DNA configuration and has the effect of impeding transcription. Exposure of UV radiated bacteria to visible sunlight is known to promote repair.

No data exists at the present time relating directly to the effects of ultraviolet radiation on bacteria under cryogenic conditions and in the absence of air and water, conditions such as would apply in interplanetary or interstellar space. It should also be pointed out that microorganisms are easily shielded against ultraviolet light. Indeed molecular clouds in the galaxy are highly effective in this respect, both in cutting out the glare of ultraviolet radiation and permitting the growth of protective mantles around bacterial particles. Thin skins of carbonized material around individual bacteria, only 0.02 μm thick, would also effectively block the damaging ultraviolet light [13].

On the whole microbiological research of the past 10 years has shown that microorganisms are remarkably space-hardy. Thermophiles are present at temperatures above boiling point in oceanic thermal vents, and as we have already pointed out entire ecologies of psychrophilic and psychrotrophic microorganisms are present in the frozen wastes of Antarctica. A formidable total mass of microbes also exists at great depths in the Earth's crust, some 8 kilometers below the surface, greater than the biomass at the surface [14]. A species of phototrophic sulfur bacterium has been recently recovered from the Black Sea that can perform photosynthesis at exceedingly low light levels, approaching near total darkness [15]. There are bacteria (e.g. *Deinococcus radiodurans*) that thrive within the cores of nuclear reactors [16]. Such bacteria perform the amazing feat of using an enzyme system to repair DNA damage, in cases where it were estimated that the DNA was experienced as many as a million breaks in its helical structure.

Most modern objections to panspermia have been based on arguments relating to cosmic ray survival [12] – it being claimed that cosmic ray exposures in space over hundreds of thousands of years would prove fatal for microorganisms. These criticisms are again highly dubious and, moreover, fail to take account of the fact that the replication power of bacteria is so great that only a minute ($\sim 10^{-21}$) survival is required at each regeneration site between periods of freeze-dried dormancy in the interstellar medium.

Ionizing radiation limits viability by dislodging electrons, causing bond breaks in the DNA and by forming reactive free radicals. The radiation doses that seriously compromise viability in cultures depend critically on the particular bacterial species, and as mentioned earlier some species such as *B. subtlis* and *D. radiodurans* are more resistant than others. In vegetative cultures, under laboratory conditions, doses equivalent to 2 megarads (2Mr) have been found to limit residual viability of *Streptococcus faeciuim* by a factor of 10^{-6} (Christensen [17]), whereas similar doses have little or no effect on cultures of *D. radiodurans* or *M. radiophilus* (Lewis [18]).

The doses of ionizing radiation received by a bacterium in interplanetary space within the solar system depends on distance from the Sun and the phase of solar activity, being highest at times near the peak of the solar sunspot cycle. In a recent NASA/LDEF (Long Duration Exposure Facility) experiment, direct exposure of spores of *B. subtilis* to unshieded solar radiation for 2107 days was found to lead to significant rates of survival (Hornek et al., [19]). The survival of common species of bacteria near the Earth's orbit for about a decade therefore seems well-attested.

The dose of cosmic rays received by a naked bacterium in a typical location in interstellar space, over a timescale of a fraction of million years, can at present be only very poorly estimated. It is possibly in the range 10-45 Megarad per million years. Doses of this order are of course higher that the doses that have been delivered to laboratory cultures where survival is well-attested. Yet the exposure conditions in space, where two successive cosmic ray ionizing events are separated by about 100 years, would be dramatically different from those pertaining to the laboratory experiments.

A low flux of ionizing radiation in space delivered over astronomical timescales to dormant freeze-dried bacteria (in the absence of H_2O and air) would perhaps bear no comparison with equivalent doses on vegetative cultures in the laboratory. The nearest terrestrial analogue might be for microbial spores that have been exposed to the natural radioactivity of rocks for geological timescales. Indeed viable cultures of bacteria have been recovered from ice drills going back 500,000 years, from isolates in amber over 25-40 million years [20, 21] and from 120 million year old material [22]. Similarly bacteria have been recovered in salt crystals from a New Mexico salt mine dated at 250 Myr (Vreeland et al., [23]). The present day dose rate of ionizing radiation on the Earth arising from natural radioactivity is in the range 0.1-1 $\mathrm{r\,yr^{-1}}$. These well-attested recoveries of dormant bacteria/spores after 10^8 yr must therefore imply tolerance to ionizing radiation with total doses in the range ~10-100Mr.

4. Interstellar Organic Molecules and Dust

Notwithstanding the remarks of the previous sections some fraction of cosmic bacteria which have no protective coatings and which are exposed remorselessly to cosmic rays and to the background of starlight in unshielded regions of interstellar space would be subject to degradation and eventual destruction. The polyaromatic hydrocarbons that are so abundant in the cosmos could have a similar origin to the organic pollutants that choke us in our cities – products of the degradation of biology, biologically generated fossil fuels in the this case, cosmic microbiology in the interstellar clouds. The theory of cosmic panspermia that we propose leads us to argue that interstellar space serves both as a graveyard of cosmic life as well as its cradle. Only the minutest fraction (less than one part in a trillion) of the interstellar bacteria needs to retain viability, in dense shielded cloudlets for instance, for panspermia to hold sway.

My own interest in panspermia began with attempts to understand the nature of cosmic dust [24]. Interstellar dust grains populate the vast open spaces between stars of the Milky Way, showing up as a cosmic fog, dense enough in many directions to blot out the light of distant stars. Remarkably these dust grains can be shown to be of a size that would be typical for a bacterium, a micrometer or less.

A fact that impressed me from the outset was that the total mass of interstellar dust in the galaxy is as large as it possibly can be if all (or nearly all) the available carbon, nitrogen and oxygen in interstellar space is condensed in the grains. The amount is about three times too large for the grains to be mainly made up of the next commonest elements, magnesium and silicon, although magnesium and silicon could of course be a component of the particles, as would hydrogen, and also many less common elements in comparatively trace quantities.

If one now asks the question: what precisely are the dust grains made of, a number of inorganic molecules composed of C, N, O in combination with hydrogen present themselves as possible candidates. These include water ice, carbon dioxide, methane, ammonia, all such materials being easily condensable into solids at temperatures typically of about 20-50 degrees Kelvin, which is the average temperature of the dust grains in space. During the decade starting from the early 1960's Fred Hoyle and I studied the properties of a wide range of inorganic grain models, comparing their electromagnetic properties against the formidable number of observations that were beginning to emerge. Such models stubbornly refused to fit the available data to anything like the precision that was required. The correspondences between predictions for assemblies of inorganic particles and the observations could be lifted to a certain moderate level of precision but never beyond that, no matter how hard one tried.

It was a milestone in our progress towards interstellar panspermia when I realized that there is another very different class of materials that can be made from the same four commonest elements - C, N, O, H, namely organic materials, possibly of a polymeric type [25]. Of course there are a vast number of organic compositions that are possible, making for a great number of further investigations that could be done. By the mid-1970's, the astronomical observations were spanning a large range in wavelength, from 30 microns in the infrared, through the near infrared, into the visible spectrum, and further into the ultraviolet. So a satisfactory theory of the nature of interstellar dust grains had by now to satisfy a very large number of observational constraints.

In 1979 Fred Hoyle and I stumbled on a result that led to many further discoveries, all of which pointed in the direction of panspermia. As already noted bacterial grains in interstellar space would be freeze-dried. Freeze drying of a bacterium would ensure that free water in the cell diffuses out of the porous cell wall leading to the development of a vacuum cavity.

The volume of the vacuum cavity for a typical bacillus amounts to about 60% of the total, and the resulting average refractive index of the entire structure is readily calculated as n = 1.16. Next we require a distribution of sizes for the bacteria, which was available in the literature for spore-forming microorganisms (see Figure 1).

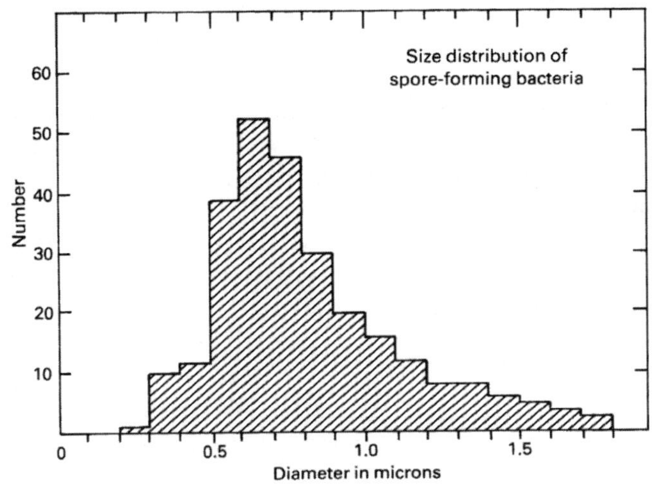

Fig. 1. Histogram of diameters of spore forming bacteria.

When the extinction of this ensemble of freeze-dried bacteria was calculated, the result compared with the astronomical data on the dimming of starlight is shown in Fig. 2.

I was indeed dumbfounded to find a result so good after so many years of failure to obtain a satisfactory fit using highly contrived inorganic models. Once the proposition is made that "interstellar grains are bacteria" there is no further room for maneuver – the model is completely defined. But for the astronomical data further in the ultraviolet another refinement was required. Added to the bacterial population we need two further components derived from biology: non-hollow viruses and/or nanobacteria contributing 29% to the total mass, and free biological aromatic molecules, which would be the most stable subunits and molecules to result from the degradation of unshielded bacteria in space. The combination of these components leads to the curves depicted in Fig. 3

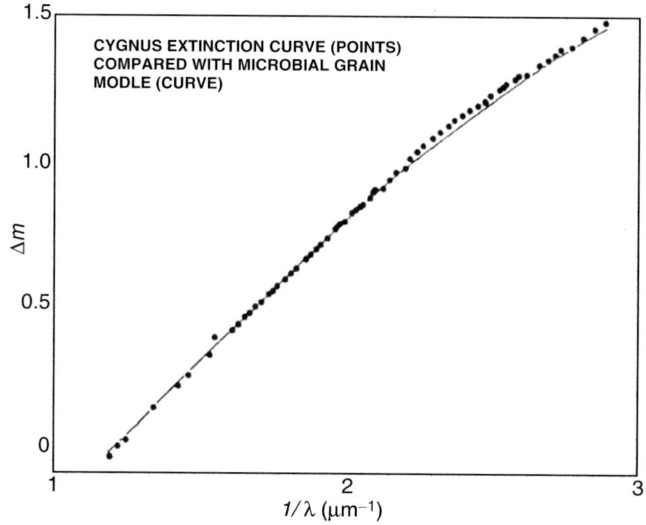

Fig. 2. Bacterial extinction over the visible spectrum compared with interstellar extinction data.

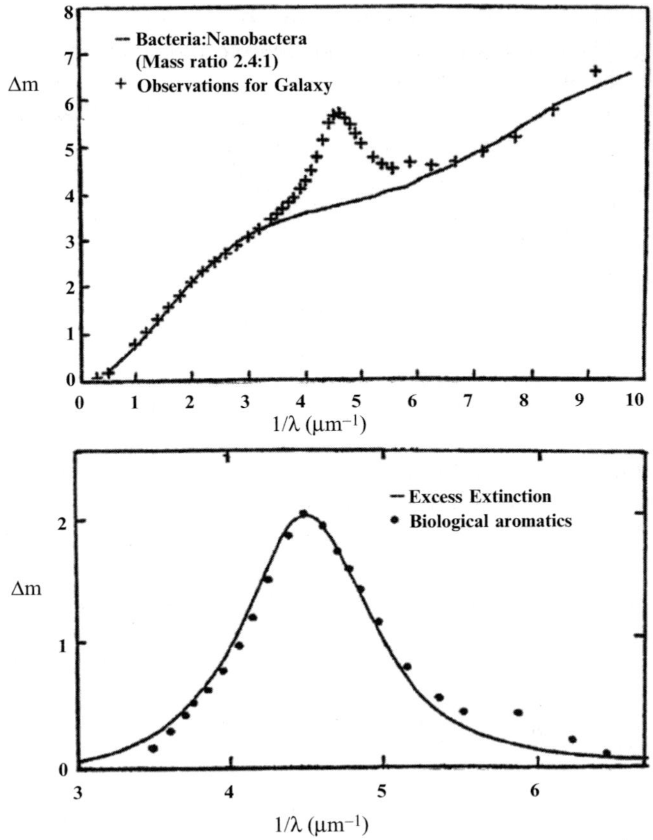

Fig. 3. **Top**: Interstellar extinction data over the 0.2-10 μm waveband compared with scattering by freeze-dried bacteria and nanobacteria. **Bottom**: Excess extinction over the mid UV band compared with absorption by an ensemble of biologically derived aromatic molecules.

Perhaps the most startling confirmation of the bacterial model followed the observations [26] of a source of infrared radiation, GC-IRS7, located near the center of our galaxy. The spectrum of this source revealed a highly detailed absorption profile extending over the 2.9-3.8 micrometre wavelength region, indicative of combined CH, OH and NH stretching modes. A laboratory spectrum of the desiccated bacterium *E. Coli*, together with a simple modeling procedure provided an exceedingly close point by point match to the astronomical data over the entire 2-4 micron waveband (Fig. 4).

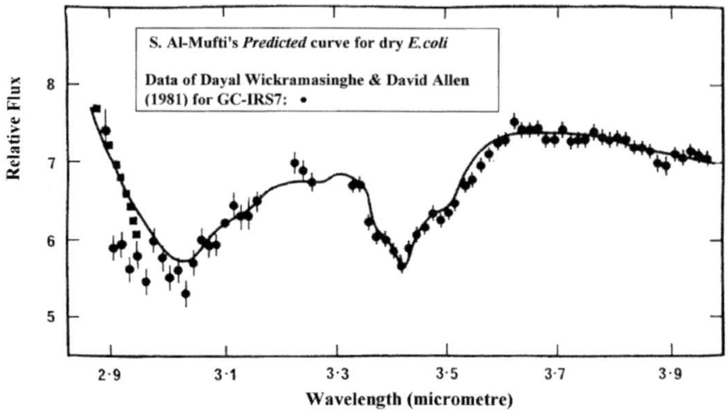

Fig. 4. Infrared spectrum of GC-IRS7 compared with bacterial model.

At this stage we found there was no alternative but to face up squarely to the conclusion that a large fraction of the interstellar dust was not merely hollow and organic, but they *must* spectroscopically be indistinguishable from freeze-dried bacterial material. In our galaxy alone the total mass of this bacterial type material had to be truly enormous, weighing a formidable 10^{33} tons.

5. Replication in Comets

By far the simplest way to produce such a vast quantity of small organic particles everywhere of the sizes of bacteria is from a bacterial template. The power of bacterial replication is immense. Given appropriate conditions for replication, a typical doubling time for bacteria would be two to three hours. A continued cascade of doublings with unlimited access to nutrients would lead to a culture that enveloped the interior of a 10km radius comet in less than a week. No abiotic process remotely matches this replication power of a biological template. Once the immense quantity of organic material in the interstellar material is appreciated, a biological origin for it becomes an almost inevitable conclusion.

An individual comet is a rather insubstantial object. But our solar system possesses so many of them, perhaps more than a hundred billion of them, that in total mass they equal the combined masses of the outer planets Uranus and Neptune, about 10^{29} grams. If all the dwarf stars in our galaxy are similarly endowed with comets, then the total mass of all the comets in our galaxy, with its 10^{11} dwarf stars, turns out to be some 10^{40} grams, which is just the amount of all the interstellar organic particles.

How would microorganisms be generated within comets, and then how could they get out of comets? We know as a matter of fact that comets do eject organic particles, typically at a rate of a million or more tons a day. This was what Comet Halley was observed to do on March 30-31, 1986. And Comet Halley went on doing just that, expelling organic particles in great bursts, for almost as long as it remained within observational range. The particles that were ejected in March 1986 were well placed to be observed in some detail. No direct tests for a biological connection had been planned, but infrared observations pointed unexpectedly in this direction. An independent analysis of dust impacting on mass spectrometers aboard the spacecraft Giotto also led to a complex organic composition that was fully consistent with the biological hypothesis [27]. Broadly similar conclusions have been shown to be valid for other comets as well, in particular Comet Hyakutake and Comet Hale-Bopp. Thus one could conclude from the astronomical data that cometary particles, just like the interstellar particles, are *spectroscopically* identical to bacteria.

In summary, the logical scheme for the operation of cometary panspermia is as follows: The dust in interstellar clouds must always contain the minutest fraction of viable bacteria (less than one in 10^{21}) that retain viability despite the harsh radiation environment of space. When a new star system (e.g. a solar system) forms from interstellar matter, comets condense in the cooler outer periphery as a prelude to planet formation. Each such comet incorporates, at the very least, a few billion viable bacteria, and these bacteria are quickly reactivated and begin to replicate in the warm interior regions of the comets, thus producing vast numbers of progeny. As a fully-fledged stellar or planetary system develops, comets that plunge into the inner regions of the system release vast quantities of bacteria. Some of the evaporated bacterial material is returned into the interstellar medium. New stars and star systems form and whole cycle continue with a positive feedback of biologically processed material.

6. Oldest Life on Earth

Along with the accumulation of astronomical evidence supporting panspermia in one form or another there has also been evidence from geology. The earliest evidence for terrestrial life has now been pushed back beyond 3.83 billion years BP, well into an epoch when we know for certain that the Earth was severely pummeled by comet and meteorite impacts [33]. This evidence comes in the form of a slight enhancement of the lighter isotope of carbon 12C relative to 13C in the oldest metamorphic rocks. The argument is that life has a slight preference for the lighter isotope of carbon and this is reflected in the carbon extracted from rocks that could date back to about 4 billion years. Whilst the early epoch of heavy bombardment would not have been conducive to prebiotic chemistry, it would nevertheless have offered ample scope and many occasions for the transfer of cometary life to Earth. It is interesting to note that this mechanism for transfering life from comets to Earth would permit some types of microbial life adapted to high pressures and subsurface conditions to become trapped in a stable way. As the impacts of comets and asteroids continued to add material to the Earth's crust in the last stages of the "late accretion epoch" a deep hot biosphere [14], such as we now have, would easily have been generated. So also could microbial life in thermal vents of a deep sea to be explained as representing a primordial habitat that accommodated the most heat resistant of the microbes that arrived from space.

7. More Evidence of Microbiology Outside the Earth

We have discussed earlier how modern microbiology has yielded a wealth of new discoveries relating to 'unearthly properties' of microorganisms. Furthermore, from recent explorations of the solar system we know that other planetary bodies besides Earth might have conditions appropriate to serve as habitats for microbial life. For instance, the Jovian satellite Europa, with growing evidence of its subsurface oceans beneath a frozen crust, provides many opportunities for a highly developed microbiota. Life may even be present in the clouds of Venus. Whilst the surface of Venus is too hot to sustain life, there is an intriguing possibility of an aerobiology thriving in the Venusian clouds. Dirk Schulze-Makuch and his colleagues have recently pointed out that 30 miles above the surface there are droplets of water and chemical tell-tale signs of life. They had expected to find high levels of carbon monoxide, but instead found hydrogen sulphide and sulphur dioxide, gases normally not found together and also carbonyl sulphide, a gas so difficult to produce by inorganic chemistry that it is generally considered to be a marker for living organisms. Likewise with new evidence of liquid water on Mars, there could be plenty of scope for microbial life in secluded subsurface niches.

According to theory described in this paper life on Earth began with the introduction of microorganisms from comets. It is clear, however, that this process could not have stopped at some distant time in the past. Comets have been with us throughout, and the Earth has continued to plough through interplanetary and cometary dust. In our view the evolution of terrestrial life is controlled and directed by the continuing input of cometary debris in the form of bacteria, fragments of bacteria, nanobacteria and smaller particles such as viruses and viroids. It is well known that viral genes sometimes come to be included in the genomes of cellular lifeforms, and that such genes could serve as potential for further evolution. Without this input of cometary genes life on Earth could not have evolved beyond the stage of a simple ancestral microbe.

There are several recent reports of genes that appear to be older, when dated by the rate of sequence variation, than the composite systems or species, in whose genomes they are included [34, 35]. Other reports show that genes required by more highly evolved species may reside without evident function in the genomes of prokaryotes [36] or viruses [37]. One cannot help but notice that these findings corroborate the concept of cosmic bacteria and cosmic genes.

8. Present-Day Tests

A direct way to test the theory of cometary panspermia is to examine a sample of cometary material under the microscope and search for signs of microbial life. Comets are literally at our doorstep and the technology to carry out the relevant microbiological experiments has been available for at least a decade. The basic procedure would involve collection of cometary material as it enters the stratosphere, with suitable precautions being taken to eliminate spurious contamination from terrestrial sources, and then examine the samples for extraterrestrial microorganisms. With a daily input of cometary debris averaging some 100 tons, the possibility of detecting infalling microbes must surely exist.

The earliest experiments to search the upper atmosphere for microorganisms were carried out using high altitude balloons in the early to mid-1960's. Although microbiological techniques available at the time were primitive compared to the present, there were already some intriguing indications of the presence of extraterrestrial microbes in air samples collected at heights of 30 km and above [38]. Positive detection of microorganisms at 39 km and a population density that increased with height pointed to a possible extraterrestrial source. Not surprisingly these early results were not taken seriously, nor were they followed up at a later date by NASA with more refined experiments as the relevant microbiological techniques evolved.

The sample return mission "Stardust", which was launched on 7 February 1999 heading to Comet Wildt-2 (rendezvous date, 2 January 2004;

return 2006) was conceived and planned before a change of attitude to panspermia took place. In the event no microbiological experiments as such were catered for. The comet dust is to be captured in a "particle catcher" filled with aerogel, a material of extremely low density. The hope is that the aerogel would act as a soft landing cushion to show down particles from an initial relative speed of 6.1 km/s to rest fairly gently, without significantly modifying original chemical structures. The thinking behind the experiment was to bring back prebiotic organic molecules. No provisions were made for the possibility that living cells might be present, so the best one might hope for when we get samples back in 2006 is the intervention of serendipity. Perhaps one might find evidence of "dead bacteria" or other clues for life in the molecules that are recovered.

The stratospheric collection experiments of the 1960's have also been resumed by the Indian Space Research Organization (ISRO) in collaboration with groups at Cardiff and Sheffield. The aim was to collect stratospheric air aseptically, and to examine it in the laboratory for signs of life [39]. The sample collection was done using a number of specially manufactured sterilized stainless steel cylinders that were evacuated to almost zero pressures and fitted with valves that could be open and shut at different heights in the atmosphere. An assembly of such cylinders was suspended in a liquid Ne environment to keep them at cryogenic temperatures, and the entire payload launched from the TATA Institute Balloon launching facility in Hyderabad, India on 20 January, 2001. As the valves of the cylinders are opened upon ground telecommand at predetermined heights, ambient air rushes in to fill the vacuum, building up high pressures within the cylinders. The valves are shut after a prescribed length of time, the cylinders hermetically sealed and parachuted back to the ground.

Back on the ground the cylinders were carefully opened and the collected air made to flow through sterile membrane filters in a contaminant free environment. Any bacteria or clumps of bacteria present in the stratosphere would then be collected on these filters.

In the first phase of this investigation evidence for the presence of clumps of viable cells were discovered in air samples collected from as high as 41 kilometres, well above the local tropopause (16 km), above which no aerosols from lower down would normally be transported [40, 41]. The detection was made using a fluorescent dyes which are only taken up by the membranes of living cells. When the isolate treated with the dye is examined under an epifluourescence microscope the picture on the left of Fig. 5 is obtained. The picture on the right is an image from an electron microscope which shows a similar structure comprised of cocci and rods. The variation with height of the distribution of such cells indicates strongly that the clumps of bacterial cells are falling from space.

Dr. Milton Wainwright of the University of Sheffield was further able to isolate a culture of two organisms: one micrococcus and one microfungus closely related to known species which must therefore have fallen from the skies. The daily input of such biological material is provisionally estimated to be in the range one third to one ton over the entire planet.

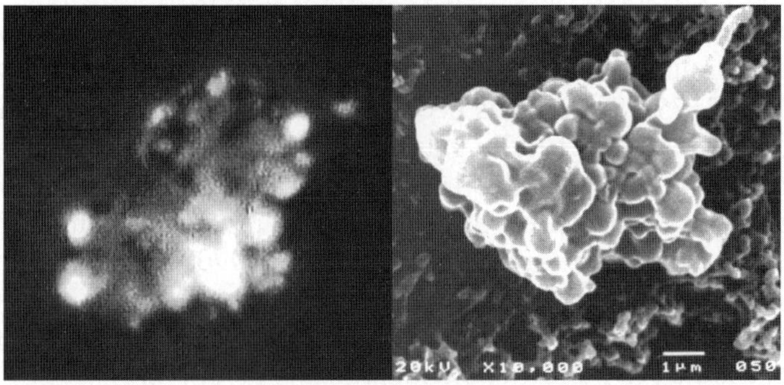

Fig. 5. **Left:** Clump of viable bacteria fluorescing in cyanine dye. **Right:** Scanning Electron Microscope picture of similar structure showing clump of cocci and a rod.

Work to verify these results are in progress, with ISRO launching a new balloon-borne cryosampler in April 2005. If these findings are confirmed panspermia would cease to be a theory, it would have become a fact.

9. Distribution of Life Beyond the Galaxy

The transfer of life across extragalactic distances requires the pre-existence of the elements C,N,O and other metals in adequate quantity, which implies access to regions where star-formation is under way. Spectroscopic studies of nearby galaxies have also shown evidence of advanced stages of nucleosynthesis and chemical evolution involving large quantities of carbonaceous material suitable as a feedstock for life, often including evidence of CO and H_2CO and other organic molecules in the gas phase.

The most efficient intergalactic transport of biological information can be achieved with microbes attached onto iron whiskers, such as are important for explaining aspects of the cosmic microwave background. Such metallic whiskers have diameters typically 0.02μm and lengths about a millimeter and they condense naturally in expanding envelopes of supernovae, as metallic vaporous cool [42]. These whiskers, along with their microbial hitch-hikers, are very strongly repelled by the infrared radiation from parent galaxies, reaching typical speeds of ~10^4 km/s in intergalactic space. Even with the minutest fraction of surviving biological "messages" attached to fast moving whiskers in this way, biology could diffuse through a radius of ~50Mpc of intergalactic space, a volume occupied by ~10^6 galaxies, in a mere Earth age, ~4.7×10^9 years.

From an observational point of view the clearest signal that may be indicative of biology is perhaps the λ2175A ultraviolet feature of interstellar dust. The λ2175A feature has been found in dust associated with both the SMC, LMC and a few nearby galaxies including M31, where observation of this band was feasible [43, 44].

The most dramatic recent discovery is the presence of the 2175A band in the lens galaxy of the gravitational lens SBS0909+532 which has a redshift of z=0.83 (Motta et al.[45]). The extinction curve for this galaxy is reproduced in Fig. 6, with the dashed curve representing a scattering background attributed to hollow bacterial grains.

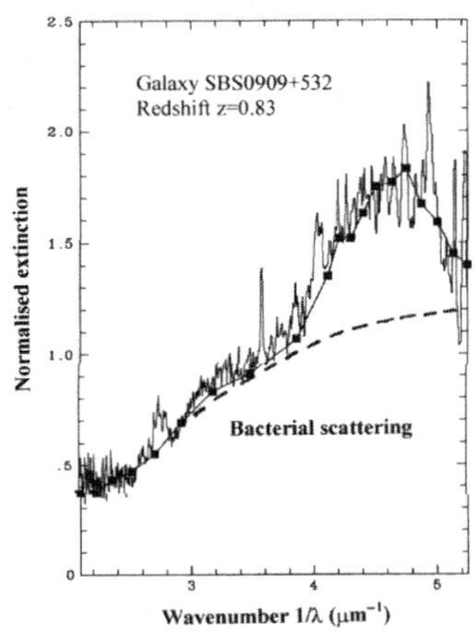

Fig. 6. The continuous line is the extinction curve for the gravitational lens galaxy SBS0909+532 excluding well-defined spectral lines due to MgII, CIII and CIV (Motta et al. [45]). The dashed curve is the scattering background attributed to bacterial particles.

The excess absorption over and above this scattering background, normalized to unity at the peak, is plotted as the points with error bars in Fig. 7. The curve is this figure shows the absorption of biological aromatic molecules similarly normalized [46].

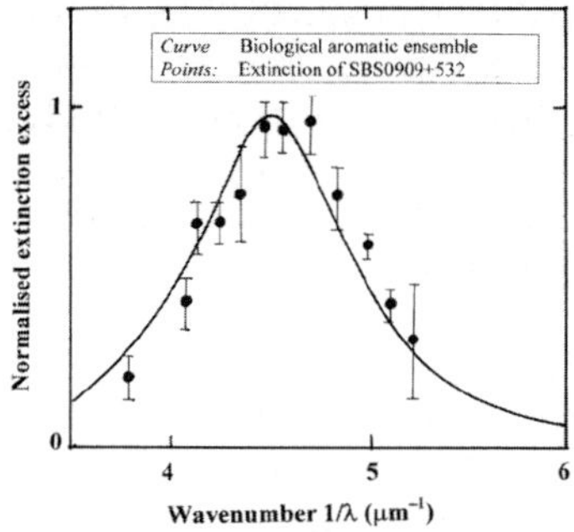

Fig. 7. The curve is the normalized absorption coefficient of an ensemble of 115 biological aromatic molecules. The points are the observations of the galaxy SBS0909-532 due to Motta et al. [45]).

The correspondence between the astronomical data and the model in Fig. 7 can be interpreted as strong evidence for biology prevailing at redshifts $z \approx 0.83$ that is to a distance $D \approx c\ z / H \sim 2.5$Gpc, assuming a Hubble constant of 100 km/s per Mpc. The 3.3μm infrared emission band and other indications of biological-type organics have also been observed in extragalactic sources (including the galaxy Antennae referred to above), and particularly starburst galaxies up to red shifts $z \sim 0.2$. All such data are consistent with the spread of microbial life encompassing a significant fraction of the radius of the observable universe.

References

1. Oparin, A.I., The Origin of Life, transl by S. Margulis, Dover. 1953; J.B.S. Haldane, The Origin of Life, Chatto and Windys. 1929
2. Miller, S.L., Science 117, 528. 1953
3. Margulis, L., As quoted by Horgan, J., in End of Science, Ch.5 (Addison Wesley, Publ), 1996
4. Hoyle, F. and Wickramasinghe, N.C., In Comets and the Origin of Life (ed. C. Ponnamperuma), D. Reidel , p. 227, 1981
5. Hoyle, F. and Wickramasinghe, N.C., Astronomical Origins of Life: Steps towards Panspermia (Kluwer Academic Publ)., 1999
6. Pasteur, L., C.R.Acad.Sci., 45, 913-916, 1857
7. Pasteur, L., C.R.Acad.Sci., 45, 1032-1036, 1857
8. von Helmholtz, H., In W. Thomson & P.G. Tait (eds) Handbuch de Theortetische Physik, Vol.1. Pt.2. Brancscheig. 1874
9. Thomson, W., British Association for the Advancement of Science, Presidential Address, London. 1871
10. Arrhenius, S., Worlds in the Making, Harper, Lond. 1908

11. Becquerel, P., Bull.Soc.Astron., 38, 393. 1924
12. Mileikowsky, C., et al., Icarus, 145, 391, 2000
13. Wickramasinghe, N.C. and Wickramasinghe, J.T., Astrophys.Sp.Sci, in press, 2003
14. Gold, T., Proc.Natl.Acad.Sci.,89, 6045-6049, 1992
15. Overmann, J., Cypionka, H. and Pfennig, N., Limnol.Oceanogr., 37 (1), 150-155, 1992
16. Secker, J., Wesson, P.S. and Lepock, J.R., Astrophys.Sp.Sci. 329, 1, 1994
17. Christensen, E.A., Acta Path. et Microbiol. Scandinavia, 61, 483, 1964
18. Lewis, N.F., J. Gen. Microbiol, 66, 29, 1971
19. Hornek, G., et al., Adv.Space Res., 14, 41, 1994
20. Cano, R.J. and Borucki, M., Science, 268, 1060, 1995
21. Lambert, L.H., et al., Int. J. Syst. Bact., 48, 511, 1998
22. Greenblatt, C.L., et al., Microbial Ecology, 38, 58, 1999
23. Vreeland, R.H., Rosenzweig, W.D. and Powers, D., Nature, 407, 897-900, 2000
24. Hoyle, F. and Wickramasinghe, N.C., The Theory of Cosmic Grains, Kluwer. 1990
25. Wickramasinghe, N.C., Nature 252, 462. 1974
26. Allen, D.A. and Wickramasinghe, D.T., Nature, 294, 239. 1981
27. Wickramasinghe, N.C., In Infrared Astronomy, p. 303, ed. By A. Mampaso et al.,
28. Claus, G., Nagy, B. and Europa, D.L., Ann. NY Acad. Sci., 108, 580.1963
29. Pflug, H.D., In Fundamental Studies and the Future of Science, ed. C. Wickramasinghe, Univ. College Cardiff Press. 1984
30. Hoover, R.B., Proc. SPIE Conference on Instruments, Methods, and Missions for the Investigation of Extraterrestrial Microorganisms, 3111, 115-136, 1997Cambridge University Press. 1993
31. Hoover, R.B., Rozanoz. A.Y., Jerman, G.A., 2003, Proc. SPIE, Vol. 5163, in press, 2003
32. McKay, D.S., et al., Science, 273, 924. 1996
33. Mojzsis, S.J. et al., Nature 384, 55. 1996
34. Kumar, S. and Blair Hedges, S., Nature, 392, 917. 1998
35. Cooper, A. and Penny, D., Science, 275, 1109. 1997
36. Bult, C.J., et al., Science, 273, 1058. 1996
37. Smith, M.C., et al., Science, 279, 1834. 1998
38. Bruch, C.W., in Airborne Microbes Symposium of the Society of Microbiology (eds P.H. Gregory and J.L.Monteith) p. 345, Cambridge University Press, 1967
39. Narlikar, J.V., Ramadurai, S., Bhargava, P., Damle, S.V., Wickramasinghe N.C., Lloyd, D., Hoyle, F. and Wallis, D.H., Proc. SPIE Conference on Instruments, Methods, and Missions for Astrobiology, Vol. 3441, 301-305, 1998
40. Harris, M.J., et al., Proc. SPIE Conf., 4495, 192, 2002
41. Wainwright, M., Wickramasinghe, N.C., Narlikar, J.V., Rajaratnam, P. and Perkins, J. Proc SPIE, Vol. 5163, in press 2003

42. Hoyle, F. and Wickramasinghe, N.C., 1988. Astrophys.Sp.Sci., 147, 245
43. Fitzpatrick, E.L., 1986. Astron. J., 92, 1068
44. Biachi, L., Clayton, G.C., Bohlin, R.C., Hutchings, J.B. and Massey, P., 1996. Astropphys.J., 471, 203
45. Motta, V., Mediavilla, E., Munoz, J.A., Falco, E., Kochanek, C.S., Arribas, S., Garcia-Lorenzo, B., Oscoz, A. and Serra-Ricart, M., 2002. Astrophys. J., in press (arXiv:astro-ph/0204130 v1 8 April)
46. Wickramasinghe, N.C., Hoyle, F. and Al-Jabori, T., 1989. Astrophys.Sp.Sci., 158, 135

Our Understanding of the Evolution of the Sun and its Death

Nami Mowlavi

INTEGRAL Science Data Center, Ecogia, 1290 Versoix, Switzerland,
Observatoire de Gen`eve, Sauverny, 1290 Versoix, Switzerland,
Nami.Mowlavi@obs.unige.ch

We summarize our understanding of the sun and of stars from modeling their structural and chemical evolution. We first recall the equations governing stellar evolution, mention the main uncertainties in stellar modeling, and show with two examples how confrontation between model predictions and observations allow to test stellar models. The case of the Sun is addressed by highlighting two recent developments. The first concerns solar neutrinos, the measurement of which validates solar models. The second deals with helioseismology that enables to probe the solar interior with high accuracy and to test its chemical composition.

1. Introduction

Stars play an important role both in the evolution of the universe and in the development of life. Not only do they provide the energy necessary to sustain life, but more fundamentally they are the basic factories in which elements that form living organisms are synthesized.

Because stars are obviously not directly accessible to experiment, our knowledge of their structural and chemical evolution in general, and of the Sun in particular, has only developed during the last century, mainly through our understanding of the laws of physics down to the subatomic level and the analysis of the light spectrum of the stars. Our knowledge of the laws of physics and of the properties of matter have enabled to model the structure of stars and their evolution with computers, while the observations, mainly of the light from stars, have enabled to improve and validate

V. Burdyuzha (ed.), The Future of Life and the Future of Our Civilization, 23–30.

model predictions. The purpose of this contribution is to summarize how we derive our knowledge about the evolution of the stars in general, and of our Sun in particular. Section 2 first describes the Sun as a standard star in the Universe. I then summarize in Sect. 3 the usage of stellar models in understanding stellar evolution, and come back more specifically to the evolution of the Sun in Sect. 4 where I compare model predictions of the solar interior with specific observations. Conclusions are finally drawn in Sect. 6. The paper is not meant to be exhaustive, but to provide to the non-astrophysicists participating in this symposium few examples of how we know about stars and the Sun.

2. The Sun, a Standard Star

The properties of a star and of its evolution are determined by only a handful of basic parameters that are its total mass and its chemical composition at star formation. Initial rotation velocity is another parameter that is taken into account since a decade or so in stellar models.

The chemical composition of the Universe after the Big Bang, from which the first generation of stars was formed, was mainly hydrogen and helium, and a little lithium and boron. Very quickly, however, the first generation of stars synthesized heavier elements in their interior during their life and polluted the interstellar medium through either strong winds or supernova explosion for the most massive stars. Successive generations of stars thereby contributed to the build up of the chemical composition of our Galaxy, from which the Sun formed 4.5 billions years ago.

Stars can be roughly divided in two categories, depending on the way they end their life. Low- and intermediate-mass stars, which comprise about 80% of all stars, end their life as white dwarfs of 0.6 to 1.2 solar masses (M_s). Those are stars with initial masses below about 8 M_s. Massive stars, on the other hand, with masses up to 150 M_s, explode at the end of their life as supernovae, leaving in most cases either a neutron star or a black hole as a remnant.

The Sun appears as a rather normal low-mass star, lying at the edge of one of the spiral arms of our Galaxy at about 8000 parsecs from its center. Its proximity to Earth makes it the most well known star, and has enabled to study its interior using neutrinos or seismology, techniques available with high accuracy only for the Sun. From a theoretical point of view, however, models of the solar interior and of its evolution are computed in a similar way as for other stars. Stellar modeling is thus first reviewed in the next section before coming back to the Sun in the section after the next.

3. Stellar Models

Stellar modeling makes usage of all our current knowledge in physics to describe the structure of a star and its evolution. The basic sets of equations governing the structural and chemical evolution of stars can be found in almost any astrophysics textbook. We just recall here the five sets of equations, summarizing where the main uncertainties still remain.

1) The *equation of hydrostatic equilibrium*. It determines the balance between the mechanical forces acting on the star, mainly gravity and pressure. Pres- sure depends on density and temperature, the relation between them being given by the equation of state. Uncertainties in the equation of state exist for extreme states of matter such as those prevailing at the end of the life of stars (supernovae, neutron stars).

2) The *conservation of mass*.

3) The *conservation of energy*. The main sources of energy in the life of a star like the Sun are gravitational contraction and nuclear fusion. Nuclear fusion occurs on timescales of billions of years (10 billion years for hydrogen burning in the Sun), and is described by *nuclear reaction rates* that depend on the density, temperature and chemical composition. Gravitational contraction (or expansion), on the other hand, acts where thermonuclear reactions are absent (star formation or intermediate stages after exhaustion of the burning nucleus in the core, for example), and has a typical timescale of millions to hundred millions years. The amount of energy liberated by a change of state in matter depends on the internal energy of matter, whose dependence on density, temperature and chemical composition is described by the *internal energy equation*. It must be mentioned that uncertainties still exist in many reaction rates, including the important reaction $^{12}C\ (\alpha,\gamma)\ ^{16}O$ which determines the amount of carbon and oxygen produced in stars.

4) The *energy transport*. Energy is transported either radiatively or, if the opacity of matter to radiation is high, convectively. The *opacity* depends on the density, temperature and chemical composition. An important consequence of convection is the transport of the ashes of thermonuclear reactions from the interior of the star where they are produced to the surface. The observation of the chemical composition at the surface of stars is a key method to test stellar model predictions. The description of convection inside stars, however, is one of the main shortcomings affecting stellar modeling. Convection is typically a three dimensional process with short timescales, and its inclusion into 1-d stellar evolution models that cover hundred thousand years is not yet feasible with current computer capacities.

5) The *nucleosynthesis* of elements at temperatures above 10^6 K, leading to the transformation of elements. As mentioned above, predictions still suffer from uncertainties in some reaction rates.

We must also add to these equations a prescription for mass loss from the surface of the star. The mass loss rate is already important during the main sequence phase for stars more massive than about $30M_s$. In lower mass stars, mass loss becomes important in the late phases when the star becomes a red giant. The mass loss prescription depends on the phase of evolution and is subject to some uncertainties.

Finally, let us say that the recent developments in stellar evolution concern the inclusion and treatment of rotation and magnetic fields in the models. We refer to Maeder and Meynet (2003) for a discussion of the relative importance of these two effects.

The above set of equations is solved numerically for each time step, starting from a cold (less than 10^6 T) and chemically homogeneous star, and ending at the formation of a white dwarf for initial masses below about $8M_s$ or at the pre-supernova stage for more massive stars. Model predictions of surface properties (luminosity, temperature, chemical composition) are confronted to observations whenever possible. We give in this section three examples of confrontation between model prediction and observation, and in the next section two more specific examples related to the Sun.

The first basic properties of a star are its luminosity and its color, from which are computed its surface temperature. Those quantities are conveniently analyzed in the so-called Hertsprung-Russel (HR) diagram that plots the luminosity of a given star against its surface temperature. The evolution of a star is represented in this diagram by characteristic paths that depend on the star's initial mass and composition. Comparing the observed locations of an ensemble of stars with those predicted by stellar models is a classical test of model predictions. An example is given by measuring the number of massive stars that are red to the number which are blue, and comparing to model predictions. We refer to Eggenberger et al. (2002) for such an analysis in open clusters at various metallicities.

Another usage of the HR diagram is the determination of the age of clusters of stars. All stars in a given cluster are born at the same time and hence have the same age and initial chemical composition. They only differ by their initial mass. Isochrones are computed in the HR diagram for different ages and compared to the distribution observed for the cluster. An example of a database of isochrones is given by Lejeune and Schaerer (2001). The isochrone that best fits the observed diagram provides the age of the cluster. Examples of the recent usage of this technique are given in, among many others, Ortolani et al., (2005); McCumber et al., (2005); Piatti et al., (2004)

The third example of comparison between observations and model predictions concerns the chemical abundances observed at the surface of stars. I have provided a review for light elements from ^{3}He to ^{19}F a few years ago in Mowlavi (2002). I will just mention here the puzzling case of fluorine. The nucleosynthesis site of this element was unknown until less than 20 years ago, unlike all other elements. Fluorine has been observed in asymptotic giant branch stars (i.e. low- and intermediate-mass stars at the end of their life) in the beginning of the nineties (Jorissen et al., 1992), and its production in those stars confirmed by model predictions (Forestini et al., 1992; Mowlavi et al., 1996). Two other production sites have also been suggested based on model predictions, supernovae (Woosley and Haxton, 1988) and Wolf-Rayet stars (Meynet and Arnould, 2000), but with so far no observational support. Recently, Cunha et al., (2003) challenged the red giant origin of fluorine based on new observations of this element at the surface of stars with low metallicities, favoring supernovae. Using a model of the chemical evolution of the Galaxy, Renda et al., (2004) actually show that all three sites may contribute to the galactic fluorine. Fluorine is a good example of the difficulties involved in both model predictions and observations.

4. Understanding Our Sun

The global evolution of the Sun as a star is rather well known. The Sun is in its main sequence phase, burning hydrogen in its center since about 4.5 billion years, and will continue to do so for another 5 billion years. A summary of its main phases of evolution until its death can be found in Mowlavi (2000). Because of its proximity, however, the interior of the Sun can be probed with techniques not applicable to other stars. I give here two examples, one related to helioseismology, and the second using solar neutrinos.

Helioseismology is a powerful tool that allows to probe the interior of the Sun with unprecedented accuracy and provides a unique test of the theory or stellar evolution. Seismology techniques start to be applied to other stars (leading to the growing field of asteroseismology), but does not lead yet to the accuracy obtained with the Sun. Helioseismology allows to measure the depth of the solar surface convection zone, the surface helium abundance, the internal rotation, the speed of sound as a function of depth down to almost the very center of the Sun, and the density distribution, though to a lesser degree than the speed of sound. I refer to Christensen-Dalsgaard (2002) for a review on the history and techniques of helioseismology. The agreement between helioseismology and model predictions was excellent. Helioseismology validated stellar model calculations. Recently, however, the chemical composition of the solar surface was revised, with much lower abundances of some heavy elements (Asplund et al., 2004). These new abundances now lead to a discrepancy in the sound speed and in the depth of the surface convective zone between model predictions and helioseismology. The measured depth of the convective zone, for example, is 0.713 solar radii from helioseismology, while the one predicted from solar models assuming the new abundance determinations is 0.726 solar radii (it is 0.714 solar radii with the old abundances). I refer to Bahcall et al., (2005a) for a discussion on the implications of the new abundances on solar models using helioseismology. The point I want to make here is that helioseismology has become so accurate that a small change in our knowledge of some basic properties of the Sun (in this case its surface abundances) leads to inconsistent results between model predictions and helioseismic measurements. The origin of the discrepancy is unknown so far. Solar models are however used to identify the cause of the mismatch. An attempt in that direction is done by Bahcall et al. (2005b) who show that the mismatch is removed if the neon abundance is increased by 0.4-0.5 dex (keeping the lower abundances for the other elements). These tests with solar models suggest that maybe the measurement of the abundance of neon at the surface of the sun is wrong.

Another tool to probe the most central regions of the Sun is solar neutrinos. Neutrinos are produced during hydrogen burning in the core of the Sun, at a rate that is very sensitive to temperature. The temperature itself is well determined by solar models. A long standing discrepancy was present between standard solar model predictions and measurements, measurements leading to a flux of neutrinos detected on Earth from the Sun which was a factor of 2 to 3 lower than predicted. Solutions were drastic: either the standard solar model is wrong, or there is a problem in our current understanding in neutrino physics. The solution came from new measurements of solar neutrinos: the solar models are right. I refer to Bahcall (2004) for a review of solar neutrinos and how it has been solved.

5. Conclusions

Our knowledge of the interior of the Sun is known to a high degree of accuracy. This has been validated with solar neutrinos and helioseismology.

The future of the Sun is globally known, but many uncertainties still affect model predictions for the phases after the main sequence. These include, for example, the mass loss rates when the sun will become a red giant, the relative abundances of carbon and oxygen that will be produced during helium burning and the structural and chemical evolution during the asymptotic giant branch phase before the Sun ends its life as a white dwarf. For a recent review on the difficulties related to this late phase, the reader is referred to Herwig (2005).

References

Asplund, M., Grevesse, N., Sauval, A. J., Allende Prieto, C., Kiselman, D., Apr. 2004. Line formation in solar granulation. IV. [O I], O I and OH lines and the photospheric O abundance. A&A417, 751–768.

Bahcall, J. N., Dec. 2004. Solar Models and Solar Neutrinos: Current Status. ArXiv High Energy Physics - Phenomenology e-prints.

Bahcall, J. N., Basu, S., Pinsonneault, M., Serenelli, A. M., Jan. 2005a. Helioseismological Implications of Recent Solar Abundance Determinations. ApJ618, 1049–1056.

Bahcall, J. N., Basu, S., Serenelli, A. M., Oct. 2005b. What Is the Neon Abundance of the Sun? ApJ631, 1281–1285.

Christensen-Dalsgaard, J., Nov. 2002. Helioseismology. Reviews of Modern Physics 74, 1073–1129.

Cunha, K., Smith, V. V., Lambert, D. L., Hinkle, K. H., Sep. 2003. Fluorine Abundances in the Large Magellanic Cloud and ! Centauri: Evidence for Neutrino Nucleosynthesis? AJ126, 1305–1311.

Eggenberger, P., Meynet, G., Maeder, A., May 2002. The blue to red supergiant ratio in young clusters at various metallicities. A&A386, 576–582.

Forestini, M., Goriely, S., Jorissen, A., Arnould, M., Jul. 1992. Fluorine production in thermal pulses on the asymptotic giant branch. A&A261, 157–163.

Herwig, F., Sep. 2005. Evolution of Asymptotic Giant Branch Stars. ARA&A43, 435–479.

Jorissen, A., Smith, V. V., Lambert, D. L., Jul. 1992. Fluorine in red giant stars - Evidence for nucleosynthesis. A&A261, 164–187.

Lejeune, T., Schaerer, D., Feb. 2001. Database of Geneva stellar evolution tracks and isochrones for (UBV)J (RI)C JHKLL'M, HST-WFPC2, Geneva and Washington photometric systems. A&A366, 538–546.

Maeder, A., Meynet, G., Dec. 2003. Stellar evolution with rotation and magnetic fields. I. The relative importance of rotational and magnetic eects. A&A411, 543–552.

McCumber, M. P., Garnett, D. R., Dufour, R. J., Sep. 2005. Stellar Populations in the Wing of the Small Magellanic Cloud from Hubble Space Telescope Photometry. AJ130, 1083–1096.

Meynet, G., Arnould, M., Mar. 2000. Synthesis of 19F in Wolf-Rayet stars. A&A355, 176–180.

Mowlavi, N., 2000. The Future of our Sun and Stars. In: The Future of the Universe and the Future of our Civilization. pp. 161–+.

Mowlavi, N., 2002. Nucleosynthesis in red giant stars. In: IAU Symposium. pp. 57–69.

Mowlavi, N., Jorissen, A., Arnould, M., Jul. 1996. Fluorine production in intermediate-mass stars. A&A311, 803–816.

Ortolani, S., Bica, E., Barbuy, B., Jul. 2005. Colour Magnitude analysis of five old open clusters. A&A437, 531–536.

Piatti, A. E., Clarira, J. J., Ahumada, A. V., May 2004. The relatively young, metal-poor and distant open cluster NGC 2324. A&A418, 979–988.

Renda, A., Fenner, Y., Gibson, B. K., Karakas, A. I., Lattanzio, J. C., Campbell, S., Chieffi, A., Cunha, K., Smith, V. V., Oct. 2004. On the origin of fluorine in the Milky Way. MNRAS354, 575–580.

Woosley, S. E., Haxton, W. C., Jul. 1988. Supernova neutrinos, neutral currents and the origin of fluorine. Nature334, 45–47.

Planetary Cosmogony: Creation of Homeland for Life and Civilization

Alexander V. Bagrov

Institute of Astronomy of Russian Academy of Sciences, Pyatnitskaya str.48, 119017 Moscow, Russia, abagrov@inasan.ru

Natural evolution of matter of the Universe comes to formation of stars of the second generation that are rich with heavy elements. Proposed by author planetary cosmogony shows that formation of planetary systems near such stars must be resulted in creation at least one planet with comfortable conditions for life, as well as in intrusion into Galaxy number of interstellar striders (loosed planets and planetesimals). The living forms appear on habitable planets from the beginning, and their origin may be connected with very complicated chemistry in interstellar gaze-and-dust nebulae. Evolution of life must overcome chemical limitations by formation of multicellular organisms and intelligence as necessary stages. At the intelligence stage there is only one mortal to its future factor - collision of habitable planet with interstellar strider.

1. Introduction

Many scientists after Drake (1961) believe that as life as intelligence in the Universe are lucky chance phenomena (Sagan, 1961, Shklovskii, 1963). Probabilities of existences in our Galaxy of solar-type stars, planetary systems, comfortable conditions on planets, origin of life and so on are still believed to be independent. Even a technology stage of intelligence is considered as lucky accident, and its durance being unpredictable. This paradigm looks very strange on background of generally accepted idea that whole universe is ruled by same natural laws. Moreover, we now

V. Burdyuzha (ed.), The Future of Life and the Future of Our Civilization, 31–39.

understand well that all natural laws and basic constants are combined just in unique superposition that allows existence of human beings (the Anthrop Principle). Therefore, the main goal of SETI investigators seems to be detection and following the chain of determined by these laws steps of the Universe evolution. Here we attempt to connect some links of the chain. Is really creation of homeland for life and civilization a destiny or a simple prize of dice play?

2. New Phenomenological Planetary Cosmogony

For more then two centuries a lot of cosmogony theories were suggested, but none of them satisfies whole set of observational data. Proposed by author hypothesis allows overcoming of main difficulties of the origin of the Solar System and can reveal history of extrasolar planetary systems. Due to the hypothesis planetary systems have to be formed during compression of protostellar nebula enriched with heavy elements at very low temperatures - about 30-60 K.

Now it is a common approach in cosmogony that stars and their planets seem to be formed from protostar nebulae (Edgeworth, 1949, Safronov, 1991). A large nebula should produce a massive star or binary one. If the nebula has rotation moment, it can prevent some matter to fall down to the central condensation (protostar) and force it to remain on orbits centered to the star. Some times later, this matter will be condensed into a number of planetesimals and planets.

When angular moment of the nebula with mass $1 \div 2\ M_{\odot}$ is large enough ($\sim 2 \cdot 10^{52}\ g \cdot cm^2 \cdot s^{-1}$) to stop accretion to the center of nebula, and its evolution ends with a disc, rotating around its center of masses. Particles and molecules in the disk will be separated by angular momentum: some of them that loss momentum will fall to the center of nebula, when the others which momentum increased, will be dragged outside. The growing central mass will form by time a star, and material that rests in the disk may form planetary system. The formation of planets takes much less time then formation of central star. The growth of a central protostar must be very slow, because only molecules without angular momentum can fall down to the center of the system. Under such conditions it may take tens and hundreds million years for central accumulation to gather enough matter to form a star.

Evolution of disk nebula takes place at very low temperature up to the time when nuclear reactions in the central protostar turn it into a common star. Therefore, gaseous component of the nebula must stop any turbulence inside it and keeps all orbits circular and complanar (Safronov, 1966). This is why planetesimals could be form on circular orbits only, as well as large planets after their combining. So evolution of the protoplanetary disk produces set of planets just on stable and comfortable circular orbits.

The Earth-type planet can be formed for less then million years. It must collect inside its body short-living isotopes of ^{26}Al as well as various molecules that were synthesized on surfaces of interstellar dust particles. Accretion of protoplanet material and its radioactivity heated and melted interiors of large planets, and this thermal energy was the first energy source for primitive anaerobic life. Later, when the central star begins its shining, new high-energy source stimulated appearance of living forms based on photosynthesis. It could take some billion years.

All large planets were formed on circular orbits, so at least one of them has to provide good conditions for life. Habitable planets can accompany only stars of the second generation with nearly solar chemical abundances. Second necessary condition for the habitable planetary system is slow rotation of a central star that is result of steady evolution of protostar/protoplanetary nebula. Simultaneously it provides circular orbits to habitable planets, so temperature conditions on their surfaces will be stable enough for life existence.

The Solar System shows initial order of its construction as well as evidences and consequences of its disaster. The order is obvious in preserved circular orbits of large planets, of the gathered to the rings satellites of giant planets and of the bodies in outer region of the Solar system fare from giant planets influence. All of these circular orbits are nearly in one plane since the beginning of the Solar System. Even Jupiter's gravitation being main conductor to the rest of the planets could not destroy it during 4.5 billion years.

On the other hand a lot of minor bodies of the Solar system have significantly elliptic orbit as well as non-zero orbit inclination relative to the plane of the large planets' orbits. All of them seem to be born by destruction of former planet Phaeton. Being by this time nearly totally melted, this planet, was spilled to billions drops by collision with some interstellar strider. The largest of these "drops" hardened as asteroids, the remnants of planet's nuclei being remained on orbits nearly the same as of parent planet. Some other drops were destroyed into tiny mouldings by internal pressure, producing compact swarm, which crossed proto-planetary disk and heaped up its snowflakes producing planetesimals with number mouldings inside. As most material of these bodies was gathered from placed on circle-like orbits particles, they obtained slightly elliptical orbits. Just that time many of half-hardened bodied collided each other like pieces of paste - without breaking out. Numerous impact craters of that epoch may be seen on the surface of atmosphere-less bodies.

The initial order of our planetary system was partially destroyed by explosion of Phaeton at the beginning of the Solar System. Since that time some element of chaos can be observed till now. The most interesting ones are comets and meteor streams. Among the number of observed comets there was no one with parabolic orbit - all of them belonged to the Solar System. Observations of comet remnants - meteors - can confirm presence mouldings from Phaeton in comet nuclei.

Chemical and mineral composition of meteorites, which may be considered as the remnants of Phaeton, shows that at beginning stages of planet formation atmospheres of planets had no free oxygen. Without it only primitive anaerobic bacteria could exist, and there are some speculations that some meteorites have traces of such cyan bacteria.

Besides that, escaped from parent stars bodies can be observed inside the Solar system when they cross it, and be identified as having large velocities. In contrast to ordinary comets nuclei of which consist of the same proto-planetary material, these bodies have to lost gas from their crust and got thick dust shells that protect interiors from heating during their encounters with stars. This is why most of interstellar vagrant bodies do not behave themselves as comets when they cross Solar System. Nevertheless they have to look like a common minor body but with parabolic orbits.

Frequency of vagrant bodies crossing Solar system may be roughly estimated by number of stars in environments of the Sun. Supposing this density to be $0.1M_\odot$ per cubic parsec and average number of escaped planets from every parent star to be 10^6, then inside Jupiter's orbit one such Moon-like object would pass once in 10^7 years with speed of 100 km/s. As the number of less massive bodies increases as power 2 of decreasing diameter, asteroid-like interstellar vagrant body with diameter 3 km may be seen every 10 years. This observation can be made occasionally only, but it will be strong evidence of reality of solar planet destruction in the past and of its cause.

Direct evidences of destruction of Phaeton planet may be obtained from meteor observations too. If the origin of dense particles from comet nuclei will be proved, it will be decisive argument for truth of the proposed scenario.

It seems interesting that estimation of probability of appearance interstellar bodies inside Jupiter's orbit leads to the probability of direct collision with the Earth of 3-km body about $3 \cdot 10^{-7}$. This value is strangely close to the estimation of time distance between most disastrous astroblemes on the Earth's surface. Besides that, there is interesting peculiarity of the huge astroblemes: no meteoritic material has been founded nor in the impact craters, nor in the external ridges, therefore, impactors that produced astroblems could not be asteroids but cometary nuclei only (A.V. Bagrov, 2001).

If pre-life chemical complex molecules are formed on surfaces of interstellar dusts, many of them may be kept save inside planetesimals which temperature must be very low and can not exceeds temperature of evaporation of most volatile molecules. Such planetesimals can play a role of containers of pre-life material or even some primitive organisms as seeds of life. Such interstellar striders may be a source of life for the newly formed planets.

3. Evolution of Universe and Predestination of Intelligence

Life is natural part of evolution of matter in the Universe. All flesh is entropy philter between high temperature of the stars and low temperature of the Universe. Being open systems, living creatures can survive by following to laws of evolution of energy-cybernetic systems.

There are three main directions of evolution of a single system: increasing of controlled energy, increasing of survives reactions and increasing of the self-structure (Hil'mi, 1972). This thesis may be illustrated by tendencies of life evolution on the Earth. So primitive species like unicellular have tiny dimensions and miserable masses that cannot allow these species to possess complex internal structure with great energy and information supply. When this main possibility of single unicellular were exhausted, the evolution created multicellular species that could keep save chemical energy and survive reactions inside them for instant counteraction under environment troubles. The multicellular species may consists of billions elementary cells combined in very complex unit. This creature can possess sophisticated information for proper behavior as well as energy to do it. The more mass creature has the more complex constitution it can be. On the other hand, there are some limiting factors for growth of living species in the nature. Say, gravity strongly restricts masses of creatures on our planet. No plant can be 200 m high, no land animal can be 20 tons weight as well as no water-living creature can be 100 tons. Moreover, growth to the limit must be paid by depression of other necessary abilities as energy under operation or information for survives reactions.

The next step of the life evolution after reaching mass limit for single creature may be combination of creatures into a group. Many fish combine into runs, birds live in flocks, among animals packs, herds tribes and so on are very common. Besides that, we know deeply specialized communities among insects. Bees, ants and termites form communities perfectly different from any previous kind of social structures. However, their communities are based on hereditary transmitting of information. It is not suitable for species that live in rapidly changeable environments.

The top of known evolution is people civilization. Our humanity manipulate now external to anyone energy so powerful that it is enough to produce global changes in climate and allowed us became dominant living form on the planet. The knowledge about world we gather and use for whole civilization is million times larger than anyone can possess as an individual. The structure of our civilization is not simple sum of its members, but includes industry, communications and other particularities of our infrastructure. It seems that possibilities for further evolution of our society are far away from visible limitations. The future of our civilization will be human very long.

The main goal of evolution is to keep save homeostasis or to adjust species themselves to the changes of the environments. Intelligence is the highest level of known life evolution. Now our civilization has energy and knowledge enough to counteract nearly any natural cataclysm of planetary scale. If we are lucky, our civilization will develop to the level that allows counteracting of any Galactic cataclysm. As for the very far future, there is one factor that may be mortal to any living forms in the Universe - it is total expansion of the Universe. Some hundred billion years later physical interactions in the Universe will be too low to provide any chemistry, and life will disappear. To counteract such changes of environment in the Universe scale is impossible. Therefore, the far predestination of inter-galaxy intelligence is to stop expansion of space in the vicinity of habitable region and creation of new Universe with proper environment.

Our Universe has the one set of basic constants that is suitable as for long existence of the Universe as for complex chemical compositions that are necessary for life. This wonderful correlation is known as anthrop principle. It means that the Universe was created just suitable for terrestrial kind of life and intelligence as a top of its evolution. The physic laws determine direction of nature evolution to the thermal equilibrium of the Universe, even at the zero temperature due to its expansion. But life is counteracting this general tendency depressing entropy in its close environments.

4. Natural Limits to Life Duration of Civilized Planets

Proposed by the author a hypothesis of origin and evolution of planetary systems shows that a number of stars of late classes seems to have their own planets. As more then half of stars in our Galaxy are believed to be low-mass dwarves, their numerous planetesimals (including moon-size planets) must be connected to central stars by weak gravity. Passages of such stars close to other stars of neighborhood lead to lose of some members of initial planetary system. Free planets become "interstellar striders" that move through Galaxy and can collide with another planet as well as another star. Therefore, the population of the Galaxy consists of two kinds of condensed matter - massive stars and comparatively low-massive planetesimals. Their gravitational interactions by the time have to equalize kinetic energies of stars and planetesimals. On the other hand, planetesimals could not pass too close to stars; otherwise, planetesimals would be destroyed by tidal forces and evaporated by star radiation. Therefore, the upper level of velocities of interstellar striders seems to be 100 - 200 km/s.

Collision of such moon-size interstellar strider with planet like Earth will lead to total destruction of the both. The history of our own Solar system has evidence of such disaster: the Main Asteroid Belt is relict of exploded planet Phaeton. The recent direct investigations of asteroids by space probes proved that asteroids are hard rocky bodies, and they can not be ancient planetesimals that did not form a planet between Mars and Jupiter due to gravitational influence of the last. Asteroid shapes and composition may be easily explained if purpose that they were formed from fractions of melted interiors of large planet like Earth when hardened.

It is obvious that threat of doom impact with any of asteroid or comet nuclei is very serious. This is why the Spaceguard Program and other such investigations are popular now. Nevertheless, asteroids inside Solar system exist for billion years on orbits that are strongly directed by large planets and may not impact the Earth. Real threat is due to comet nuclei and their analogues - interstellar striders. Impact craters on the Moon surface as well as on surfaces of Mercury, Mars and known asteroids have nearly the same age of 4 billion years. The younger impact craters on the Earth (astroblems) with hundred kilometers in diameter seem to be produced by comet nuclei. At least some of them (but may be all of them) were small interstellar striders, as no meteorite material was found inside these astroblems.

We cannot calculate exactly probability of such collisions, but very rough estimations show that total disastrous Phaeton-type collision may be once in 40 billion years, and impacts that lead to planetary life disaster - once in 100 million years. Besides that, impact of interstellar strider to the central star of habitable system will induce a terrible X-ray burst that can kill all living creatures. These estimations put strong limits on lifetime for life and for intelligence in particular. It may be much less then is necessary for civilization to develop itself enough to overcome potential threat of doom collision with interstellar strider.

If any civilization would be lucky enough to overcome treat of interstellar striders impacts and outlast period of mortal dependence of possible disaster from space, it will escape natural limits to life duration and live long.

5. Conclusions

Habitable planets seem to be formed as members of planetary systems near single solar-type stars with very slow rotation. At least one planet of planetary system can possess conditions suitable for life. Appearance of life on planets can take some hundreds millions years, and its evolution to intelligence can take 4-5 billion years.

Social stage of intelligence life is necessary for advanced evolution of living matter. The strongest limit to its existence is extraterrestrial natural factor - valuable probability of mortal impact by interstellar strider. Up today our humanity cannot prevent such collision, nor overcome following disaster. When Civilization does it, it will be safe for billion years.

Simultaneous evolution of the Universe and intelligence may be ended only by two ways: they will be ruin both or intelligence will change habitable part of the Universe to the stage with properties good enough for life.

References

Bagrov A.V., (2001) Astronomical problems of the Earth protection against dangerous asteroids. / Proceedings of IV ISTC Scientific Advisory Committee Seminar on "Basic Science in ISTC Activities", Academgorodok, Novosibirsk // Novosibirsk: Budker Inst. of Nuclear Physics SB RAS, 2001, p. 285-291.

Drake F.D., (1962) Intelligent Life in Space. Macmillan, New York, 128 pp.

Edgeworth K.E., (1949) The origin and evolution of the Solar System // Mon. Not. Roy. Astron. Soc. 1949. V. 109. p. 600-609.

Hil'mi G.F., (1972) Chaos and Life. (in: "Habitable Universe", Konstantinov B.P. & Pekelis V.D, eds.). Moscow: Nauka Pabl., 1972. - p.33-49. (in Russian).

Safronov V.S., (1991) Kuiper prize lecture: Some problems in the formation of the planets. // Icarus. 1991. Vol. 94, N2. p. 260-271.

Safronov V.S., (1966) Accumulation of minor bodies at the external boundary of planetary system. // Astron. Vestnik, 1966. - V. 30, p. 291-298 (in Russian).
Sagan C., (1961) On the origine and planetary distribution of life. Radiation Res. 15, 174-192 (1961).
Shklovskii I.S., (1963) Is communication possiblewith intelligent beings on other planets? Interstellar communications, A.G.W.Cameron, ed., W.A. Benjamin, Inc., New York, 1963, 5-16.

Impact Phenomena: In the Laboratory, on the Earth, and in the Solar System

Jacek Leliwa-Kopystyński

Warsaw University, Institute of Geophysics, ul. Pasteura 7, 02-093 Warszawa, Poland, and Space Research Center of Polish Academy of Sciences, ul. Bartycka 18A, 00-716 Warszawa, Poland, jkopyst@mimuw.edu.pl

First, the fundamental data concerning classification of the Solar System bodies and their origin are gathered. Next the impacts/collisions leading to different outcomes are discussed. The energy criterion related to the strength-dominated and gravity-dominated impact shattering is presented. Laboratory experiments and their relation to impacts in space are discussed. A review of the impact phenomena in the Solar System is presented. Finally, the most interesting impact structures on the Earth are listed, with regard to the biological effects of the impacts.

1. Introduction: An Overview of the Solar System

Our planetary system exists for about 4.6 billions years. It has been formed from a primordial gaseous nebula composed mostly of hydrogen and helium. On the evolutionary path of the nebula the Sun was formed first. The remaining matter of the disk around the Sun was partly dispersed and partly it became the building material for the planets, the satellites, as well as numerous minor bodies. The planets were formed by the accretion processes governed mostly by gravity of the Sun and by gravity of the locally largest body that is a planetary embryo. Here 'locally' means 'at an appropriate interval of distance' from the Sun. The unit of distance in the Solar System (SS) is the astronomical unit, 1 AU = 1.5×10^8 km, that is equal to the mean Sun-Earth distance.

The numerous minor bodies existing in the SS are the debris that have remained from the epoch of formation of the planets. The minor bodies

V. Burdyuzha (ed.), The Future of Life and the Future of Our Civilization, 41–54.

are: (i) Classic asteroids, orbiting mostly in the belt 1.6 – 5.2 AU spreading in between Mars and Jupiter. (ii) The centaurs that are in between Saturn and Neptune, 9.5 – 30 AU. (iii) The Trans Neptunian Objects (TNOs) that are orbiting beyond the orbit of Neptune. To the last group belong the Kuiper Belt Objects (KBOs) at the solar distances spanning 30 – 200 AU. The presently known number of the asteroids is of the order of 10^5; the number of known KBOs is of the order of 10^3. The ratio 10^2 reflects to the possibilities of the asteroids and the KBOs detection rather than the true populations of both groups. The sizes of observable asteroids are, in general, at least hundreds of meters. The KBOs that can be discovered have the size of hundreds of kilometers. So, only some of them, sufficiently large, are known at present; the planet Pluto with the radius 1200 km is considered as the largest known KBO.

Comets, another group of minor bodies, are not mentioned as yet. The eccentricities of the cometary orbits are on average much greater then the eccentricities of other SS objects. So, the orbits of the comets are very elongated and therefore comets can be found at any distance from the Sun. The typical sizes of the cometary nuclei are estimated to be between hundreds of meters and a few kilometers. Due to their small size comets are observed at distances from the Sun no larger than a few astronomical units. The hypothetical reservoir of comets, the so called Oort Cloud (OC), extends spherically up to 10^4 – 10^5 AU from the Sun. The estimated number of comets in the OC is 10^{12}. However, their total mass is estimated to be no larger than a few Earth masses.

The formations of the SS bodies happened over the large interval of distances from the Sun, so in very different thermodynamic conditions. The bodies that formed nearer to the Sun are composed of the materials that condensed at higher temperatures, i.e. from minerals. We call them rocky bodies. Their elemental compositions are mainly Si, Mg, Fe and O. The rocky bodies are the terrestrial planets (Mercury, Venus, Earth, Mars), the asteroids, and some of the satellites (the largest of them are the Jovian satellite Io and our Moon). Large amount of the rocky material was accumulated together forming the cores of the future giant planets. The volatile elements or compounds namely H_2, He, H_2O, CH_4, NH_3 have condensed on these cores forming the gaseous giant planets Jupiter, Saturn, Uranus, and Neptune.

The remaining rocky and icy materials were swept out of the solar nebula or they formed the numerous medium-sized and small bodies. Those are most of the satellites of the giant planets, the TNO, and the comets. All of them are called icy bodies nevertheless they are composed both of ices and of minerals in proportion by mass roughly 1:1 (by volume 3:1). Soon after the formation of the planetary system the collisions of the bodies were much more frequent than they are now. So, the epoch of some first few hundred millions of years is called the Big Bombardment Epoch (BBE). Then, the number of bodies being the potential impactors decreased considerably, the probability of collisions dropped, and the BBE finished. In the present SS the asteroids and the comets are also potential impactors. However, they are rather the sources of the 'impactors of the second order'. The latter are produced in two different ways:

The comets lose their matter due to solar irradiation. The volatile material sublimes and the boulders are released into space forming a meteorite stream following the parent comet's orbit.

The asteroids sometimes collide with each other, so the fragments of different sizes are produced. The largest fragments can form a family of genetically connected asteroids. The small solid state debris become potential future meteorites.

In the large impact events the impactors are the comets and the asteroids themselves but rather not the debris produced from them by sublimation or collisions.

2. Impacts and Collisions

Let us assume that the colliding bodies are spherical. Some convenient symbols and terminology are:

M, R, ρ – mass, radius and mean density of the body being bombarded (the target).

m_p, d – mass of the impactor (the projectile) and its radius. By definition $m_p \leq M$.

m_l – mass of the largest fragment of the target after the impact (after the collision).

v – velocity of the impactor relative to the target at the moment of collision.

v_{esc} – escape velocity from the target (that is equal to the free fall velocity on that target).

$G = 6.672 \times 10^{-11}$ N m^2 kg^{-2} – the gravity constant.

$E = m_p v^2/2$ – energy of the impact, expressed in J or in kilotons of trinitrotoluene (TNT).

Energy of explosion of 1 kg of TNT is equal to 10^6 cal = 4.186×10^6 J. This unit is commonly applied to large impacts. The energy of the atomic bomb blasted over Hiroshima was about 20 kt. So, it was equivalent to the energy of an impact of the meteorite with mass 4.19×10^5 kg hitting the Earth surface with velocity 20 km s^{-1}. The radius of this meteorite would be as small as only 3.2 m (stony meteorite with density 3000 kg m^{-3}) or only 2.3 m (iron meteorite with density 8600 kg m^{-3}).

There is a convenient relation for the impact events:

$$v_{esc} = (2GM/R)^{1/2} = (8\pi G\rho/3)^{1/2} R = (\rho/1789)^{1/2} R \text{ [m s}^{-1}\text{]} \quad (1)$$

Here R is in km and ρ is in kg m^{-3}. The escape velocity v_{esc} depends weakly on density. The latter is close to 1789 kg m^{-3} for the giant planets and for many of their satellites. So, roughly speaking for these bodies the free fall velocity in meters per second is equal to their radii in kilometers. For many rocky bodies (the terrestrial planets – Mercury, Venus, Earth, and Mars; Moon, Io, the low-porosity asteroids) there is roughly v_{esc} [m s^{-1}] $\approx 1.5\ R$ [km]. The impact velocity of a SS impactor onto a SS target is the combination of v_{esc} and the orbital velocities of both the target and the impactor and thus $v > v_{esc}$. For the Earth one usually assumes $v_{esc} = 20$ km s^{-1} as a reasonable estimate.

The words 'impact' and 'collision' are frequently used as synonyms. However, their meaning is somewhat different. Let us adopt for convenience a definition of an impact as an event that can produce a crater on the target body but it cannot destroy it. On the other hand, a collision means an encounter of two bodies with comparable masses and sizes. Collisions can lead to break-up of the target. In the planetary sciences the term 'giant impact' is also used for collisions of the large bodies.

We are predominantly interested in the fate of the target but not of the impactor. So the crucial parameter is neither the impactor size nor its density but the impact energy. Typical ranges of ρ and v can be specified easily. The density of the targets (as well as of the impactors) is about 1000 kg m^{-3} or 3000 kg m^{-3} for the icy and for the rocky bodies, respectively. The velocity attainable in the laboratory for the studies of impact events is usually from about 100 m s^{-1} to about 10 km s^{-1}. However, the velocities of the order of mm s^{-1} are used as well to study growth of fluffy aggregates. The typical impact velocities in the present-day planetary system are from a few km s^{-1} to several tens km s^{-1}.

The following classification of the impact events in the SS can be introduced:

Rebound of the impactor. This is related mostly to low-velocity impacts onto small targets (with neglected gravity), e.g. the stone-on-stone impact with the velocity of the order of 100 m s^{-1}.

Impact that leads to sticking of a projectile in a target. This class is relevant to the very beginning of the SS. At that time much of the solid material had been in the form of the small fluffy aggregates. Their mutual gravity was negligible and their relative velocities were small. For solid-solid (metal-metal) impact sticking experiments (Leliwa-Kopystyński et al., 1984 and Leliwa-Kopystyński, 1984).

Impact that leads to cratering of a target. The mass loss of the target is negligible and therefore mass of the largest post-impact fragment $m_l \approx M$. The post-impact mass of the target may even increase by 'consuming' the impactor and therefore the target may grow (accretion process). Roughly, the upper limit of the mass of the impactor that produces a crater but does not destroy the target is $m_p \leq (10^{-6} - 10^{-3})M$. For large craters on small SS objects this estimation was given (Thomas, 1999).

Sub-catastrophic (intermediate) range. The mass of the largest fragment is $0.5M \leq m_l < M$ while the mass of impactor is in the interval $(10^{-6} - 10^{-3})$ $M \leq m_p \leq 0.1M$. (Jach et al., 1994, Svetsov, 2005).

Collision that leads to a catastrophic break-up of a target (disruption, shattering). The post-collision mass of the largest fragment is less than half of the initial mass of the target $m_l \leq 0.5M$ while $m_p \approx (0.1 - 1)M$. The term 'giant impact' is commonly in use (Love and Ahrens, 1996).

The above ranges of mass ratio m_p/M are very tentative. Note that the ratio $m_p/M \propto (r/R)^3$. So, for the impactors causing a crater $r < 0.01R$ and for those causing shattering $r > 0.1R$, approximately. The term 'catastrophe' is related to the post-collisional mass of the target. Indeed, the target body is considered to have survived if more than half of its mass is not fragmented. However, from the point of view of the target surface (in particular the potential biosphere on the surface) a disaster can occur in the cratering impact regime. For example, the Earth impacted by a kilometre-sized asteroid or comet (m_p/M of an order of 10^{-12}) would suffer a disaster catastrophic for the biosphere, see Section 6.

3. Energy Criterion for Shattering of the Target: The Distinction between Strength-Dominated and Gravity-Dominated Regimes

There is an essential qualitative difference between the small-scale and the large-scale impacts. The small-scale events concern the situation when the gravitational interaction between the impactor and the target can be neglected. This is evidently the case in the experiments performed in the Earth-based laboratories. However, gravity of the Earth influences the tracks of the individual impactor as well as the surface distribution of the impactor blanket. The effects of impacts and collisions in the planetary system depend both on the mechanical strength Y of the target as well as on its self-gravity. The strength is expressed in the units of pressure [N m^{-2}] that represents the durability against the pressure-due or tension-due destruction or, equivalently, it is expressed in the units of specific energy per unit volume [J m^{-3}] that is needed for destruction.

There are two regimes of target destruction. They are the strength-dominated regime, when M is small and the gravity-dominated regime when M is large. Let us estimate the border point between them. The gravitational energy accumulated in a spherically-symmetric body is:

$$E_G = \frac{3}{5}\alpha\frac{GM^2}{R} \tag{2}$$

For the body with density increasing towards the center the coefficient α is $\alpha > 1$. For the Earth $\alpha = 1.13$. For the uniform bodies (most of the asteroids, small moons, cometary nuclei are supposed to be uniform) we have $\alpha = 1$. The energy needed for shattering the body against its selfgravity is βE_G. Here $\beta = 1$ for shattering and dispersing infinitesimally small grains to infinity and $\beta < 1$ when the post impact debris with finite sizes remain. The mechanical (strength) energy needed for disruption of this body is approximately:

$$E_Y = \frac{4\pi R^3}{3}Y, \quad \text{or} \quad E_Y = MQ, \tag{3}$$

here: $Q = Y/\rho$ is the specific energy for shuttering of the target per unit mass [J kg^{-1}]. The limiting target radius $R_{Y/G}$ between the strength- and the gravity-dominated regimes is given by the condition $\beta E_G = E_Y$ and therefore

$$R_{Y/G} = \left(\frac{5}{4\pi\alpha\beta G}\frac{Y}{\rho^2}\right)^{1/2} \tag{4}$$

This formula for $\alpha = \beta = 1$ gives the estimates: $R_{Y/G} = 77$ km for the porous icy mineral targets ($Y = 10^6$ N m^{-2}, $\rho = 1000$ kg m^{-3}) and the similar result $R_{Y/G} = 81$ km for the rocky targets ($Y = 10^7$ N m^{-2}, $\rho = 3000$ kg m^{-3}).

4. Laboratory Experiments: their Parameters and Expected Outcome

Apart from the limitation relating to the scale the most important difference between the Earth-based laboratory studies and the events in space is that the Earth gravity cannot be removed. So gravity is always acting on the flux of the impact ejection. Its influence on the crater form is less important.

The laboratory conditions restrict several parameters related to the target and impactor properties as well as to geometrical conditions:

- Mass regime, typical: M from grams to kilograms, m_p from micrograms to grams.
- Target properties:
  (i) Geometry: flat ('infinite half-space target') or spherical;
  (ii) Structure: uniform or layered; consolidated (e.g. rock) or loose (e.g. sand);
  (iii) Porosity: nonporous or porous.
- Impactor: spherical or cylindrical.
- Impactor trajectory: horizontal (the most typical), vertical, or inclined.
- Impact: head-on (perpendicular) or oblique. Studies of oblique impacts are far less typical.
- Impact velocity regime: from as low as mm s^{-1} in grain-grain sticking experiments related to the early stages of planetary embryos formation to as high as ~10 km s^{-1} (hypervelocity regime) aimed at studies of the present-day impacts in the SS).
- Impact energy regime: up to 10^3 J, approximately. However, the explosions performed in the field tests are used to simulate much more energetic impacts.

Expected output from laboratory experiments:

- Crater's morphology, therefore crater's depth H, diameter D, and its shape (Arakawa et al., 2000, 2002). For example, it was found that in icy targets an axial cross-section of the crater changes when the impactor energy increases Evolution is from bowl-shaped to trumpet-shaped or the other way round depending on the target material (H_2O ice or CO_2 ice). The intermediate shape is simply conical one (Burchell et al., 2005).
- Scaling laws for crater's depth H, diameter D, and volume V versus impactor energy $mv^2/2$ (Burchell et al., 2005) or impactor momentum mv.

- In most cases the material of an impactor is not critically important. However, the response of the very low density targets to an impact strongly depends on the projectile density. High-density impactor produces narrow and deep carrot-type craters in low-density target (Kadono, 1999).
- Criteria for catastrophic disruption of the targets. In particular measurements of specific energy of disruption (Arakawa et al., 2002).
- Studies of the distributions of the elevation angle of ejecta by number and by mass (Burchell et al., 1998, Arakawa et al., 2002).
- Studies of cumulative distributions of ejecta by number and by mass.

Extrapolation of the small-scale laboratory results to the planetary scale is crucial (Holsapple 1994, Housen and Holsapple, 1999, 2003). However, it is not simple to do since (i) extrapolation should span several orders of magnitude of impactor energy, and (ii) in the laboratory we have neither a possibility to switch-off the Earth gravity nor to change its value. The craters' morphology on the laboratory targets and on the SS bodies can differ considerably. In particular, the central peaks are very common in lunar craters but they are rather not observed in the laboratory tests. This is easy to explain. In the laboratory, where targets are made from brittle material, the slow post-impact rheological behavior of the targets cannot be observed. On the SS target bodies the solid-state highly viscous rheology can play an important role.

The unique 'laboratory in space experiment' was performed 4 July 2005 when an copper impactor with the mass 370 kg separated from the NASA Deep Impact mission spacecraft and hit the surface of the comet 9P/Temple 1 with relative velocity 10.2 km s^{-1}. The impact energy was 1.9 $\times 10^{10}$ J. Brightness of the comet increased from magnitude 9.3 before impact to about magnitude 6 after impact. At the time of preparation of this paper (mid-July 2005) the data concerning the impact crater are not yet available.

5. The Consequences of Impacts in the Solar System

On the surfaces of the most of the atmosphere-free bodies of the SS the impact craters are ubiquitous. This applies most of the satellites, including our Moon the asteroids, and Mercury. It is possible to gather information about the geological history of the surfaces from the density of the craters. The surfaces densely covered (saturated) by craters of very different sizes are geologically old. On the other hand, the surfaces with low density of impact craters are young. Jupiter satellites Io and Europa are good examples. On Io the process of resurfacing is fast due to intensive volcanic activity of this satellite. The volcanic deposits spread over the globe and they obscure older structures. So, the impact craters on Io are practically absent. The icy surface of Europa is also renewed due to outflows of liquid (contaminated water) from the sub-surface regions. Water freezes and covers the older relief of the surface. The impact craters on Europa are not numerous and their age is estimated to be as young as only tens of millions of years. In conclusion, studies of the craters' size-density relationships and craters saturation allow to estimate the geological age of the surface. Surfaces of the bodies without atmosphere are reached by the grains of space matter from the full spectrum of sizes. So, the studies of size distribution of the craters give estimates concerning the energy and size spectra of the impactors in that region of the SS in which the target body resides.

Impact-due modification of the surfaces of the bodies with substantial atmosphere (Venus, Earth, Mars, Saturnian satellite Titan) is radically different. At first, small impactors are burning in the atmospheres and they do not reach the surfaces. So, only the traces of the sufficiently large impactors that have not disintegrated during the passage through the atmosphere are imprinted on the surfaces. Secondly, the impact craters are degraded due to mechanical and chemical weathering, and due to various geological processes.

The weathering processes are the wind-weathering (on all bodies with atmospheres), the water-weathering (confirmed for sure only on the Earth and on Mars, but in the past) the chemical-weathering (on all planets where the components of the atmosphere reacts with these of the crust). The plate tectonics, a process that is ascertained for sure only on the Earth, is a phenomenon leading to the disappearance of those craters that were formed on the downgoing edges of the tectonic plates. Volcanic activity producing the lava flows, leads to covering of the elder surface. The lunar basalt lava fields and the icy (cryogenic) lava on the icy satellites stand as examples

All impact structures underwent degradation due to the relaxation processes. In particular, the crater depth H decreases exponentially with time

$$H = H_0 \exp(-t/t_{rel}), \qquad \text{where } t_{rel} = 8\eta / (Dg\rho) \qquad (5)$$

is the relaxation time (Melosh, 1989). Here D is crater diameter, g is surface gravity, ρ is density, and η is viscosity. The viscosity coefficient η depends on material and on temperature. So, its reasonable range spans several orders of magnitude, $\eta \sim 10^{(22 \pm 5)}$.

In Section 3 it is shown that the sufficiently energetic impacts lead to shattering of the target body. The giant impact hypothesis is used for a plausible explanation of several facts observed in the SS. The consequences of such large impact (a catastrophic collision) can be manifold, as discussed below.

(i) *Asteroid families.* The orbital elements of some groups of asteroids are nearly identical. Such groups are called the asteroid families. There is well-documented hypothesis that the members of a family are the fragments of collisional shattering of a common parent body. More than 20 families are identified. According to Bendjoya and Zappala (2002) the most numerous are the following ones: 8 Flora (>600 identified members), 24 Themis (>500), 221 Eos (480), 44 Nysa (380), 158 Koronis (300), and 4 Vesta (230). The analysis of dispersion of the orbital parameters allows to estimate the age of a family. The age of the Flora family is, for example, significantly less than 10^9 years (Nesvorny, 2002).

(ii) *Origin of the Moon* and peculiarities of the Earth-Moon system. In our SS the Earth-Moon system is exceptional as the mass ratio Moon/Earth = 1/81 is three orders of magnitude higher that any other planet/satellite mass ratio (the Pluto-Charon system belongs to the class of binary KBOs but it is not a regular planet-satellite system). According to the present state of knowledge the most plausible scenario concerning the origin of the Moon is the giant impact hypothesis formulated by in 1986 (Benz, Slattery and Cameron, 1986, Cameron 1997, Benz and Asphuag, 1999). An impactor with mass equal to about 1/10 of the present mass of the Earth has collided with the proto-Earth. Several particular values of the impactor/target mass ratio, collision velocity, and collision angle have been studied. After the giant collision the collider has disintegrated and large amount of the mass, mostly from the proto-Earth rocky mantle (but not from the proto-Earth iron core) has been thrown away. Some portion of that mass escaped and some has accumulated on the orbit around the Earth forming the Moon. The mean density of the Moon, 3344 kg m^{-3}, indicates that its composition is mainly rocky but not irony. It is one of the arguments that the Moon was formed from the proto-Earth mantle material.

(iii) *High inclination of the axis of rotation of Uranus.* Tilt of the Uranus rotation axis with respect to the line perpendicular to its orbital plane is equal–82°. The minus sign indicate that the Uranus' rotation is retrograde. An unresolved question arises: Was Uranus high inclination caused by an impact of a late-arriving giant planetesimal? If not, then another mechanism must be responsible. But such a mechanism is not known as yet.

(iv) *Origin of Miranda*, the Uranian satellite geologically very strange. It is possible that a catastrophic impact broke down the pristine proto-Miranda. If so, the present Miranda is a product of re-accumulation on the orbit of the collisional debris, including the large fragments, some 100 km across. During re-accumulation the fragments hit each other with velocities of some dozens of m s^{-1}. Perhaps the famous Miranda coronae structures are related to those giant collisions.

At the end of this section it is necessary to mention an example of spectacular impact events that begun 18 July 1994 when comet Shoemaker-Levy 9 (SL9) struck Jupiter's atmosphere had velocity ~60 km s^{-1}. The comet was previously disrupted into several fragments by the Jovian tidal forces. The fragments dispersed along the comet's orbit. More than 20 observed fragments penetrated into the huge Jovian atmosphere within a few days. The sizes of fragments were estimated to be from hundreds of meters to a few kilometers. The traces of impacts on the atmosphere (with sizes comparable to size of the Moon) lasted for several months.

6. Impacts Onto the Earth

It is estimated that the Earth gathers each year $(3 \pm 1.5) \times 10^7$ kg of extraterrestrial material (Hutchison, R., 2004). Most of it are the micrometeorites that are burning in the uppermost layers of the atmosphere. The numbers of meteoritic falls with mass larger than 0.1 kg is no more than 17000 per year. Of that, no more than 5000 falls per year have mass larger than 1 kg (less than one falls with mass over 1 kg per year per 10^5 km^2). In general, the cumulative meteorite flux rate at the top of the atmosphere follows the power law: The number of meteorites with radius larger than r falling in the unit of time on the unit of surface of a planet is

$$n(>\mathrm{r}) = a\, r^{-b} \tag{6}$$

The coefficients a and b depend on the region of the SS in which the impacted planet resides and on mass of the planet. According to Ward (2002), for the bolides and asteroids reaching the Earth atmosphere the coefficients are $a = 3.89 \times 10^{-14}$ $[\mathrm{m}^{1/3}\,\mathrm{yr}^{-1}]$ and $b = 7/3$. So, formula (6) gives one impactor with $r > 50$ m in 464 years, one with $r > 5$ m in 2.2 years, and twenty with $r > 1$ m annually.

The impactor flux and the sizes of impactors at the top of the atmosphere are larger than they are when reaching the Earth surface. Of course, only a meteorite that reaches the surface can produce a crater. In Table 1 (Leliwa-Kopystynski and Burchell, 2004) are listed the largest and/or the most interesting impact structures known on the Earth. During the last billion years the events that produce hundred kilometer size craters happen with average frequency one per 10^8 years. (During the BBE impact frequency was much larger). The impactor sizes are estimated to be 1 – 2 km but they are certainly no larger than a few kilometers. However, their influence on the biosphere could be catastrophic, see species extinction data in Table 1. The Tunguska-scale events happened once per 10^3 years. The lands on the Earth take only 1/4 of its total surface; of that 1/5 is densely populated. So, we estimate that the probability of devastation of a human settlement by a Tunguska-like event is 1/20000 per year.

The problem of hazard due to asteroid impacts on the Earth was widely discussed in the last decades. This problem concerns several aspects, namely: estimation of the probability of an impact (celestial mechanics applied to the asteroids crossing the Earth orbit), impact prevention (human technical capabilities), impact-due disaster modeling (crater formation and scale of surface damage), and finally the pre- and after-impact disaster logistics.

Table 1. Selected impact structures identified on the Earth.

Impact site	Diameter D km	Age 10^6 years	Remarks
Sudbury, Canada	200	1850	
Chicxulub, Yucatan, Mexico	180	64.98	Extinction of 38.5% of genera (67% of species).
Vredefort, South Africa	140	1970	
Manicouagan, Quebec, Canada	100	214	Extinction of 34.1% of genera (62% of species).
Popigai, Russia	100	35	Extinction of 11.4% of genera (25.5% of species).
Acraman, South Australia	90	>570	
Puchezh-Katunki, Russia	80	220	Extinction of 20.1% of genera (43% of species).
Kara, Russia	65	73	Twin structure of Kara and Ust-Kara has a combined diameter ~120 km.
Beaver head, Montana, USA	60	~600	
Chesapeake Bay, Atlantic shelf, close to the West USA	80	35	Impact happened on the sea. Extinction of some species.
Mjølnir, Barents Sea	40	142	Impactor size 1 – 2 km. Impact happened in a 400 m deep sea.
Aourtounga, Sahara, north of Chad	17	A few hundreds	Impactor size 1 – 2 km.
Arizona, USA: Berlinguer crater	1.6	0.056	The best preserved and the youngest crater on the Earth. Impactor size 10 – 100 m.
Tunguska event, Russia: no crater, only forest devastation	----	Event in 1908 yr.	Disintegration of an (icy?) impactor above devastated area.

Acknowledgments

The author is grateful to organizers of the Symposium 'The Future of Life and our Civilization' (Frankfurt, May 2005), particularly to Prof. V. Burdyuzha and to Prof. C. Gros for their kind invitation. This work is partially supported by the grant No. 2P03D 01025 provided by the Polish Ministry of Scientific Research and Information Technology.

References

Arakawa, M., M. Higa, J. Leliwa-Kopystynski, and N. Maeno, 2000. Impact cratering of granular mixture targets made of H_2O ice – CO_2 ice – pyrophylite. *Planetary and Space Science*, 48, 1437-1446.

Arakawa, M., J. Leliwa-Kopystynski, and N. Maeno, 2002. Impact experiments on porous icy-silicate cylindrical blocks and the implication for disruption and accumulation of small icy bodies. *Icarus*, 158, 516-531.

Bendjoya, Ph., and V. Zappala, 2002. Asteroid family identification. In: *Asteroids III.* W. F. Bottke Jr., A. Cellino, P. Paolicchi, and R. P. Binzel (Eds.), University of Arizona Press. pp. 613-631.

Benz, W., W. L. Slattery, and A. G. W. Cameron, 1986. The origin of the Moon and the single impact hypothesis V. *Icarus*, 66, 515-535.

Benz, W., and E. Asphuag, 1999. Catastrophic disruption revised. *Icarus*, 142, 5-20.

Burchell, M. J., W. Brooke-Thomas, J. Leliwa-Kopystyński, and J. C. Zarnecki, 1998. Hypervelocity impact experiments on solid CO_2 targets. *Icarus*, 131, 210-222, 1998.

Burchell, M. J., J. Leliwa-Kopystynski, and M. Arakawa, 2005. Cratering of icy targets by different impactors: laboratory experiments and implications for cratering in the solar system. *Icarus*, in press.

Cameron, A. G. W., 1997. The origin of the Moon and the single impact hypothesis V. *Icarus*, 126, 126-137.

Holsapple, K. A., 1994. Catastrophic disruption and cratering of solar system bodies: a review and new results. *Planet. Space Sci.*, 42, 1067-1078.

Housen, K. R., K. A. Holsapple, 1999. Scale effects in strength-dominated collisions of rocky asteroids. *Icarus*, 142, 21-33.

Housen, K. R., K. A. Holsapple, 2003. Impact cratering on porous asteroids. *Icarus*, 163, 102-119.

Hutchison, R., 2004. *Meteorites*. Series: Cambridge Planetary Science (No. 2). Cambridge University Press. 520 pp.

Jach, K., J. Leliwa-Kopystyński, M. Mroczkowski, R. Swierczynski, and P. Wolanski, 1994. Free particle modelling of the hypervelocity asteroids collisions with the Earth. *Planet. Space Sci.*, 42, 1123-1137.

Kadono, T., 1999. Hypervelocity impact into low density material and cometary outburst. *Planet. Space Sci.*, 47, 305-308.

Leliwa-Kopystyński, J., 1984. Sticking experiments and non-gravitational component of the mechanism of the growth of planets. J. *Journal de Physique*, Colloque C8, supp. au no. 11, tome 45, pp. C8-109-112.

Leliwa-Kopystynski, J., T. Taniguchi, K. Kondo and A. Sawaoka, 1984. Sticking in moderate velocity oblique impact: application to planetology. *Icarus*, **57**, 280-293.

Leliwa-Kopystynski, J., and M. Burchell, 2004. Impact cratering of icy and rocky targets in planetary sciences and in the laboratory. In: *Cratering in Marine Environments and on Ice*. Eds: H. Dypvig, M. Burchell, and P. Clayes. Springer Verlag series *Impact Studies*. pp. 223-249.

Love, S. G., and T. J. Ahrens, 1996. Catastrophic Impacts on Gravity Dominated Asteroids. *Icarus*, 124, 141-155.

Melosh, H. J., 1989. *Impact cratering*. Series: *Oxford Monograph on Geology and Geophysics*. No. 11. Oxford University Press, 245 pp.

Nesvorny, D., A. Morbidelli, D. Vokrouhlicky, W. F. Bottke, and M. Broz, 2002. The Flora family: a case of dynamically dispersed collisional swarm? *Icarus* 157, 155-172.

Svetsov, V. V., 2005. Numerical Simulation of very large impacts on the Earth. *Planet. Space. Sci.*, in press.

Thomas, P. C., 1999. Large craters on small objects: occurrence, morphology, and effects. *Icarus*, 142, 86-96.

Ward, S. N., 2002. Planetary cratering: A probabilistic approach. *J. Geoph. Res.*, 107, No. E4, 7-1 – 7-12.

Further Reading:

De Pater, I., and J. J. Lissauer, 2001. *Planetary Sciences*. Cambridge University Press. 528 pp. The fundamental book concerning planetology.

Gehrels, T., editor, 1994. *Hazards due to comets and asteroids*. University of Arizona Press. 1300 pp.

Melosh, H. J., 1989. *Impact cratering*. Series: *Oxford Monograph on Geology and Geophysics*. No. 11. Oxford University Press, 245 pp. The fundamental book concerning impact cratering.

II

THE ORIGIN OF LIFE

The Structural Regularities of Encoding of the Genetic Information in DNA Chromosomes

Anatolyj Gupal

V.M. Glushkov Institute of Cybernetics, Kyiv, Ac.Glushkov str. 40, Ukraine, gupal_anatol@yahoo.com

We determined the new complementary principles of encoding bases by one chain in DNA chromosomes of human genome and other investigated genomes. On the basis of obtained statistical data one can draw a conclusion that there exist strict rules of forming DNA structure valid for all species. The obtained results will significantly improve our present view of encoding genetic information.

In 2003 the International Project "Human Genome" was declared completed and immediately a new project ENCODE (Encyclopedia of DNA Elements) started the same year. At present several hundreds of genomes of viruses, bacteria, animals and plants are determined. The current human genome sequence contains 2.85 billion nucleotides interrupted by only 341 gaps. It covers ~99% of the euchromatic genome and is accurate to an error rate of ~ event per 100,000 bases. Notably, the human genome seems to encode only 20.000 – 25.000 protein-coding genes [1-3].

Although DNA is relatively simply and well understood chemically, the human genome's structure is extraordinarily complex and its function is poorly understood. Only 1 – 2% of its bases encode proteins, and the full complement of protein-coding sequences still remains to be established. Even less is known about the function of the roughly half of the genome that consists of highly repetitive sequences or of the remaining non-coding, non-repetitive DNA.

Two-chain spiral of DNA is written in a four-letter alphabet composed of the four bases adenine (A), cytosine (C), guanine (G) and thymine (T). It is known that C – G, A – T are complementary pairs of bases connecting two chains of DNA. The statement of the problem is to develop and

V. Burdyuzha (ed.), The Future of Life and the Future of Our Civilization, 57–62.

confirm fundamental principles in encoding complementary bases A and T and also C and G by one chain in DNA chromosomes of human genome and other investigated organisms; the relationships for letter pairs, a triplet of tree adjacent bases etc. On the basis of obtained statistical data one can draw a conclusion that there exist strict rules of forming DNA structure valid for all species. The obtained results will significantly improve our present view of encoding genetic information and also data on DNA replication process as well as formation of compact shape of chromosomes packing in the kernels of cells of organism.

1. Complementary Principles for Bases

The statement of the problem is to determine the new fundamental complementary principles in encoding bases by one chain in DNA chromosomes of human genome and other investigated organisms. As was reported in [4-5] we discovered that for each chain of DNA chromosomes the following complementary relationships hold. The frequency of the letter A is equal to the frequency of the letter T, and the frequency of the letter C is equal to the frequency of the letter G:

$$\frac{n(A)}{n} = \frac{n(T)}{n}, \quad \frac{n(C)}{n} = \frac{n(G)}{n}, \tag{1}$$

where: $n(j)$, $j \in \{A, C, G, T\}$ is the number of letter j, n is the length of chromosome (Table 1).

Table1. Frequencies of the letters A, C, G, T.

Chromosome	Human genome			
	A	C	G	T
1	0,291	0,209	0,209	0,292
2	0,299	0,201	0.201	0,299
3	0,301	0,198	0,198	0,302
4	0,309	0,191	0,191	0,309
5	0,302	0,197	0,198	0,303
6	0,302	0,198	0,198	0,302
7	0,296	0,204	0,204	0,297
8	0,299	0,201	0,201	0,299
9	0,293	0,207	0,207	0,293
10	0,292	0,208	0,208	0,292
11	0,292	0,208	0,208	0,292
12	0,296	0,204	0,204	0,296
13	0,307	0,193	0,193	0,308
14	0,294	0,204	0,205	0,297
15	0,289	0,211	0,211	0,289

16	0,275	0,223	0,224	0,277
17	0,272	0,228	0,227	0,273
18	0,301	0,199	0,199	0,301
19	0,258	0,242	0,242	0,259
20	0,278	0,220	0,221	0,280
21	0,297	0,204	0,205	0,294
22	0,261	0,240	0,240	0,260
X	0,302	0,197	0,197	0,303
Y	0,299	0,199	0,200	0,301

2. Complementary Principles for Letter Pairs

The frequencies of bases pairs are calculated by the relation:

$$\frac{n(ij)}{n(i)} \tag{2}$$

in which: $n(ij)$ is the number of pairs in the chromosome chain. Remarkable peculiarity of frequencies of bases pairs is that for all chromosomes of human genome the following relationships hold

$$\frac{n(AA)}{n(A)} = \frac{n(TT)}{n(T)}, \frac{n(CC)}{n(C)} = \frac{n(GG)}{n(G)}. \tag{3}$$

Complementary principles for letter pairs are (with accounting gaps in human genome as well as accuracy of sequencing)

$$\begin{aligned} &n(AC) \sim n(GT),\ n(AG) \sim n(CT) \\ &n(TC) \sim n(GA),\ n(TG) \sim n(CA) \\ &n(AA) \sim n(TT),\ n(CC) \sim n(GG). \end{aligned} \tag{4}$$

Complementary relations for triplets of three bases are (Table 2)

$$n(ijk) \sim n(\bar{k}\,\bar{j}\,\bar{i}) \tag{5}$$

where: $i, j, k \in \{A, C, G, T\}$, $\bar{A} = T$, $\bar{C} = G$, $\bar{T} = A$, $\bar{G} = C$, $(\bar{k}\,\bar{j}\,\bar{i})$ is anticodon.

Table 2. Number of triplets in chromosome 6 in human genome.

codon	number	codon	number	codon	number	codon	number
AAA	6742017	TTT	6744661	CAG	3216761	CTG	3217346
AAC	2509339	GTT	2507886	CCA	2932409	TGG	2932367
AAG	3412539	CTT	3407422	CCC	1980135	GGG	1986846
AAT	4419198	ATT	4420523	CCG	394680	CGG	396760
ACA	3417383	TGT	3417331	CGA	341096	TCG	340572
ACC	1872766	GGT	1869465	CGC	345302	GCG	346653
ACG	391422	GGT	390169	CTA	2226977	TAG	2227635
ACT	2735979	AGT	2734072	CTC	2680818	GAG	2686241
AGA	3741389	TCT	3735896	GAA	3394901	TTC	3388807
AGC	2242727	GCT	2239440	GAC	1533503	GTC	1532047
AGG	2824985	CCT	2821248	GCA	2330699	TGC	2327157
ATA	3684661	TAT	3682369	GCC	1793026	GGC	1794632
ATC	2260505	GAT	2265164	GGA	2490014	TCC	2482545
ATG	3129388	CAT	3128346	GTA	1962626	TAC	1966011
CAA	3229842	TTG	3228944	TAA	3716329	TTA	3718080
CAC	2408697	GTG	2408478	TCA	3303155	TGA	3307301

3. Complementary Property of Sequences Consisting of Identical Letters (Bases)

Table 3 shows data about the number of sequences, consisting of identical letters A, C, G, T in the chromosome 2 of human genome. For instance, A20 is the sequence of 20 identical letters A; T10 is the sequence of 10 identical letters T; correspondingly n (A20) is the number of such sequences occurring in chromosome 2, etc. We can draw a conclusion that the following relationships hold

$$n(A...A) \sim n(T...T),\ n(C...C) \sim n(G...G)\,. \tag{6}$$

Table 3. Human genom Chromosome 2.

Number n	A	T	C	G
1	32 885 475	32 885 555	26 877 090	26 900 089
2	8 802 666	8 823 505	6 695 063	6 700 669
3	3 452 571	3 465 217	1 730 704	1 730 727
4	1 330 971	1 335 874	416 239	417 181
5	502 189	505 290	96 319	96 463
10	8 179	8 255	131	131
15	2 525	2 604	8	10

18	1 389	1 434	3	1
20	1 044	1 022	1	1
24	635	608	1	1
40	18	15	0	0
50	3	3	0	0
62	1	1	0	0

These relationships were confirmed for the rest chromosomes of human genome. Note that the number of different variants of sequences consisting of 20 letters is $4^{20} = 2^{40} \sim 10^{12}$, and for 50 letters is $4^{50} = 2^{100} \sim 10^{30}$. It is clear, that probability of accidental revealing that relationships (6) hold is infinitesimal small.

The obtained results will significantly improve our present view of encoding genetic information. On the basis of obtained statistical data one can draw a conclusion that there exist strict rules of forming DNA structure, valid for all species. In our opinion, this result can contribute to the theory of evolution.

The genome is a dynamic structure, continually subjected to modification by the forces of evolution. The genomic variation seen in humans represents only a small glimpse through the larger window of evolution, where hundreds of millions of years of trial-and-error efforts have created today's biosphere of animal, plant and microbial species. A complete elucidation of genome function requires a parallel understanding of the sequence differences across species and the fundamental processes that have sculpted their genomes into the modern-day forms.

In the framework of abovementioned problems, we performed comparative analysis of human genome with other organisms (approximately 70 genomes): chimpanzee, mouse, rat, chicken, Tetraodon, C.elegans, Drosophila melanogaster, Arabidopsis, bacterium, yeasts for identification of structural elements of genomes.

References

1. The International Human Genome Sequencing Consortium. Initial sequencing and analysis of the human genome. (2001) *Nature* 409, 860-921.
2. Collins F.S. et al., (2003) A Vision on the Future genomics Research. *Nature* 422, 835-847.

3. The International Human Genome Sequencing Consortium. Finishing the euchromatic sequence of the human genome (2004) *Nature* 431, 931-945.
4. Sergienko I.V., Gupal A.M., Vagis A.A. (2005) Complementary principles in encoding bases by one chain in DNA chromosomes. *J. Autom. Inform. Scien.* 4, 153-157.
5. Gupal A.M., Vagis A.A. (2005) Complementary principles of bases in DNA chromosomes. *J. Autom. Inform. Scien.* 5, 153-157.

The Origin of Life on the Earth: As a Natural Bioreactor Might Arise

Mark Nussinov and Veniamin Maron

Janush Korchak 11/41, Kiriat Norday, 42495 Nethanya. Israel, mark80@bezeqint.net
Moscow Academy of Oil and Gas, Leninskiy prospect 65, Moscow, Russia maronvi@gubkin.ru

On the base of evolution characteristics a model of self-organization of primitive nanosize biological structures on the early Earth was worked out. Evolution characteristics testify to primary of nucleic acids and secondary of polypeptides. This model describes a scenario of appearance of the life phase before a cell on thin clayey kernels of regolith of the early Earth near 4.5 billions years ago. At the same time first liquid water has arisen on surface of the Earth. The appearance of ribonucleoproteinic virus complexes is connected with their synthesis in interlayer spaces of clayey kernels of regolith (here a natural bioreactor worked probably). Similar clays were detected in meteorites (Al'ende and others). Nitrous bases of nucleic acids, aminoacids, glucoses, phosphates and other compounds were in a water solution surrounded a kernel of regolith. In the result of absorption of these compounds with next polymerization in gaps between layers of a clayey kernel a RNA molecule together with a D-ribosa was firstly synthesized by abiotic way. The RNA synthesis in similar conditions was recently demonstrated in a laboratory (New Scientists, N 10, 2003). Subsequently RNA molecules selected L-aminoacids that were in the stereochemical correlation with D-ribosa and condensed them in short polypeptide chains. Thus, homohelicity of modern biopolymers has arisen. Our researches are shown that appearance of the biot on the Earth was a result of symbiosis of living and lifeless nature. Probably the same symbiosis is going on now.

V. Burdyuzha (ed.), The Future of Life and the Future of Our Civilization, 63–64.

References

M. Nussinov, A. Vekhov "The formation of regoliths on the early Earth" 1978, Nature 275, 19.

M. Nussinov, V. Maron, S. Santoli "Self-organization of the Universe and Life", 1999, Verba-Press, Jerusalem, Israel.

Volcanoes and Life: Life Arises Everywhere Volcanoes Appear

Oleksandr S. Potashko

Marine Geology and Sedimetory Ore Formation Section of National Academy of Sciences of Ukraine, Kiev, Ukraine, potashko@mail.ru

A volcanic activity was discovered at the summer of 2005 on Enceladus (a Saturn's satellite). Earlier the activity of volcanoes on Io was detected. On the Earth volcanism is a usual phenomenon. Each planet comes a period of crust formation and a thickness of this crust was growing during geological evolution. The volcanic activity begins when an intensity of a crust to be thrown down. In the initial stage of any planet evolution a furious volcanism occurs. Probably life may be produced in vents of undersea volcanoes in their chimneys. This bioniche may include a biodiversity from microbe mats to crabs. A typical life layer is 2-3 cm. The chimney of an underwater volcano in Black sea was covered by "soft tissue". Underwater volcanoes "smokers" may exist several hundred thousand years. In this period of time the common principles of life formation work.

Besides, we connect a volcanic activity with climatic cycles. Nuclear winter and nuclear summer (even spring) could take place on our planet. We may reconstruct a life crisis scenario based on ice core data from Antarctica and Greenland. Life exists as long as there is a mechanism of its formation of simplest forms of inorganic substances. Life changes up to now. Now we have new forms of viruses, microorganisms and so on. It will be always. A volcanic criterion for search of life beyond Solar system may be useful also.

V. Burdyuzha (ed.), The Future of Life and the Future of Our Civilization, 65–66.

References

O. Potashko "Volcanoes activity as transformers from mineral to organic life" Proceedings III European Workshop on ExoAstrobiology Mars: Search for Life, Madrid, Spain, (ESA SP-545, 2004), ESA p. 171-172.

L. Mukhin "Evolution of Organic Compounds in Volcanic Regions" 1974, Nature 251, 50-51.

Lessons of Life

Christian de Duve

Christian de Duve Institute of Cellular Pathology, ICP 75.50, Avenue Hippocrate 75, B-1200 Brussels, Belgium, deduve@icp.ucl.ac.be

The origin and evolution of life are ruled by chemical determinism and a frequently optimizing process of natural selection. As a result, there is more necessity and less chance than is often stated.

The chemical evolution leading to cellular life on Earth almost four billion years ago likely passed through a stage where RNA alone performed all of the functions of the modern macromolecules RNA, DNA and protein. However the so-called RNA world was itself too complex to evolve directly from organic molecules found on the prebiotic Earth. More likely, the RNA world emerged from and was supported by a primitive sort of metabolism fueled by the bonds in sulfur-containing compounds called thioesters.

V. Burdyuzha (ed.), The Future of Life and the Future of Our Civilization, 67.

Human Evolution: Retrodictions and Predictions

David R. Begun

Department of Anthropology, University of Toronto, Toronto, Ontario, M5S 3G3, Canada, bedun@chass.utoronto.ca

Human Evolution is a contentious topic. Understanding the human evolutionary past is complex enough; predicting the future of human evolution is nearly impossible. However, we can reconstruct events that led to the evolution of characteristics that have contributed to our success, and may hasten our extinction.

1. Introduction

Primates, including humans, are vision dominant mammals, largely arboreal, with unusually high levels of limb and dietary flexibility. Among primates, hominoids are the evolutionary group that includes humans and apes. All hominoids except humans practice a form of arboreal locomotion in which the body is positioned below rather than on top of a branch (suspension), and their brains are as large, or larger, than in any other primate. In hominids, the group that includes the great apes (orangutans, gorillas, chimpanzees and bonobos) and humans, there is a general slowing or delay in life history (lifespan, age at menarche, age of dental eruption and skeletal maturation, etc.), and a dramatic increase in brain and body size. All hominids are capable of extracting embedded, concealed or otherwise protected resources from the environment with a level of efficiency not generally seen in other primates, and all engage in intensive, prolonged, and complex forms of social interaction. Humans, of course, take all of these attributes to the extreme. If great apes are the gifted members of the primate community, humans are the super geniuses. We excel in information acquisition, processing and retrieval, and we are distinct from

V. Burdyuzha (ed.), The Future of Life and the Future of Our Civilization, 69–81.

other hominids in being bipedal, permitting the development of greatly enhanced manipulative capabilities of the hands.

All of these attributes make humans the most impressive, and most dangerous, animal on the planet. In this chapter I will outline the major events in the evolutionary history of the primates that help explain the origins of the most important attributes that make us human. I will then consider the future course of human evolution, including the inevitability of extinction.

2. The Primate Fossil Record and the Origin of the Hominoidea

2.1 Early Primates

There is some debate about the age of the earliest primates, so I will begin my survey with the earliest easily recognizable primates, the adapiformes and the tarsiiformes. Adapiformes are primates mainly of Eocene age (about 56 to 35 million years ago, or Ma), although a few taxa persist into the late Miocene (roughly 10 Ma) [1]. Although very diverse in morphology and adaptation, they generally resemble living strepsirhines (lemurs, lorises and their kin). Tarsiiformes are also mainly Eocene, with a few taxa extending the range to about 20 Ma. Again, while highly diverse, tarsiiformes resemble living tarsiers, which belong to the haplorhines, to which monkeys, apes and humans belong as well [2].

Adapiformes are among the earliest primates to show characteristic features of living primates. Adapiformes had skulls with reduced snouts and eyes facing forward, reflecting the increasing importance of vision over the sense of smell. Their brains were small by modern primate standards but larger than most other mammals of similar size living at the same time. Their skeletons provide evidence of a somewhat mobile and powerfully grasping limb structure, and a long and flexible vertebral column.

Adapiformes were probably fairly agile in the trees, and had begun to develop enhanced eye-hand coordination, perhaps to improve their mobility in the challenging arboreal environment, or perhaps to enhance their ability to use the hands to acquire food items. Most adapiformes were fruit or leaf eaters.

Table1. Classification of the taxa described in this chapter.

- Primates
  - Strepsirhines
    - Adapiformes
    - Lemuriformes (lemurs and kin)
    - Lorisiformes (lorises and kin)
  - Haplorhines
    - Tarsiiformes (Tarsius)
    - Anthropoidea (monkeys, apes and humans)
      - Platyrrhini (New World monkeys)
      - Catarrhini (Old World monkeys, apes and humans)
        - Cercopithecoidea (Old World monkeys)
        - Hominoidea (apes and humans)
          - Hylobatidae (gibbons and siamangs)
          - Hominidae (great apes and humans)
            - Homininae (African apes and humans)
            - Ponginae (orangutan)

Tarsiiformes are also modern primate-like, but were generally smaller than adapiformes. Unlike most adapiformes, many tarsiiformes were insect or small vertebrate eaters, and may have used their large eyes and enhanced eye-hand coordination to capture prey.

Both of these groups of early primates appear to have dominated the arboreal environment at many localities. The features that define them as primates are the precursors to many uniquely human attributes, such as our large brains, highly manipulative hands, and mobile limbs. Paleoprimatologists disagree on which of these early primate groups is most closely related to anthropoids (New and Old World monkeys, apes and humans) [3-4]. There were many more events in the evolutionary history of the primates that led, by chance of course, to the origin of higher primates including humans.

2.2 Early Anthropoids

The oldest anthropoids are thought to date to the beginning of the middle Eocene, about 50 Ma [3]. The earliest anthropoids are recognized mainly by their dentition, which resembles living anthropoids in that the molar teeth are broader and have lower, more rounded cusps than in tarsiers or most strepsirhines. While the earliest anthropoids are more tarsier-like, eventually the mandibles would become more strongly built and fuse in the midline. These features distinguish modern anthropoids from other primates.

From the earliest anthropoids the two major groups of living anthropoids, the Platyrrhini or New World monkeys, and the Catarrhini or Old World monkeys, apes, and humans, would emerge. The origins of the New World monkeys are somewhat mysterious, as they live today only in Central and South America, which were not connected to North America at the time primates first appear in South America. But it is the origins of the catarrhines that are of special interest here. One of the best collections of fossil anthropoids comes from the famous Fayum deposits of Egypt [5]. Among these fossils are some of the earliest catarrhines.

2.3 Early Catarrhines

The most advanced of the Fayum anthropoids is the early catarrhine Aegyptopithecus, which had powerful jaws and low, rounded molars, a brain larger than would be typical in a strepsirhine and some further reduction in the snout. Aegyptopithecus is still quite primitive, and in many ways is intermediate between strepsirhines and more advanced catarrhines. Its limb structure, for example, while flexible as in living strepsirhines, lacks the increased mobility and overall elongation (especially the forelimb) present in more modern catarrhines. Over time, later early catarrhines, such as Pliopithecus from Europe, would develop more modern catarrhine characteristics, including a short snout, larger brain and long, highly mobile limbs [6].

2.4 Early Hominoids

Hominoids first appear in the fossil record in the early Miocene, at about 20 Ma, though a few specimens may date back to the Oligocene, about 26 Ma [7]. The best known of the protohominoids is Proconsul, present at many sites in Kenya. Proconsul has all the attributes of modern catarrhines, but only a few hominoid characteristics, including the absence of a tail (or the presence of a coccyx), and subtle indications of enhanced limb mobility [8-12]. Proconsul had powerful, grasping hands and feet, and some indications of encephalization (brain size increase) compared to most monkeys [13-16]. Proconsul probably ate soft, ripe fruits and moved through the environment as do living monkeys, on the top of branches. A protohominoid similar to Proconsul probably moved into Eurasia about 17 Ma, where the first specimens more closely resembling modern hominids appear. Several taxa are known with more powerful jaws, large teeth with thick enamel, but limbs still largely like those of Proconsul. The jaws and teeth may have permitted these new species to exploit a broader range of dietary resources [6].

3. The Evolution of Hominids

3.1 Early Hominids

By about 13 Ma hominoids appear with the hallmark of modern hominoids, highly mobile forelimbs capable of supporting body mass below branches. This, combined with the ability to exploit a wider range of resources that comes from having more powerful jaws, leads to the earliest hominids. The best known of the early hominids are Dryopithecus from Europe and Sivapithecus from South Asia. Dryopithecus is probably closely related to African apes and humans, while Sivapithecus is likely to be a close relative of the orangutan [17-18]. Both share many attributed with living great apes, including large bodies and brains, slower life histories, elongated faces with large front teeth and many other detailed resemblances to orangutans in the case of Sivapithecus and to African apes in the case of Dryopithecus [16 - 19]. The postcranial skeleton is better known in Dryopithecus, and includes typical hominoid features such as long arms and short legs, short, stiff backs, broad thoraxes, extremely mobile shoulders and elbows, and very powerfully grasping hands and feet [9].

The success of Dryopithecus, Sivapithecus and other late Miocene hominids was probably due at least in part to their enhanced cognitive abilities and flexible adaptations. During the late Miocene the Earth's climate was becoming more variable and unpredictable [20]. While most other mammals became increasingly specialized to exploit ever changing niches, hominids developed flexible strategies to confront ecological changes. With larger brains come more complex forms of behavior, including complex feeding techniques, more complex social interactions and enhanced communicative abilities. As these capacities become better developed, selection acts to increase their efficiency and effectiveness, leading to feedback loops or arms races resulting from competition among members of the population (figure 1) [21-24]. These developments set the stage for the immense increase in brain size that comes with the origin of the genus Homo.

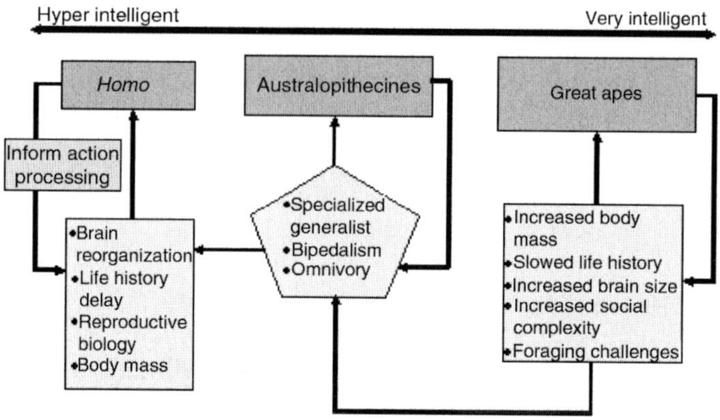

Fig. 1. Feedback loops among a variety of factors that contribute to the evolution of higher intelligence.

3.2 Early Humans

Dryopithecus, or a closely related species, is probably an ancestral to African apes and humans. While there is almost no fossil record of the African apes, many human taxa appear in the fossil record before the appearance of Homo at about 2.5 Ma. The best known of these is Australopithecus. Australopithecines are chimpanzee sized fossil humans with brains only slightly larger than those of chimpanzees, and possible evidence of some degree of cortical reorganization [16, 25 - 29]. There is no direct evidence of tool use in australopithecines, though most presume that they had, at least, the tool using capacities of living chimpanzees. Chimpanzees use stones as tools and make tools from perishable materials that would not preserve in the fossil record [30 - 32].

The most important differences between chimpanzees and australopithecines are that australopithecines are bipedal and their jaws and teeth are massive and designed to exploit very hard and/or tough food items. Bipedalism, along with a persisting ability to use arboreal resources, allowed australopithecines to travel greater distances between forest patches, increasing their potential daily ranges. A massive masticatory apparatus probably allowed australopithecines to broaden their resource base, even beyond that of contemporaneous great apes.

3.3 Homo

Within the successful radiation of australopithecines there is the ancestor of the genus Homo, although there is little agreement on the identity of this ancestor. The earliest specimens of Homo are known from Ethiopia, Kenya and Malawi and are about 2.5 Ma, which is also the age of the earliest identifiable stone tools [33 - 37].

The earliest specimens of Homo are fragmentary, but about 2 Ma more complete crania are known. These show that early Homo (Homo habilis and Homo rudolfensis) had significantly larger brains than australopithecines, and clearer evidence of reorganization. The later includes more marked cerebral asymmetries such as the development of a "Broca's cap", a bulge on the left frontal lobe in a region involved in spoken language in modern humans [38 - 39]. We can not know if enlargement in this region in early Homo indicates spoken language competence or some other capacity that preceded language. Broca's cap is adjacent on the cerebral cortex to the motor cortex region controlling movements of the arm and hand. It is possible that enlargement in this region is related initially to increasing manual dexterity or grip variety and strength, and was only secondarily co-opted for language production [40]. The simultaneous appearance of stone tools, Broca's cap and early Homo is probably not coincidental.

3.4 Later Homo

The evolution of Homo is complex and includes many species over time, but for the purposes of this discussion it can be summarized by a few major trends. Homo erectus and Homo ergaster appear in Asia and Africa, respectively, between about 1.6 to 1.8 Ma, with larger brain and body masses than early Homo [40 - 42]. By about 0.9 Ma Homo erectus ranged from Europe to China in the north and from the Mediterranean to South Africa in the south. By about 0.8 Ma new species appear, the best known of which is Homo heidelbergensis. Neandertals first appear about 0.25 Ma and Homo sapiens sapiens by at least 0.13 Ma and possibly as early as 0.2 Ma [42 - 45].

During the transition from early Homo to modern humans, the main trends include fairly steady increases in brain size, reduction in the size of the jaws and teeth, geographic variability in body size, and, with modern humans, a general decrease in skeletal robustness.

3.5 The Biology/Technology Transition

The rate of increase in brain size begins to slow by about 0.5 Ma (figure 2). At the same time, thousands of localities are known around the world with millions of paleolithic artifacts. The rate of change in the complexity of these technologies, and in other aspects of the culture of prehistoric humans (settlement patterns, land use, hunting strategies, geographic distribution, etc.), is very different (figure 2).

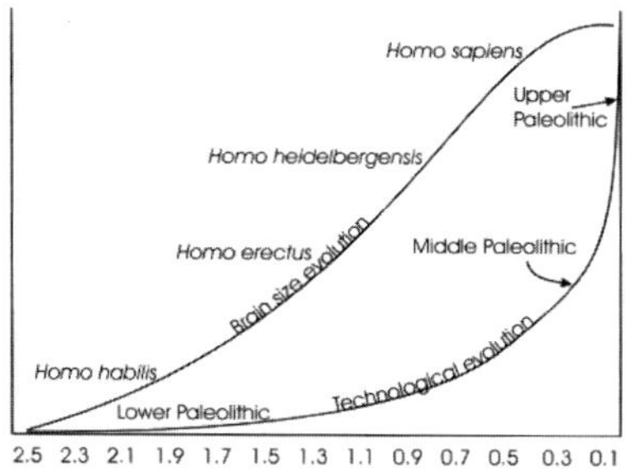

Fig. 2. Comparison of rates of change in biological and technological evolution. X axis units are millions of years.

Early stone tool technologies remain relatively unchanged for about 1 Ma. By about 1 Ma the pace of technological evolution accelerates. This rate of acceleration increases over time, reaching dramatic rates of change after about 0.2 Ma. The lower paleolithic time period, associated with Homo habilis, Homo erectus and their contemporaries, lasts about 2 Ma. The middle paleolithic, associated with the Neandertals, lasts about 0.2 Ma. The upper paleolithic, associated with modern humans, lasts about 0.035 [46]. In other words, each of the major archeological periods associated with prehistoric humans is about one order of magnitude shorter than the preceding period. Since the end of the paleolithic and the beginning of the agricultural revolution, about 12,000 years ago, the pace of technological change increased to unimaginable rates in comparison to the rate of cultural evolution during the period in which our brains evolved. It could be argued that the pace of technological change is reaching a point beyond our biological capacities to manage it.

3.6 Genetics of Human Evolution

It has taken about 7 Ma for humans and chimpanzees to accumulate about 1 to 2% genetic divergence, based on most measures, although there are between 20 and 40 million base pair differences between the two species [47]. Of these, it has recently been estimated that only about 70,000 have resulted in adaptive changes in protein coding sequence regions [47]. A few specific differences related to brain evolution have recently been identified. A mutation in a gene affecting myosin expression is thought by one group of researchers to have caused an overall reduction in the size and power of the jaws and muscles of mastication in humans, which they correlate to increases in brain size [48]. A number of other promising discoveries of genetic differences between apes and humans that may be correlated to important differences between the two have also been documented [49].

4. The Future "Evolution" of Humans

4.1 Beyond Biological Evolution?

It is often asserted that humans have moved beyond biological evolution, given the buffering effects of technology. Technology protects humans from the rigors of the natural environment, improves our minor imperfections, and permits individuals who would have died in the pre-industrial era to live productive lives. It has even been suggested that technology works against human evolution by allowing mutations that would lead to reduced fitness to be maintained in the population. This naïve view fails to consider the continuous effects of mutation and the fact that even the most deleterious alleles are maintained at low levels in all populations by simple Mendelian processes.

It is difficult to quantify the rate of evolution in current populations of humans. The fossil record provides ample evidence that humans, even with sophisticated technology, were still responding to environmental stresses with evolutionary innovations. Neandertals cared for their infirm and used technology to protect themselves from the elements, but they clearly changed over time [43, 46]. Homo floresiensis, the recently discovered diminutive fossil human from Indonesia, underwent dramatic morphological changes from its putative Homo erectus ancestor, despite a sophisticated technology. These changes, including a marked decrease in body size and a spectacular reduction in brain size, mirror those documented in classic cases of island biogeography, and show that humans respond to isolation in a manner similar to other mammals. Since the agricultural revolution, humans have become much less robust skeletally, and there is a trend toward reduction or loss of the last molar. But the time scale on which evolution operates makes it difficult to document ongoing biological evolution in humans.

Some current activities could conceivably lead to a reduction in genetic diversity within modern humans, precipitating either a speciation event, or more likely, a catastrophic event and possible extinction. Genetic screening during pregnancy has the potential to reduce genetic diversity by artificial selection of desirable characteristics in offspring. Dramatic declines in genetic diversity could leave humanity susceptible to pandemic outbreaks that could lead to extinction. Currently, screening consists largely of detecting genes or gene by-products that indicate the presence of genetic disorders. It is possible in the future that screening will evolve to the point where parents might have the ability to "customize" their offspring, allowing only those with a specific desired set of features (sex, height, weight, skin tone, IQ, etc.) continue in utero. While this may lead to a reduction in genetic diversity over time, mutation and recombination will continue to produce diversity, and recessive genes will persist in all populations. It seems unlikely to me that the number of genes that could be screened in the future will be significant relative to the total number of genes in the human genome.

Cloning, the ultimate in gene diversity reducing technology, if it becomes widespread, would place humanity in a precarious position. Even if human cloning becomes feasible and common, despite technical and ethical issues, it again seems unlikely to me that this form of reproductive technology would become any more widespread across all human populations than are current widely available technologies.

5. The One Confident Prediction about Human Evolution

All species eventually become extinct, at least among multicellular organisms. There are really two forms of extinction. Populations of a species may be isolated from other populations of the same species and, over time, evolve into a new species. The old species may persist for a time along with the new species. Or, a species may disappear without descendents. Either way, once the old species no longer exists, even if there is genetic continuity between it and the new species, it is considered to have become extinct.

Humans will surely follow one of these paths. It would be arrogant, and dangerous, to think otherwise. Ironically, the idea that we humans can think ourselves out of extinction makes it tempting to continue the policies and behaviors that demonstrate a callous disregard for the sustainable future of life on earth, and may precipitate our extinction. We can act to maximize the likelihood of the first option of extinction, that which leaves descendents. But we will become extinct either way. This is a natural phenomenon, and, as Shakespeare suggests, nothing to fear:

"Be cheerful, sir. Our revels now are ended. These our actors, as I foretold you, were all spirits and are melted into air, into thin air; And, like the baseless fabric of this vision, the cloud-capp'd towers, the gorgeous palaces, the solemn temples, the great globe itself, yea, all which it inherit, shall dissolve, and, like this insubstantial pageant faded, leave not a wrack behind. We are such stuff as dreams are made on, and our little life is rounded with a sleep."

Shakespeare, The Tempest.

Acknowledgments

I am grateful to Vladimir Burdyuzha and Claudius Gros for inviting me to participate in the Frankfurt symposium "The Future of Life and of Our Civilization - Predictions and Modeling" in May, 2005. This work was supported by funds from the Volkswagen Foundation and NSERC.

References

1. D. L. Gebo, in The Primate Fossil Record. W. Hartwig, Ed. (Cambridge University Press., Cambridge, 2002) pp. 21-43.
2. G. F. Gunnell, K. D. Rose, in The Primate Fossil Record. W. Hartwig, Ed. (Cambridge University Press., Cambridge, 2002) pp. 45-82.
3. K. C. Beard, in The Primate Fossil Record. W. Hartwig, Ed. (Cambridge University Press., Cambridge, 2002) pp. 133-149.
4. M. Dagosto, in The Primate Fossil Record. W. Hartwig, Ed. (Cambridge University Press., Cambridge, 2002) pp. 125-132.
5. D. T. Rasmussen, in The Primate Fossil Record. W. Hartwig, Ed. (Cambridge University Press., Cambridge, 2002) pp. 203-220.
6. D. R. Begun, in The Primate Fossil Record. W. Hartwig, Ed. (Cambridge University Press., Cambridge, 2002) pp. 221-240.
7. T. Harrison, in The Primate Fossil Record W. C. Hartwig, Ed. (Cambridge Univ. Press, Cambridge, 2002) pp. 311-338.
8. K. C. Beard, M. F. Teaford, A. Walker, Folia. Primatol. 47, 97-118 (1986).
9. D. R. Begun, Scientific American 289, 74-83. (2003).

10. M. D. Rose, in Function, Phylogeny and Fossils: Miocene Hominoid Evolution and Adaptations. D. R. Begun, C. V. Ward, M. D. Rose, Eds. (Plenum Publishing Co., New York, 1997) pp. 79-100.
11. A. Walker, in Function, Phylogeny and Fossils: Miocene Hominoid Evolution and Adaptations. D. R. Begun, C. V. Ward, M. D. Rose, Eds. (Plenum Publishing Co., New York, 1997) pp. 209-224.
12. C. V. Ward, in Function, Phylogeny and Fossils: Miocene Hominoid Evolution and Adaptations. D. R. Begun, C. V. Ward, M. D. Rose, Eds. (Plenum Publishing Co., New York, 1997) pp. 101-130.
13. D. Falk, in New Interpretations of Ape and Human Ancestry R. L. Ciochon, R. S. Corruccini, Eds. (Plenum Press, New York, 1983) pp. 239-248.
14. A. C. Walker, D. Falk, R. Smith, M. F. Pickford, Nature 305, 525-527 (1983).
15. D. R. Begun, M. F. Teaford, A. Walker, J. Hum. Evol. 26, 89-165 (1994).
16. D. R. Begun, L. Kordos, in The Evolution of Thought: Evolutionary Origins of Great Ape Intelligence A. E. Russon, D. R. Begun, Eds. (Cambridge University Press, Cambridge, 2004) pp. 260-279.
17. J. Kelley, in The Primate Fossil Record. W. Hartwig, Ed. (Cambridge University Press., Cambridge, 2002) pp. 369-384.
18. D. R. Begun, in The Primate Fossil Record. W. Hartwig, Ed. (Cambridge University Press., Cambridge, 2002) pp. 339-368.
19. J. Kelley, in The Evolution of Thought: Evolutionary Origins of Great Ape Intelligence A. E. Russon, D. R. Begun, Eds. (Cambridge University Press, Cambridge, 2004) pp. 280-297.
20. R. Potts, in The Evolution of Thought: Evolutionary Origins of Great Ape Intelligence A. E. Russon, D. R. Begun, Eds. (Cambridge University Press, Cambridge, 2004) pp. 237-259.
21. G. Yamakoshi, in The Evolution of Thought: Evolutionary Origins of Great Ape Intelligence A. E. Russon, D. R. Begun, Eds. (Cambridge University Press, Cambridge, 2004) pp. 140-171.
22. C. V. Ward, M. Flinn, D. R. Begun, in The Evolution of Thought: Evolutionary Origins of Great Ape Intelligence A. E. Russon, D. R. Begun, Eds. (Cambridge University Press, Cambridge, 2004) pp. 335-349.
23. A. E. Russon, D. R. Begun, in The Evolution of Thought: Evolutionary Origins of Great Ape Intelligence. A. E. Russon, D. R. Begun, Eds. (Cambridge University Press, Cambridge, 2004) pp. 353-368.
24. S. T. Parker, in The Evolution of Thought: Evolutionary Origins of Great Ape Intelligence A. E. Russon, D. R. Begun, Eds. (Cambridge University Press, Cambridge, 2004) pp. 45-60.
25. J. Kappelman, Journal of Human Evolution 30, 243-276 (1996).
26. R. L. Holloway, in Origins of the Human Brain J.-P. Changeux, J. Chavaillon, Eds. (Clarendon Press, Oxford, 1995) pp. 42-60.
27. H. M. McHenry, in The Evolution of the "Robust" Australopithecines F. E. Grine, Ed. (Aldine de Gruyter, New York, 1988) pp. 133-148.
28. H. M. McHenry, Am. J. Phys. Anthropol. 87, 407-430 (1992).

29. P. V. Tobias, in Origins of the Human Brain J.-P. Changeux, J. Chavaillon, Eds. (Clarendon Press, Oxford, 1995) pp. 61-83.
30. C. Boesch, M. Tomasello, Curr. Anthropol. 39, 591-614 (1998).
31. A. Kortlandt, E. Holzhaus, Primates 28, 473-496 (1987).
32. J. Mercader, M. Panger, C. Boesch, Science 296, 1452-1455 (2002).
33. A. Hill, S. Ward, A. Deino, G. Curtis, R. Drake, Nature 355, 719-722 (1992).
34. J. W. K. Harris, G. L. Isaac, Z. M. Kaufulu, in Koobi Fora. Research project. Plio-Pleistocene Archaeology G. L. Isaac, B. Isaac, Eds. (Clarendon Press, Oxford, 1997), vol. 5, pp. 115-223.
35. W. H. Kimbel, et al., J. Hum. Evol. 31, 549-561 (1996).
36. H. Roche, et al., Nature 399, 57-60 (1999).
37. F. Schrenk, T. G. Bromage, C. G. Betzler, U. Ring, Y. M. Juwayeyi, Nature 365, 833-836 (1993).
38. P. V. Tobias, The Brain in Hominid Evolution (Columbia University Press, New York, 1971).
39. P. V. Tobias, Olduvai Gorge. The Skulls, Endocasts and Teeth of Homo Habilis. (Press Syndicate of the Uniiversity of Cambridge, Cambridge, 1991), vol. IV.
40. D. R. Begun, A. Walker, in The Nariokotome Homo erectus Skeleton A. Walker, R. Leakey, Eds. (Harvard University Press, Cambridge, MA, 1993) pp. 326-358.
41. C. B. Ruff, A. Walker, in The Nariokotome Homo erectus Skeleton A. Walker, R. Leakey, Eds. (Harvard University Press, Cambridge, MA, 1993) pp. 234-265.
42. H. Dunsworth, A. Walker, in The Primate Fossil Record. W. Hartwig, Ed. (Cambridge University Press., Cambridge, 2002) pp. 419-435.
43. F. H. Smith, in The Primate Fossil Record. W. Hartwig, Ed. (Cambridge University Press., Cambridge, 2002) pp. 437-456.
44. I. McDougall, F. H. Brown, J. G. Fleagle, Nature 433, 733-736 (2005).
45. T. D. White, et al., Nature 423, 742-747 (2003).
46. R. G. Klein, The Human Career: Human Biological and Cultural Origins. (University of Chicago Press., Chicago, ed. 2., 1999).
47. S. B. Carroll, Nature 422, 849-857 (2003).
48. H. H. Stedman, Kozyak, B. W., Nelson, A., Thesier D. M., Su, L. T., Low, D. W., Bridges, C. R., Sharger, J. B., Minugh-Purvis, N., Mitchell, M. A., Nature 428, 415-418 (2004).
49. W. Enard, et al., Science 296, 340-343 (2002).

Man's Place in Nature: Human and Chimpanzee Behavior are Compared

Toshisada Nishida

The zoological department of Kyoto University, Kitashirakawa-Oiwakecho Sakyo, Kyoto, Kyoto 606-8502, Japan, nishida@jinrui.zool.kyoto-u.ac.jp

Chimpanzees are the closest human relatives. By studying them in the natural habitat and finding out similarities in morphology, behavior and psychology, we can know the biological background of humans which specify the limitation of future human life and social institutions.

V. Burdyuzha (ed.), The Future of Life and the Future of Our Civilization, 83.

Creative Processes in Natural and Artificial Systems (Third Signal System of Man)

Abraham Goldberg

Pinchas Lavon 9/63, Kiriat Nordau, 42701 Netanya, Israel, abhrzhv@gmail.com

1. Introduction

All the inanimate and living nature from viruses to a Man and his social formations, from elementary particles to galaxies is permanently changing, adapting, evolving *creating* new and new objects, forms and interrelations. They are inherent in matter *creative processes* (CP) – methods of its existence and evolution. CP isn't a philosophic conception but *physical reality* such as conservation laws, the laws of mechanics and thermodynamics, Relativity principles. This allows proposing creative methods of natural systems research and artificial systems building.

Any inanimate and living system *preserves* itself existing, surviving and developing under influences of surrounding world using its parameters, knowledge or structure. If they turn out to be insufficient then comes a crisis – "*pre-creative*" state *forcing* the system to change its parameters, knowledge and structure.

V. Burdyuzha (ed.), The Future of Life and the Future of Our Civilization, 85–94.

2. Creative Processes

CP is the process of *accidental changes* in a sphere of structure, physical and/or intellectual parameters of open *complex*[1] material System, which are *self-determining* (i.e. decreasing the sphere of changes) if a certain next change gives useful effect for System existence and/or evolution. CP is radically different from "habitual" feedback processes (and from Darwin's process of accidental changes and natural selection) because CP feedback acts not in changing itself but in the *sphere* of changing. This accelerates CP in geometrical progression *concentrating-directing* System adaptation and evolution. (Existing notions identify a creation only with human activity but it's impossible to substitute the term "creative process" for more habitual "self-organizing process of accidental search – by trial-and-error" because the latter doesn't reveal its purpose – the creation of a new and its methodology – by feedback in the sphere of accidental search.)

CP is self-*arising* and self-*developing* process. It doesn't need in any external creator or control. (System changes without any outside interference can be only accidental.) CP stops for a while (and its results are fastened) on reaching not the best but only the *first* result (solution) *sufficient* in its concrete surroundings and situation. (This accelerates adaptation in addition excluding tests of other its variants leading to "combinations explosions"[2]. "Unreasonable" nature "uses" only sufficient results. CP revives if the situation goes out of limits admissible for the System. Thus CP preserves up to pre-creative state System stability and ensures System changeability, "purposefulness" and "fine tuning" of its parameters and constants.

CP acts "according to circumstances", it doesn't have any tendency or direction set from outside. Just CP is "driving force", creator building up new Systems and destroying old ones – out of necessity that it determines itself, acting against dissipative processes. CP ensures the actual rapid na-

---

[1] The complex system (CS) is an object of any nature possessing "system properties", which any its part doesn't have by any method of division. It's qualitative difference from "simple" systems transforming CS into original "thing in itself" with own physical regularities, metrics, constants, uncertainty spheres, functional tendency with cause-effect freedom of choice, and incompletely predicted behavior. All real material systems are CS. Their presentation as "simple" systems is possible only in primitive or limited research levels.

[2] It's proved that a multitude of sufficient results is on an average just as good as best results multitude in system resources saving such as energy, volume, spectrum. A "wise" person looks for the best solution and pays this in combinations explosions, which he overcomes by his heuristics.

ture adaptation and evolution in spite of frequent cosmic and terrestrial catastrophes that is infeasible for only Darwin's process.

An algorithm of System changing can be reversible but its concrete realization is determined by irreversible CP. (Transformations of ice into water and back don't preserve the arrangement of ice crystals. Reversible equations of quantum theory determine probability densities of particles spatial-time arrangement accidental values but not the values themselves determining by irreversible CP. Travels in time ("time machines") are impossible in principle because our world evolves irreversibly.)

Changes of System structure – quantity-quality transformations (QQT) can take place in CP course, and they often have spasmodic development: instability – splash of accidental process appears in the System and then CP determination selects the one of possible attractors-results, which the development of new structure aspires to. It can be both more and less complicated than preceding one if will turn to be sufficient for new conditions. The System can come down in QQT process to its previous evolutionary level, and then develop in other way. (Such events are known with bacteria survived in extreme conditions. This can take place in critical stages of human societies, e.g. after Roman Empire destruction under barbarians onset.)

CP develops on the base of System preceding structure, parameters and knowledge and thus it has *genetic* ("hereditary") and *creative* components: Creative one is based on genetic component and then, in its turn, improves genetic one for changed conditions. CP accidental part turns CP into irreversible and unformulated (as a whole) process but it allows System to find solutions and acquire knowledge *don't logically* following from already available, i.e. *cognize*, work out *illogical judgments* (conclusions) and *hypotheses*. (The creation or cognition of a new is essentially illogical process: logic is always based on axioms, it can create and prove theorems but not hypotheses and axioms.).

It's essential for scientific and practical CP application that CP results are predicted incompletely but the pre-creative state can be both predictable and consciously organized. (This actually takes place in art of war, teaching methodic, sportsmen training, the creation of conflict situations, "shock therapy" in medicine, politics, economics, science, etc.) Well-known phenomena of physiological and psychological adaptation processes – so-called stresses are CP too.

Systems can interact with surroundings by means of:

A. One-way (from surroundings to System) energy-material connections, where energy-material CP can create local – in place and time – System self-organization (synergetic, resonance circuits, catalysts, waves, clouds);

B. Two-way energy-material connections, where CP can create local – in place only – self-organization (auto-generator, atom, molecule, planet system, star, galaxy);[3]

C. Two-way energy-material and *information* connections, where energy-material and *intellectual* CP together ensure illocal System self-organization and adaptation, cognition of surroundings, self-cognition (all natural and artificial living beings).

All the processes of matter changing including transition and establishment processes go by means of CP on nuclear-atomic and molecular levels, in the evolution of stars and galaxies, on the stages of living beings origin and evolution. System formation, specifically self-organization of connections between System elements takes place uniformly: the elements have some properties-abilities for connection (valence, electric charge, "color", gravitation, molecules and particles for exchange, signals, etc); connections accidental nature is determined in a certain (one of possible) result of CP connection. CP also "processes" accidental factors in evolution (and history) in accordance with the concrete situation, where the System exists.

Theories (models) created in the course of cognition usually have some boundaries of application. CP exist in the entire universe and determine all the stages of its evolution. That's why CP can be the base and instrument of research and even a forecasting.

3. Natural CP

Two simple examples explain CP action in inanimate matter: As the distance between atoms decreases their common electronic envelope arises unstable at the beginning and then stable – by sufficiently small distances. Nuclei repulsion prevents further atoms approach, and finally a new System – molecule is created. Possibly this takes place after several oscillations relatively stability boundaries; Cosmic objects interacting by gravitation are characterized by some sufficient limits of mutual distances and velocities, where a "self-creating" stable System arises.

Self-organized infinitely varied world can't be a result of an enormous exceptional accidental fluctuation (or a result of single creation act) but the permanent creation of vast number of interacting CP, which could have other results possibly existing in other universe formed after big bang.

[3] Systems A and B researches by I. Prigogine's school established the self-organization property inherent in matter (but didn't point out its creative mechanism).

What's more, the hypothesis of sole unique fluctuation is "unnatural" – it corresponds to time-limited infernal (hopeless) picture of matter development.

Oxygen-less life was already on the Earth three billion years ago because atmospheric oxygen was then almost absent. The life could develop so that quite different beings including intellectual ones would live on the Earth now. But one of CP results was then the appearance of plants, which in their turn formed atmospheric oxygen giving the origin of all the present-day oxygen fauna development.

Catalysis selects directions of changing in inanimate matter. It's supplemented in living matter with *fermentative* catalysis based on structurally molecular selection that partly turns into an exchange with *signals* both inside the System and with other Systems by means of switching elements "action (phenomenon) ß à signal" starting or registering signals. Ensuring this structures appear accidentally (with the probability depending on conditions in different parts of the universe) but then CP *quickly fastens* them concentrating development in this direction (that can be seemed as biological, even anthrop purposefulness). Living beings step by step *select* their accidental changes accelerating their evolution. There is a number of terrestrial and cosmic models of structures potentially suitable for Life origin but the transition from such suitability to reality is possible only by original CP work in terrestrial conditions.

The appearance of signals adds *intellectual* CP (ICP) to energy-material one and *marks* the appearance of living matter. The latter accelerates its adaptation and evolution by means of genetic mechanism (epigenetic regulation, RNA molecular system, suppression and "stammer" of genes action, etc), *functional* and *sex* dimorphism. Each of them includes vanguard (creative) part creating and approving changes and conservative (genetic) part fixing useful changes. ICP creates the new quality – *creative intellect* (CI) including *relations* to surrounding world (emotions, self-identification, studying with cognition and understanding, events prediction) and forming them *originally* using both inherited and *individually* accumulated experience. Four steps of CI hierarchical ascending ladder can be selected: "The simplest step" (ICP takes place only between surroundings and System, where internal control is absent); "Sub-consciousness" (ICP is inside System too, generalizations and consciously uncontrolled integral actions are possible); "Consciousness" (System can formulate its knowledge and exchange information with other Systems in a language in common, is capable to collective actions using signals, to cognition and understanding); "Super-consciousness" (the System acquires such individual knowledge, which it can't already pass to others by means of language and notions in common). The simplest step is widespread. Its examples are market price formation (without any understanding between sellers), elemental transport or messages movement in decentralized road or communication networks, the employment of limited common reserve for independent individuals.

A number of suppositions appeared lately about so-called "distributed brain" guiding the survival of associations (species) such as ants, lemmings, birds of passage, etc, and even about the existence of Supreme Mind ("Matrix") guiding mankind and being a source of various paranormal phenomena. But CI steps from the simplest one on early evolution stages (such as ant-hills) to Sub-consciousness and Consciousness (people) give the possibility to do well without Supreme Mind and distributed brain hypotheses. Any "distributed" brain is still centralized guidance but natural (and artificial) orchestra can play without any conductor: CI steps ensure such distributed *decentralized* guidance by means of surroundings itself (on the simplest step) and several signal systems on higher steps[4].

It's of vital importance for life and science that more complex world with the greater number of dimensions (e.g. the universe) can be cognized by the being with lesser dimensions number (e.g. by a Man or an intellectual machine) on the base of two conditions: only *indirect* cognition is pos-

[4] Living beings knowledge are directly passed to others by means of the First and the Second signal systems in all mediums excepting electromagnetic and gravitation fields. There are however non-observed authentically-repeatedly paranormal phenomena going out of these signal systems limits. It is hypothetical Third signal system (TSS) acting through electromagnetic bio-field: Electrical pulses move in brain and nervous system forming spatially distributed electromagnetic waves. Their combinations are complex and individual but they can be weakly different by similar sense for individuals with similar nervous systems parameters. This creates on principle telepathy possibilities. The "fine tuning" of neurons connections structure fulfilled by natural ICP can be the main means of TSS transmission and reception.

The bio-field exists some time after its source destruction that can on principle explain paranormal phenomena including "déjà vu", i.e. individual's intellectual imprints can be preserved and perceived by others for some time. TSS is apparently also manifested, when extreme actions of a Man or animals are precluded with stresses in their nervous systems – bio-field splashes influencing objects of actions, when they still didn't perceive them by their First or Second signal systems. TSS apparently acts stronger by decrease of consciousness withstanding role. The tendency of living beings evolution is further increase of TSS relative role between people and between a Man and intellectual machine substituting a part of Second signal system actions. Lately TSS existence was corroborated experimentally by the possibility of brain parts magnetic stimulation (TMS). But magnetic field can influence only electromagnetic field in any organism that hence exists there and takes part in the work of brain.

sible – according to changes in dimensions of cognizing System; the cognition must be *creative* – by hypotheses creation and verification. Creative understanding is the creation of *own standards* of images and ideas. Not only information quantity but also its *sense* is essential for intelligent being. Individual CI, his (its) intellectual capabilities are determined relatively to the concrete sphere of knowledge and against the background of precursors and opponents.

Living beings brain isn't a computer controlled by "Matrix" or only survival tool but is the sphere of ICP action that doesn't need in any outside control. Individual thoughts and consciousness are brain sensations of connections between neurons established in the moment. Observing "simultaneousness" of inventions (thoughts) creation is the result of Systems *society* evolution. The appearance of intelligent beings is so inevitable as ICP appearance. That's why the search for extraterrestrial life and appeals to it are reasonable by signals – ICP manifestations but not only code or graphic symbols, which can differ from terrestrial ones in conventionalities, spectrum, even logic. (The absence of contacts can also be connected with small probability of civilizations, which are interested and have possibilities for contacts and also with *non-stationary* cosmic *gravitation fields* deflecting narrow beam (due to energy possibilities) radiations. The estimation taking this into consideration showed that possible distances are only in a few hundreds light years, where such civilizations can be absent.)

By minimum (1-2%) genomes difference the main human qualitative difference from anthropoids are grown by several leaps-QQT of creative capabilities under the influence of connections increase between neurons and changed surroundings. A Man changes Life role from an object of evolution for evolution creator that can himself create artificial creators. The evolution from bio-molecules to cells went very slowly – only by Darwin's law – in the first 2-2.5 billion years, ICP appearance accelerated evolution to a great extent.

ICP determines the evolution of social relations too. Social ICP act on conscious level but their inner causes, springs and perspectives are not always realized in human societies being as if in social "sub-consciousness". Their ICP have specific QQT (political, scientific, technical, economic, religious revolutions) and are influenced through individual consciousness (PR-technologies, mass-media, bringing up, terrorism, etc). All social decisions from municipal and group levels to state and world politics go out of logic frames to the sphere of hypotheses, suppositions, preferences, intuition, etc. The reasonable conscious direction of social ICP is vital society problem so as optimum combination of democracy and leadership, i.e. stabilizing and vanguard (quite often accidental now) social ICP components.

The vital example of extreme necessity in reasonable social ICP for modern world is the struggle against Pan-Islamic expansion and terrorism. It's useless for this both expansion reconciliation (by yielding, political "correctness", liberalism, multiculturism) and wars of the recent past owing to the absence of a certain area, where enemy forces and control are stationed. The global dispersed expansion and terrorism in today's conditions of so far not united civilized world require equivalent creative methods of struggle against them. The methods can include the reasonable regional counteraction guidance, the creation of restrictions for availability and actions of immigrants from countries – expansion sources (as terrorism is impossible without them), the creation in such countries of economic conditions, which decrease demographic pressure, the need in expansion and "attractiveness" of terrorism for its potential doers. "Post-industrial" information society is the result of world economy, technology, politics and information surroundings globalization. Its perspective is Single (united): Administration taking care of all Earth inhabitances and resources; Intellect creatively organizing knowledge exchange and new knowledge acquirement.

Religions are intelligent beings prerogatives as a meaning of their life, work and intentions, a substitution of things unknown for them, their refuge in cruel surrounding world and a justification of their actions in the world. CP somewhat "reconciles" religions and science: Creator of our world gave to the world also the freedom of creative development; intelligent beings arisen in its course play in it an active and responsible role. (They made up many more or less universal world models including anthrop principle but the models have the deficiency of principle – they are metaphysical, CP and evolution are absent in them).

4. Artificial CP

A Man-designer or programmer of artificial intellect systems (AIS) introduces into AIS from outside and beforehand all their algorithms, criteria and purposes. Thus AIS are incapable to original creation by any their software or hardware improvement. They will never be intelligent beings and even living ones. To build *creative* artificial intellect systems (CAIS) is necessary to introduce into AIS *artificial* ICP thus building *creative programs* and *computers*.

Since it's practically impossible to design CAIS not on logical elements introduced by designer, artificial ICP is based on what all real data, algorithms, criteria, purposes are the functions, which so as their derivatives don't have infinite meanings and can be described by polynomials (power, harmonic, code, gauss ones – depending on problem specific features). CP accidental part (pseudo accidental – practically) changes the polynomial coefficients parameters. Their creative changing reflects changes of functions. Initial changes spheres and steps, the list of changing parameters and their initial meanings, the algorithm of artificial accidental process, etc are subjects of similar CP of more high levels. CAIS designer gradually-"asymptotically" brings nearer these levels to own intellectual level. Fig. 1 shows artificial CP or ICP fundamental scheme-algorithm with feedback in the sphere of changes in each parameter or level.

An accidental change inside the sphere of changes	An element of parameters or structure	System existence improvement or aggravation	Changing of the sphere of changes

Fig. 1.

Real control systems usually are functions of many interconnected parameters. Designer's heuristics are used in AIS in such case. CAIS automatically takes these dependencies into consideration repeatedly reiterating Fig. 1 procedure for different parameters. (Artificial ICP accidental part comes out directly – by random numbers or relatively – in the course of interaction with external accidental information. For example, computer viruses act like this. Their creation and the struggle against them are vital examples of artificial ICP.)

Self-reproducing CAIS of the future, where both knowledge and structure evolve, are capable to *surpass* intellectual levels and capabilities of their starting artificial ICP designers.

5. Some Results

CP gave the base to propose hypotheses and models of:

Sporadically arising QQT-big bangs – the explosions of extremely large black holes on reaching critical meanings of their masses with the formation of diverging and widening Met-galaxies groups, where each Met-galaxy has own types of substance and vacuum, physical laws and constants. The hypothesis explains observing course of our Met-galaxy widening, helps to understand "dark" substance composition, establishes the limits of Relativity principles action, eliminate the necessity in "dark energy" hypothesis, shows the possibility of the Universe endless renovation.

Origin and development of life forms, intellect, signal systems, individual and group consciousness and mentality, information society, extraterrestrial civilizations. This seems not less important than e.g. the determination of concrete life origin processes on the Earth and African or Asian ways of human evolution, the more so that all these processes most likely took place by several different creative ways.

CAIS are already built in the spheres of communication, control, image recognition, and they solved a number of problems beyond AIS capabilities[5]. CAIS applications in different spheres are in detail shown in [2]. They include: Decentralized situate control in large industrial, transport, power, communication systems; Source-independent image recognition using *adapting standards*; The adaptation of transmitting messages for their sufficient understanding by the concrete natural or artificial recipient; Adaptive individual patient treatment – in addition to the treatment of "an average-statistical patient for an average-statistical illness"; The search with adaptation of both its algorithm and purpose in scientific researches and processing of data arrays; The prediction of partner, consumer, adversary reactions by means of *adapting hypotheses*; The application in global Internet-surroundings for the search for information, information noise reduction, struggle against interference.

References

1. Abraham Goldberg, Felix Skorochod. Creative processes of information society. (In Russian) www.elektron.2000.narod.ru.
2. Abraham Goldberg. Natural and Artificial Creative Processes in Nature, Science and Engineering. Tel-Aviv, Pilies Studio, 2002, 169 pp.

[5] It's important to note that CAIS doesn't supplement and doesn't abolish existing AIS instruments (methods) but is the means of their choice and adaptation in concrete cases and conditions.

III

CONSERVATION OF LIFE

Human Alteration of Evolutionary Processes

John Cairns, Jr.

Department of Biological Sciences, Virginia Polytechnic Institute and State University, Blacksburg, Virginia 2406, USA, jcairns@vt.edu

Soviet climatologist Budyko has remarked: "temperature and rainfall are the two major variables of life on Earth." Human society is changing both of these phenomena markedly, along with many other key variables that affect evolutionary processes. A major risk is that the tempo (or rate of human-induced environmental change) may proceed more rapidly than the ability of scientists to understand, predict, or make any long-term changes that might reduce the severity of the consequences. Increasing evidence indicates that the general public and its leaders (i.e. policy makers and politicians) fail to grasp the full implications of a planet in which the types and rate of environmental change differ substantively from the climate records of the past 5 million years.

1. Introduction

Almost every human activity has some effect upon natural systems. When the human population was small and spread thinly over the planet, as it was for most of the 160,000 years the human species has inhabited the planet, adverse effects were localized and comparatively small. In short, the resilience of natural systems was not exceeded and, as a consequence, the impact on evolutionary processes was much less than it is today. Currently, however, effects are global and intense; illustrative examples follow.

1. Human population increased fourfold in the 20th century (Speth, 2004).The doubling time occurred within the life span of a single individual—a new phenomenon.
2. Affluence has increased even more because the global economic output has increased approximately twenty fold.

V. Burdyuzha (ed.), The Future of Life and the Future of Our Civilization, 97–104.

3. Humankind has become a major evolutionary selective force.
4. Perpetual economic growth is, arguably, the major paradigm for human society.
5. Species impoverishment (i.e. loss of biodiversity) and the consequent loss of valuable genetic information due to invasive species and habitat destruction and alteration, together with an increase in ubiquitous persistent toxic substances, has alarmed the scientific community for decades.
6. Over-harvesting, especially of marine fisheries, has made sustainable use of natural resources problematic.
7. Climate change has already become a major factor that is impairing ecosystems globally.

2. Love of Nature and Catastrophes

Two major factors may diminish or stop damage to the 30+ million other life forms with which humankind shares the planet:(a) a love for and an ethical responsibility for the well being of other life forms and (b) fear of the consequences if humankind continues unsustainable practices (Cairns, 2004 a,b).

Concern about natural systems and the environment became widespread during the latter part of the 20th century, which resulted in the first Earth Day in 1970 and the 1972 United Nations Conference on the Human Environment (The Stockholm Conference). The latter resulted in the United Nation's Environmental Program. However, the failure to implement any of these protective measures resulted in continued environmental degradation, although some notable successes were achieved. Even environmental catastrophes in the late 20th and early 21st centuries have resulted in a focus on symptoms rather than causes. Some ecological catastrophes (e.g. thinning and disappearance of the Artic ice shelf) receive little or no attention from the "popular" news media. Other ecological events receive significantly more attention, such as the sea level rise at the Pacific Ocean island country of Tuvalu (Brown, 2001-2002) and the displacement of the Inuits (Native Americans in Alaska) covered by US Senator John McCain's global warming hearings in the US Senate. Of course, the most dramatic catastrophe was the tsunami in late 2004, which caused massive loss of human life. Persuasive evidence indicates that the loss of protection from massive wave action increased tsunami damage substantially. This lack of protection would not have happened if the mangrove forests and coral reefs had not been damaged previously by human actions (e.g.

Silverstein, 2005; Sharma, 2005). Ecological catastrophes are most likely to occur in areas or nations with significant ecological deficits (i.e. natural capital has been lost at a rate greater than the replacement rate) and will almost certainly have a major effect upon evolutionary processes, which, in turn, will have both long- and short-term effects upon human society. Finally, global warming and other human induced ecological changes that will affect evolutionary processes will result in severe consequences to human society.

One example is the suddenly warming climate, which is likely to be a serious threat to political stability (Schwartz and Randall, 2003). The "Pentagon Report" (Schwartz and Randall, 2003) describes an extreme scenario whose effects might be less than described or even worse because of interactions between subcomponents of the global systems. Effects on evolutionary processes are probable, regardless of the way the scenario unfolds.

3. Evidence for Human Alteration of Evolutionary Processes

An excellent summary of the alteration of evolutionary processes is available through the US National Academy of Sciences (Myers and Knoll, 2000), which provides abundantly referenced evidence that alterations have occurred and are likely to continue if present trends persist. Significant alteration of evolutionary processes will have major effects, mostly unfavorable, upon the dynamics of human society and humankind's quest for sustainable use of the planet. Dixon and Adams (2003) speculate on what a post-human society might entail (these two authors consulted thirteen advisors with impressive credentials on evolutionary processes).
Habitat fragmentation, now a global phenomenon, is another alteration that could cause a major disruption of evolutionary processes (e.g. Templeton et al., 2000).

Attesting to evolutionary alterations with massive documentation seems superfluous. Who can contemplate the massive recent alterations humans have made in the biosphere and conclude that these alterations have no effect upon evolutionary processes? Those persons would have to deny such evidence as the development of resistance to antibiotics in some disease organisms and the continual need to develop new pesticides to control pests. Why do policy makers not regard this paucity of readily available information as major evidence of the detrimental effects to human society of altering evolutionary processes?

4. Denial, Anti-Science, and Special Interest Lobbying

One controversial explanation of the ineffectiveness of the environmental movement in the United States is that no prominent national leader has stated publicly and forcefully the detrimental consequences of present environmental trends. Leadership may fear alarming the general public or being labeled an extreme environmentalist (e.g. Shellenberger and Nordhaus, 2005). Although many laud the efforts of pioneers in the environmental field, some (Shellenberger and Nordhaus, 2005) believe that modern environmentalism is no longer capable of coping with the serious ecological crises of the world. For example, efforts to reduce global warming over at least two decades have not resulted in unsustainable practices being replaced by sustainable practices.

In contrast, Ehrlich and Ehrlich (2005) assert that, despite their belief that The New York Times Science Section has led the journalistic profession in reporting the consensus of the scientific community on the issues of climate change, the seriousness of the overall environmental situation has never been adequately covered by the media. Even though The New York Times has printed articles (Editorial, 20 January, 2005) on the human impact on the planet, no explicit statement about the seriousness of the impact has been forthcoming.

The well-known American religious leader Martin Luther King, Jr. stated: "A time comes when silence is betrayal... Nor does the human spirit move without great difficulty against the apathy of conformist thought, within one's own bosom and in the surrounding world" (Quote of the Week from Sojourners online newsletter, Wednesday, 19 January, 2005). How can the silence continue when the processes, including evolutionary, of Earth's biological life support system are being seriously disrupted by human activities? Earth's life support system has favored the human species for approximately 160,000 years, but the 30+ million species with which humans share the planet are not concerned with the fate of Homo sapiens. The other species are not committed to maintaining the life support system on the behalf of humans, even though conditions they produce now are beneficial to humans.

Speth (2004) believes that three factors are responsible for humankind's failure to respond to global threats: (a) the collective power of the forces that produced this situation will not be adequately changed by half-measures, (b) the far-reaching complex responses required make redirecting the global agenda inherently difficult, and (c) global politics impede the development of a suitable global agenda. However, Speth believes the transition to sustainability can be made.

Gelbspan (2004), a recipient of the Pulitzer Prize, focuses on the consequences of global warming, which he feels is causing the planet to fall apart piece by piece in the face of persistent and pathological denial. Since Gelbspan is a journalist, his charge that the media has failed to make the connection between climate change and other events, such as altered rainfall patterns, is very persuasive. Gelbspan also feels another major failure of the media is ignoring the ferocious battles between the fossil fuel industry lobby and credentialed scientists who have made the study of global warming a major part of their professional careers. He uses as an example (pp. xii, xiii) the assault on the character and scientific integrity of Dr. Benjamin Santer, a world-class climate modeler at the US Lawrence Livermore National Laboratory. Associated Press Special Correspondent Hanley (2005) remarks that the US delegation to a global conference on disasters wanted to purge a UN action plan of its references to climate change as a potential cause of future natural calamities. Clayton (2005) describes the fate of George Zeliger, a whistle blower (a person who makes a public disclosure of corruption or wrongdoing). Orr (2004) has written a very disturbing analysis of the effect of politics (especially when disguised as patriotism) on the environment.

The relevance of these incidences to human alteration of evolutionary processes is that the scientific process must be allowed to flourish and must not be suppressed when it appears to conflict with political or economic ideologies or matters of faith. The scientific process, including peer review, has been very successful in discrediting faulty hypotheses, but it does so by rigorous testing of them and their supporting data.

Wiener (2005) describes a situation in which 20 of the largest chemical companies in the US have developed a campaign to discredit two historians who studied the attempts of industries to conceal links between their products and cancer. This situation is unusual in that the companies have subpoenaed and deposed (in courts of law) the five academicians who recommended that the University of California Press publish the book Deceit and Denial: The Deadly Politics of Industrial Pollution by David Markowitz and David Rosner. Intimidating qualified reviewer strikes at the heart of the scientific process. In another somewhat similar situation, the British Government's chief scientific advisor, Chief Scientist Sir David King, has claimed there have been attempts to discredit him because of his attempts to call attention to the threat of global warming (Conner, 2005).

In the United States, arguably one of the scientific leaders in the world, the assault on science has three major components:(a) discredit scientists whose views differ from the dominant political or economic ideology and religious faith, (b) attempt to intimidate scientists and other academicians by litigation, which is both time consuming and expensive, (c) attempt to discredit scientific theories by implying they are merely educated guesses rather than carefully constructed frameworks for understanding a substantial body of evidence (e.g. Editorial, January 23, The New York Times, 2005).Theories supported by mainstream science are the most useful scientific theories. Attacks on the theory of evolution in the United States are increasing and persistent and are especially significant when they are against the texts used in the school system. If science is discredited in the educational system, understanding the effects of humans upon evolutionary processes will be markedly hampered. Fortunately, many scientifically advanced countries accept evolutionary theory, and both teaching and research can proceed in a systematic way in keeping with the processes of science.

Many Christians view evolution as God's means of creation, and the theory of evolution is taught in Catholic schools and many other Christian schools. Christian fundamentalists and creationists are a very politically active sub-set of all Christians, but their energy and fervor in promoting their beliefs have made teaching evolution a major issue in the United States. Sustainable use of the planet requires that humankind have a better understanding of evolutionary processes. Achieving this goal requires that the processes of science not be disrupted, especially in the education of future scientists.

5. Conclusions

The quest for sustainable use of the planet by Homo sapiens requires a mutualistic relationship between human society and natural systems. Disrupting evolutionary processes that facilitate this relationship will almost certainly have adverse, possibly fatal, effects upon human society. Another way to envision the quest for sustainability is avoiding a post-human world (Cairns, 2005). Lest this seem too fanciful, it is well to remember that Homo sapiens has only inhabited Earth for approximately 160,000 years out of an estimated 4.5 billion years that the planet has probably existed. In addition, the greatest anthropogenic damage has occurred in the last 200 years.

If ecological tipping points are reached or exceeded, disequilibrium will result. Regrettably, the only certain way to find an ecological tipping point is to reach it or exceed it, because no laboratory experiments are suitable for such large temporal and spatial spans. McCarthy (2005) discusses a report that estimates the climate change tipping point at 2° centigrade above the average world temperature prevailing in 1750 (before the Industrial

Revolution). Since that time, human production of greenhouse gases, such as carbon dioxide, has markedly influenced the retention of the sun's heat in the atmosphere.

Speth (2003) believes that globalization is one of the profound phenomena in the present era that has affected the environmental, economic, and social aspects of the nations of the world. Because globalization involves so many political and economic systems, mid-course corrections of these powerful trends will be exceedingly difficult, but not impossible, to alter. To achieve this goal, a mutualistic co-evolution of human society and natural systems is necessary (Cairns, in press). If humankind fails in this undertaking, evolutionary processes will continue, although many other species will probably be driven to extinction. Failure would also suggest that intelligence, as humans define it, did not provide the long-term survival value it was thought to have. I believe if intelligence is used to select sustainable practices, it will have proven to have long-term survival value.

Acknowledgments

The first handwritten draft of this paper was typed by Karen Cairns, who also made the changes are necessary for the third draft. Darla Donald provided her usual skilled editorial assistance. I am grateful to Vladimir Burdyuzha for reading my paper at the conference since I was unable to attend for health reasons.

References

Brown, L. R. 2001-2002. Rising sea level forcing evacuation of island country. Earth Policy News, Washington, DC.news@earth-policy.org

Cairns, J., Jr. 2004a. Ecological tipping points: A major challenge for the experimental sciences. Asian Journal of Experimental Sciences 18(1,2):1-16.

Cairns, J., Jr. 2004b. Coping with ecological catastrophes: Crossing major thresholds. Ethics in Science and Environmental Politics http://www.intres.com/atricles/esep/2004/E56.pdf

Cairns, J., Jr. 2005. Avoiding a posthuman world. Science and Society 3(1):13-24.

Cairns, J., Jr. In press. Sustainable coevolution. International Journal of Sustainable able Development and World Ecology.

Clayton, M. 2005. Hard job of blowing the whistle gets harder. The Christian Science Monitor 20 Jan http://www.csmonitor.com/2005/0120/p13s02-sten.html

Conner, S. 2005. Americans are trying to discredit me, claims chief scientist. Independent News 17 Jan http://www.news.independent.co.uk/low_res/story.jsp?story=601497&host=3&dir=58

Dixon, D. and J. Adams. 2003. The Future is Wild. Firefly Books, Toronto, Canada
Editorial. 2005. At the limits of air and water. The New York Times. January 20.
Editorial. The crafty attacks on evolution. The New York Times. January 23.
Ehrlich, P. and A. Ehrlich. 2005. The Times and the environment. Letters to the editor on The Media and The Environment. Pacific Conservation Biology 10(1):2.
Gelbspan, R. 2004. Boiling Point. Basic Books, New York, NY.
Hanley, C. J. 2005. U. S. seeks to scuttle conference text linking climate change to disasters. San Francisco Chronicle, 19 January.
McCarthy, M. 2005. Climate change: Report warns point of no return may be reached in 10 years, leading to droughts, agricultural failure, and water shortages. The Independent, 24 January.
Myers, N. and A. H. Knoll (ed). 2000. The Future of Evolution. Proceedings of the National Academy of Sciences 98(10): 5389-5479.
Orr, D. W. 2004. Patriotism, Politics, and the Environment in an Age of Terror.
Schwartz, P. and P. Randall. 2003. An Abrupt Climate Change Scenario and its Implications for United States National Security. Full PDF report (917 kb) available for download at http://www.ems.org/climate/pentagon.climate.change.html.
Sharma, D. 2005. Outside view: Tsunami, mangroves and economy. (UPI Outside View Commentator). Washington Times http://www.washtimes.com/upi-breaking/20050109-105932-8248r.htm.
Shellenberger, M. and T. Nordhaus. 2005. Death of environmentalism: Global warming politics in a post-environmental world. Grist Magazine, 13 January http://www.grist.org/cgi-bin/printthis.p1.
Silverstein, D. 2005. A tidal wave of lessons learned- an ENN Commentary. Environmental News Network http://www.enn.com/today_PF.html?id = 6926.
Speth, J. G. 2003. Worlds Apart: Globalization and the Environment. Island Press, Washington, DC.
Speth, J. G. 2004. Red Sky at Morning: America and the Crisis of the Global Environment. The Yale University Press, New Haven, CT.
Templeton, A. R., R. J. Robertson, J. Brissen, and J. Strasburg. 2000. Disrupting evolutionary processes: The effect of habitat fragmentation in the Missouri Ozarks. Proceedings National Academy of Sciences 98(10):5426-5432.
Wiener, J. 2005. Chemicals, cancer, and history. The Nation 7 February. http://www.thenation.com/doc.mhtml?i=20050207+s=weiner

The Danger of Ruling Models in a World of Natural Changes and Shifts

Nils-Axel Mörner

Paleogeophysics & Geodynamics, Stockholm, Sweden, morner@pog.su.se

Aristotle's presented the first global model; his model of the planetary system. It was totally wrong. Still, it ruled the world for 1800 years until Copernicus presented an observationally based solution. To leave observational reality behind and to hang on to models and model predictions seem utterly dangerous and basically unscientific. Still, we are today victims of many ruling models. The proclamation that nuclear waste can be stored in the bedrock for hundreds of thousands of years under "full safety" and "no problems" is not only against common sense, it also violates basic geological knowledge. IPCC's climate-modeling now totally roles the entire world. Still, it is based on very shaky ground including errors, falsifications and misinterpretations. Sea level, for example, is by no means in a rising mode and we can free the world from the condemnation to become flooded in the near future. In about 40 years we will be in a new Solar Minimum and are hence likely to experience a new "Little Ice Age". All this reveals the danger of ruling models, and calls for a return to basic observational facts. Scientific integrity has become vital.

1. Introduction

In true natural science, we have always worked with a basic three-parted scheme, viz. Observation – Interpretation – Conclusion. In case of more unified pictures, we talk about a chain of: Hypothesis – Theory – Paradigm. This is our scientific base; so it has been and so it ought to be.

In recent years of computer modeling, a new and very dangerous scheme has entered the scientific scene, viz. Idea – Models – "the Truth".

V. Burdyuzha (ed.), The Future of Life and the Future of Our Civilization, 105–114.

Modeling is a powerful tool assisting us in our search for connections and interacting variables. It should never grow to become a subject of itself. A few cases will be discussed below.

2. The First Model Ever Presented

In the Ionic settlement with the cities of Ephesos, Miletos and Kos (today's SW Turkey), a wonderful, free, natural philosophy flourished. In their understanding of the planetary system, the Sun was where is should be, i.e. in the center (Fig. 1), no questions about that. With Socrates, Platoon and especially Aristotle's things changed. The Earth was placed in the center and the Sun was proclaimed to move around the Earth.

Aristotle's presented a unified model – the ever first model – of the planetary and celestial mechanics. Everything was explained by movements of the planetary and celestial bodies along 56 independent circular paths. No one was to object to this masterly "final solution" (later updated by Ptolemaios ~170 BC).

Some clever persons objected – e.g. Anaxagoras and Aristarkos – but they were rapidly overruled by the master's model.

Aristotle's–Ptolemaios model was adopted by the Church because the Earth was in the center, where the Church wanted it to be.

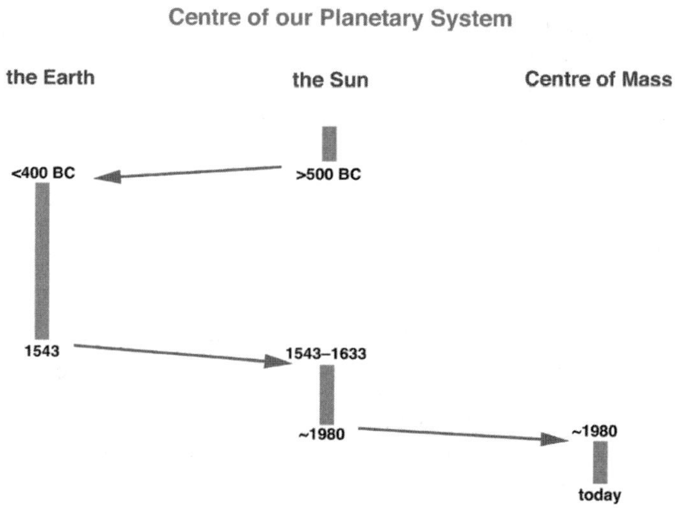

Fig. 1. Changing opinion of the center of our planetary system during 2700 years. It took some 1800 years before the ruling model of Aristotle s could be dismissed by the observationally based results of Copernicus in 1543.

It took about 1800 until reality caught up with the model illusion and Nicolaus Copernicus, in 1543, presented his outstanding observational facts proving that the Sun was in the center and the planets, including the Earth, were forced to circle around the Sun (Fig. 1). Still, the Church refused to accept the truth. Giordano Bruno was burned to death in 1600, and Galileo Galilei had to deny the facts in 1633.

3. The Word "Cosmos"

Pythagoras, the famous Greece mathematician (born around 550 BC), was the first person to use the word "cosmos". This word did not refer to the Universe, however, but to our own surroundings, i.e. nature. The Greek word referred to beauty, order and meaning. Therefore, according to Pythagoras, we have to study nature and natural phenomena. When we are doing this, we will – without really noticing it – become "cosmios", i.e. ordered in our souls.

To be ordered in the soul, what a wonderful thought. Where do we see it today? Hardly it is among politicians. Maybe it is among some scientists.

And the study of nature and natural sciences was the way to achieve this order in our souls; cosmos. The more important it is that our natural science will not be contaminated by "ruling models" but remains open for observations and unbound curiosity.

4. From Intellectualism to Empirism

The Renascence meant a quite thorough re-vitalisation of art and science after the long medieval period dominated by religion and mysticism (Fig. 2). In the last 500 years we see remarkable achievements; first dominated by intellectualism (thinking, calculation, new ideas) and, since about 1750, dominated by empirism and rationalism (observations, measurements, experiments). Carl von Linné may be regarded as the front person in the opening of nature for observations, interpretations and systematization. Much of the foundation of modern sciences (chemistry, physics, biology, geology, climatology) was lied down in the second half of the 18th century and the first half of the 19th century (Fig. 2). In the last decades, we may see the appearance and uncontrolled growth of modelling, and we may feel a growing hope of, at least, some intellectualism and observational-based common sense.

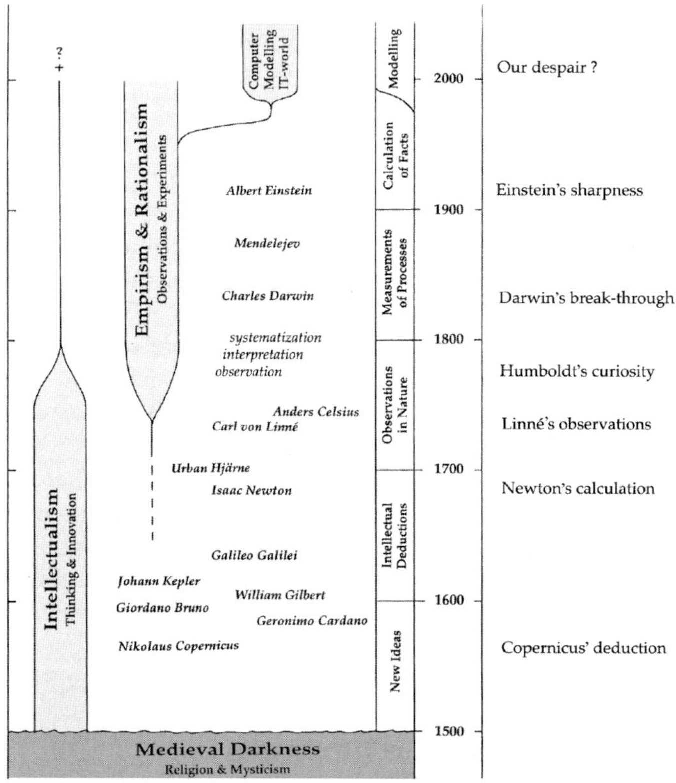

Fig. 2. Scientific changes in the last 500 years (generalized, of course) with the main steps from intellectualism to empirism & rationalism and, in the last time, a switch to our present computer modeling and IT world (modified from Morner, 2000).

With all the beautiful examples of deduction, calculation, observation, curiosity, interpretation, systematization and sharpness in the last 500 years, we have all possibilities to cope with our own despair (Fig. 2).

5. Nuclear Waste Problems

The Swedish and Finnish nuclear energy agencies proudly proclaim that nuclear waste can be stored in the bedrock for hundreds of thousands of years under "full safety" and "no problems". Such a statement is, of course, against common sense. It is to extend predictions "in absurdum" (Mörner, 2001). All efforts and all arguments are devoted to the support of the idea that a closed repository 500 m down in the bedrock will remain intact for a period of, at least, 100,000 thousand years. Again model scenarios take over and leave observational facts and common sense behind.

The truth is that the concept of a closed bedrock repository violates basic geological knowledge. There is no safety to be guaranteed for those immense time periods (Mörner, 2001). Whilst Sweden today is characterized by a low to moderately-low seismic activity, it was, during the deglacial phase, characterized by a very high to super-high seismic activity, in frequency as well as in magnitude, which is well established by recent observational facts (Mörner, 2003, 2004). In such an environment – to be repeated at future glaciation periods – there can be no safe repository in the bedrock. Therefore, it is an urgent need of a total reconsideration of the mode of handling the high-level nuclear waste (Mörner, 2001; Cronhjort and Mörner, 2004).

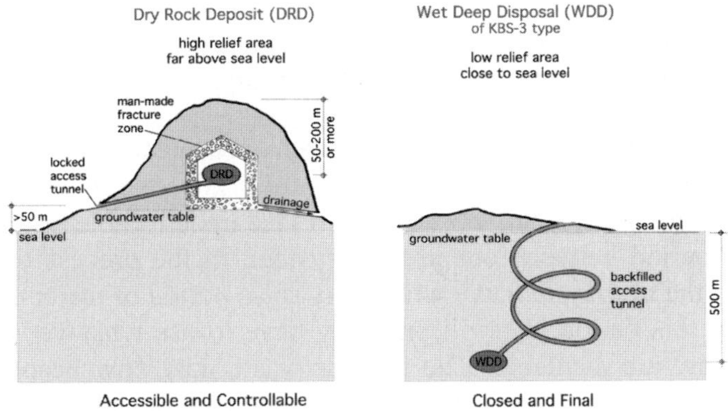

Fig. 3. Comparison between a dry DRD repository and a wet WDD repository of KBS-3 type (from Cronhjort and Mörner, 2004).

Instead of a closed repository below the groundwater level, we propose a dry rock deposit (DRD) as illustrated in Fig. 3 and further discussed elsewhere (Mörner, 2001; Cronhjort and Mörner, 2004). A DRD-repository is, of course, well locked preventing unwanted intrusion, at the same time, however, the waste remains accessible for reparation, transmutation and even utilisation. Artificial fracture zones keep the repository dry. At the same time, these fracture zones offer increased seismic protection.

The high-level nuclear "waste" contains near 96% remaining energy. Is this a waste, a dangerous plutonium source or a future energy reserve? Even if today's technology does not see a safe methodology for further energy production, this will probably not be the case in the near future.

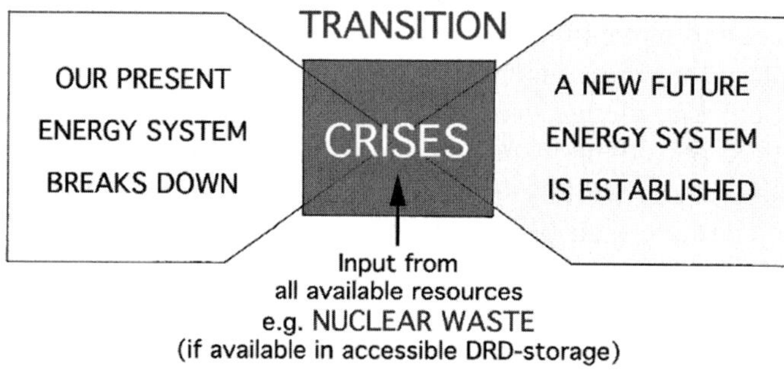

Fig. 4. In the near future, our present energy system will be insufficient and break down. At the transition period to a new future energy system, we will have to pass serious crises, when we are likely to be forced to vacuum-clean the planet of all available resources. The nuclear waste, still containing some 96% of its energy, may then become a vital resource for sustainability and survival provided it has been stored in an accessible way in a DRD repository.

We know that our present energy systems are becoming insufficient. If we, within the next 100 years (at the very most 200 years), will not be in a totally new energy system, the world will face unbelievable problems. The transition period between our present system, in the process of breaking down, and the new system, we will surely see a period of major crises (Fig. 4). During this period, we will probably have to use whatever energy resources there are available. The nuclear waste may now become a truly saving resource. Therefore, precisely just therefore, we must keep our nuclear waste accessible. This accessibility is only available in a DRD-repository, however (Fig. 3).

6. The Global Warming Issue

IPCC's climate modeling now totally rules the entire world. Still, it is based on very shaky ground including errors, falsifications and misinterpretations. Sea level, for example, is by no means in a rising mode.

Climate is becoming increasingly warmer we hear almost every day. This is what has become known as "Global Warming". The idea of IPCC (2001) is that there is a linear relationship between CO_2 increase in the atmosphere and global temperature. The fact, however, is that temperature has constantly gone up and down. From 1850 to 1970, we see an almost linear relationship with Solar variability; not CO_2. For the last 30 years, our data sets are so contaminated by personal interpretations and personal choices that it is almost impossible to sort up the mess in reliable and unreliable data.

In the IPCC-scenario, we will face a rapidly increasing temperature in the near future that will cause an opening of the Arctic Basin (ACIA, 2004). Such a view implies that we neglect the solar influence (Morner, 2005b).

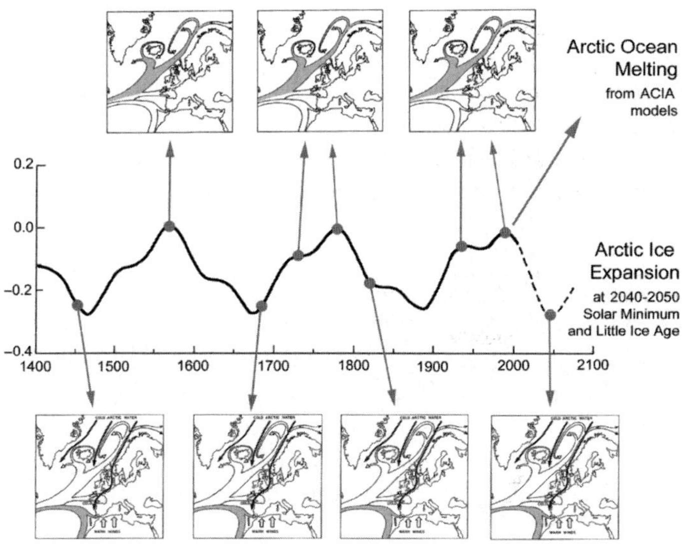

Fig. 5. The main solar cycle in the last 600 years with observed ocean circulation pattern at maxima and minima, and the expected extension into the future (from Morner, 2005b). At Solar minima NW Europe, the North Atlantic and the Arctic have experienced cold phases known as Little Ice Ages. By year 2040-2050, a new Solar Minimum is to be expected, and with a new cold phase over the Arctic and NW Europe.

The fact is that the climatic changes during the last 600 years include cold periods around 1450, 1690 and 1815 that correlate with periods of Solar Minima (the Sporer, Maunder and Dalton Solar Minima). The driving cyclic solar forces can easily be extrapolated into the future (Fig. 5). This would call for a new cold period or "Little Ice Age" to occur at around 2040-2050 in totally contrary to the IPCC-scenario. The Solar influence is simply kept out of the Global Warming concept. It is high time to bring the Sun back into the center.

In the global warming concept, it has been constantly claimed that there will be a causal rise in sea level; a rise that already is in the accelerating mode, in the near future to cause extensive and disastrous flooding of low-lying coastal areas and islands. Is this facts or fiction, what lies behind this idea, and, especially, what do the true international sea level specialists think (Morner, 2005)?

The recording and understanding of past changes in sea level, and its relation to other variables (climate, glacial volume, gravity potential variations, rotational changes, ocean current variability, evaporation/ precipitation changes, etc.) is the key to sound estimates of future changes in sea level (Morner, 2004b). The international organizations hosting the true specialists on sea level changes are to be found with the INQUA commission on sea level changes and the IGCP special projects on sea level changes. When I was president of the INQUA Commission on Sea Level Changes and Coastal Evolution, 1999-2003, we paid special attention just to this question; i.e. the proposed rise in sea level and its relation to observational reality. We discussed the issue at five international meetings and by Web-networking (INQUA, 2000). Our best estimate for the next century was +10 cm +10 cm (INQUA, 2000, Morner, 2004b), later revised by me to +5 cm +15 cm (Morner, 2004b, 2005a, 2005b).

It is true that sea level rose in the order of 10-11 cm from 1850 to 1940 as a function of solar variability and related changes in global temperature and glacial volume. From 1940 to 1970, it stopped rising, maybe even fell a little. In the last 10-15 years, we see no true signs of any rise or, especially, accelerating rise (as claimed by IPCC), only a variability around zero (Morner, 2004b, 2005b). This is illustrated in Fig. 6.

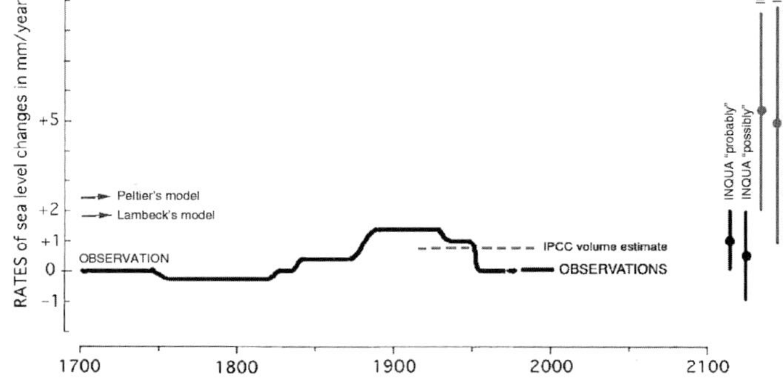

Fig. 6. Observed sea level changes in the past 300 years and estimated changes by year 2100 (from Morner 2004b).

From 2000 to the present, we have run a special international sea level project in the Maldives (Morner et al., 2004) including six field sessions and numerous radiocarbon dates. Our record for the last 1200 years is given in Fig. 7. There are no signs of any on-going sea level rise. It seems all to be a myth. The same result is obtained if one examines other regions; e.g. the famous records of Tuvalu and Venice and satellite altimetry (Morner, 2004b, 2005a, 2005b).

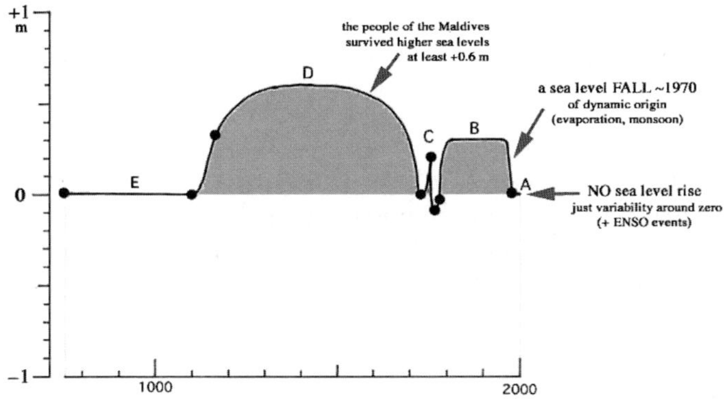

Fig. 7. The sea level record of the past 1200 years from the Maldives including a significant sea level fall in the 1970s and a lack of signs of any ongoing rise (Morner et al., 2004; Morner, 2995b).

In conclusion; observational data do not support the sea level rise scenario. On the contrary, they seriously contradict it. Therefore, we should free the world from the condemnation of becoming extensively flooded in the near future. Furthermore, in about 40 years, we will be in a new Solar Minimum with related cold period.

7. Future Perspectives

Scientific progress has always been driven by hard work, sharpness and unbound curiosity. This is our true scientific resource, and it must be the driving forces also in the future. This calls for increased independence of individual scientists and scientific organizations. Ruling models must not take over as guiding tool. Even ruling scientific paradigm must be questioned and tested. I find it significant that Swedish science never has been at a stronger front-position than during the "Period of Freedom" in the 18th century after the fall of the autocratic kings. Freedom, indeed, is the true ground for creativity. Integrity – besides knowledge, of course – is the tool for further achievements.

Finally, scientists are not bound by boundaries (national, ethnic, religious, social, etc.). On the contrary, scientists consider opinions and presentations, and make their contacts and firm links on those grounds alone. This scientific way of collaboration is another basic resource for the future. It is also the core-idea of our project "The Future of Life and the Future of Our Civilization".

Acknowledgments

I am indebted to Professor Vladimir Burdyuzha for inviting me to the organization committee of the symposium on "The Future of Life and the Future of our Civilization" and I acknowledge an excellent meeting in Frankfurt organized by Professor Claudius Gros.

References

ACIA, 2004. Impact of a Warm Arctic: Arctic Climate Impact Assessment. ACIA Overview Report. Cambridge Univ. Press. Also: http://amap.no/acia/

Cronhjort, B. and Mörner, N.-A., 2004. A question of dry vs. wet. The case for Dry Rock Disposal of nuclear waste. Radwaste Solutions, May/June-04, p. 44-47.

INQUA, 2000. Sea Level Changes and Coastal Evolution. www.pog.su.se.

IPCC, 2001. Climate Change. Cambridge Univ. Press.

Mörner, N.-A., 2000. From Intellectualism to Empirism. In: Giuseppe Toaldo and is time, Contrib. Univ. Padova, 33, 635-642.

Mörner, N.-A., 2001. In absurdum: long-term predictions and nuclear waste handling. Engineering Geol., 61, 75-82.

Mörner, N.-A., 2003. Paleoseismicity of Sweden. A noval paradigm. Stockholm University, P&G-print, 320 pp

Mörner, N.-A., 2004a. Active Faults in Fennoscandia, especially Sweden: primary structures and secondary effects. Tectonophysics, 380, 139-157.

Mörner, N.-A., 2004b. Estimating future sea level changes. Global Planet. Change, 40, 49-54.

Mörner, N.-A., 2005a. Facts or Fiction? House of Lords, Economic Affairs Committee, Report, 1-6.

Mörner,-A., 2005b. Sea level changes and crustal movements with special aspects on the Mediterranean. Z. Geomorph. N.F., Suppl.vol. 137, 91-102.

Mörner, N.-A., Tooley, M. and Possnert, G., 2004. New perspectives for the future of the Maldives. Global Planet. Change, 40, 177-182.

Ecological Limits of the Growth of Civilization

Kim S. Losev

Geological Department of Moscow State University, Leninskie gori, 119992 Moscow, Russia, kim@viniti.ru

Rome Club at the second part of the XXth century developed Thomas Malthus ideas and discussed limits of the growth as the limit of resources. Now it is obvious that the main limit of the growth is ecological, which is going to determine the other limits. Theory of biotic regulation of environment demonstrated it. The idea of biotic regulation of environment was formulated by Vladimir Vernadsky at the first part of the XXth century. He wrote that «living substance in biosphere plays the main active role and can't be compared by it is strength and continuation with any geological force. It determines all the main chemical regularity in biosphere». And: «life itself creates the environment of life».

Vladimir Timofeev-Resovsky determined the concept of «biotic regulation of environment». In 1968 he noted that "… the biosphere of the Earth – a gigantic living factory, which transforms the energy and substance on the surface of our planet – forms both the equilibrium composition of the atmosphere and the composition of solutions in natural waters, and through the atmosphere it forms the energy budget of our planet. It also affects the climate. One should remember an important role in the global water cycle of evaporation of water over vegetation cover. Hence, the biosphere of the Earth forms the human environment. And the careless attitude to it, violation of its correct functioning will mean not only a violation of food resources for humans and a number of raw materials necessary for people but also a violation of the gas and water environment of humans. Finally, people without the biosphere or with a poorly functioning biosphere cannot exist on the Earth».

V. G. Gorshkov from St. Petersburg in 1990s carried out difficult and somewhat thankless work on the empirical generalization of available information and as a result formulated theory of biotic regulation of the envi-

V. Burdyuzha (ed.), The Future of Life and the Future of Our Civilization, 115–119.

ronment. One can understand why this work is difficult. It is thankless because, like every new paradigm, it is first unaccepted and rejected. But gradually this theory became accepted by most of the scientific community and not only in Russia but also throughout the world. It is enough to quote from Steffen and Tyson (2001) book which was the basis of the Amsterdam Declaration, the final document of the important international conference held on July, 2001 "Challenges of Changing Earth" dedicated to the results and accomplishment of the four largest international scientific programs. They wrote: "The Earth is system that life itself helps to control. Biological processes interact strongly with physical and chemical processes to create the planetary environment, but biology plays much stronger role in keeping Earth's environment within habitable limits than it was previously thought". The theory of biotic regulation of the environment was published in English in three books (Gorshkov, 1995; Gorshkov et al., 2000; Kondratiev et al., 2004).

Theory of biotic regulation of the environment is quite different to Gaia hypothesis. **Biota cannot be a globally correlated system**, as J. Lovelock (1982) supposed. In the absence of the competitive interaction, it is impossible in such a system to distinguish between progressive changes and regressive ones, which are equiprobable within it. Therefore changes in such a system will weaken its regulated nature, that is, lead to degradation. This process is equivalent to accumulation of irregularity (entropy) in closed systems. With the assumed existence of such a system of life, it would not be able to evolve, because all changes in it should be prohibited. **Only one mechanism is capable to preserve the regularity and at the same time to ensure the evolution. It can be a stabilizing selection, which acts through a competitive interaction of a totality of homogeneous communities**.

Thus organized in sets numerous correlated communities biota is the main actor in the biosphere and the basic mechanism ensuring the regulation and dynamic stability of the environment, including the climatic stability. Preservation of natural biota in sufficient volume is the key problem of preservation of life and civilization stability. Another words it is problem of what part of global ecosystem humankind civilization can use for own growth. This part of the global ecosystem calls carrying ecological capacity or carrying economic capacity. This is the key problem of ecology.

There is enough evidence that illustrates how life, materialized through total sum of natural organisms (biota), regulates of the substances fluxes, build and organizes own environment using solar energy. A complicated interaction of biota with the environment leads to form intercorrelated communities of organisms – biogeocenoses or local ecosystems, which are elementary units of the biochemical cycle and elementary cells of the environmental regulation. In biogeocenoses, both synthesis and decomposition of organic are accomplished by great number (hundreds of thousands) of independent elements. For instance, synthesis in a tree is realized through needles or leaves (which average 200 thousands), which act independent of each other and even compete for the solar radiation flux, and decomposition is realized by microorganisms (billions of species) and by fungi hyphae in soils.

Because of huge amounts of microorganisms and fungi hyphae in soils, the organic decomposition is characterized by low natural fluctuations and, as a rule, does not introduce any disturbance into biogeocenoses (local ecosystem). But usually natural systems include large moving animals like elk or deer consuming vegetative organic and seemingly breaking the good closed cycle of biogens. One way to prevent an irreversible increase of openness is to reduce the share of consumption by these animals. Experimental data on numerous natural ecosystems (landscapes) show that the share of consumption by large animals does not exceed 1% of net primary biologic production (Fig. 1). The feeding territory of large animals includes a multitude of biogeocenoses, and the density of their mass per unit area of the ecosystem (landscape) is constant through a time period of decade years. In the mid-latitude forests it constitutes 2-3 kg/ha.

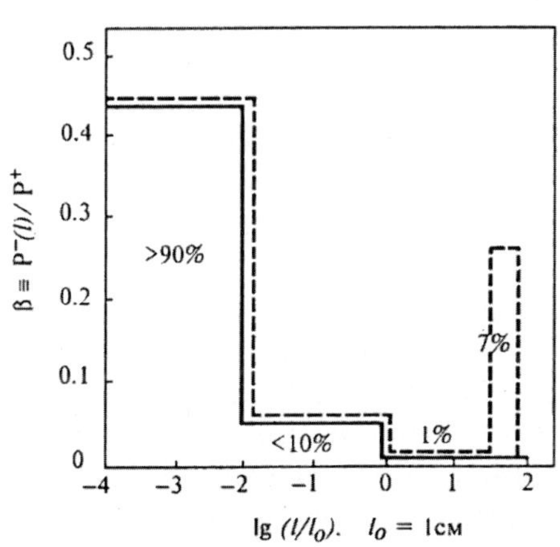

Fig. 1.

Correlation of organisms (producers and consumers) in the biogeocenoses is provided by the flux of substances and energy. It is at a maximum when the size of biogeocenoses is a minimum. The size of biogeocenoses ranges between 1 cm and dozens of meters but not more than the canopy projection of higher plants in the forests. Quasi-homogeneous, correlated communities – biogeocenoses – form populations – ecosystems. The competitive interaction of biogeocenoses in them provides the elimination of those biogeocenoses, which cannot any more regulate and stabilize the environment, since biogeocenoses, like any correlated systems, can disintegrate.

In anthropogenic system the environment-forming factor is the human activity aimed at providing favorable conditions for existing of only one species – *Homo sapiens*, at a given moment and in a short-range perspective. In anthropogenic systems all natural balances, processes, and regularities are actively violated. Therefore such systems are **not** the subject of «joint creation by nature and man» or "co-evolution" of nature and society. In fact, human creates his systems on the ruins of biogeocenoses and ecosystems when he oversteps the limits of its carrying ecological capacity. Therefore in the anthropogenic spatial formations immediately appear various problems connected with the environmental changes. In solving them, humans do not eliminate the basic cause of the problem, but reduce the damage caused by the economic activity, to an acceptable level established subjectively, proceeding from short-term interests, economic and financial possibilities. Ecologically sustainable development, in principle, is impossible in such anthropogenic systems after exceeding carrying economic capacity.

The carrying economic capacity or the limit to disturbance of natural ecosystems (landscapes) is demonstrated in Figure 1. Figure 1 is a graphic expression of the law of distribution of energy fluxes among the consumers (geterothroph organisms) of different sizes in biota, obtained from an empirical generalization of all available data of observations carried out in different natural ecosystems. It shows an acceptable level of consumption of net primary biological production by humans for conditions of preserved environmental stability. Humans may use in their interests not more than 1 per cent of net primary production of global or local ecosystems. It can be recalculated into an admissible economic capacity, which will constitute 1 TW power capacity, or into a value of the mastered area, which ranges within 20-30 per cent of land surface. The theory of biotic regulation has solved the problem about which Holdgate (1994) wrote - that on the problem of carrying capacity of ecosystems "many scientists have broken their teeth". The present-day civilization use near 10% net primary biologic production, use 10TW power capacity and disturbed natural ecosystem on 60% of land. It means that ecological limit overstepped and it has led to hard global ecological crisis and destabilizes life on Earth at all scales – from molecular to global. Thus first limit has overstepped, later will come the others.

The above does not imply an appeal to refuse to create anthropogenic systems, which is unavoidable in conditions of the present civilization. But creating them, it is necessary to have an idea of what man is really doing, of the limits to admissible destruction of the mechanism for the biotic regulation and stabilization of the environment in each ecosystem (landscape) and in the biosphere (geographical environment) on the whole. With these limits taken into account one can speak about the possibility of ecological stability of life within certain way of development.

In the light of the above, the notion of **"ecology" can be defined as follows: it is the science that studies the laws and mechanisms, which provide the stability of life and environment on the Earth**. It is based on the theory of biotic regulation where the key problem is the carrying capacity of ecosystems. Humans have left behind long traversed path, repeatedly violating local ecosystems resulting in local ecological and then resource, social and political crises, before in the twentieth century they overstepped the limits of carrying capacity of the global ecosystem. But still not all historians consider the way of human's development from this point of view. Only V. Vernadsky (1944) wrote that humans cannot independently build his history regardless of the laws of biosphere. It is necessary to build a new history of Civilization, which would be agreed with laws of biosphere.

References

Gorshkov V.G. Physical and Biological bases of Life Stability. (1994) Berlin: Springer-Verlag. 340 pp.

Gorshkov V.G., Gorshkov V.V., Makarieva A.M. (2000) Biotic Regulation of Environment: Key issue of Global Change. Chichester: Springer-Praxis Pbl. 367 pp.

Holdgate M.W. Ecology, Development and Global Policy. J. Appl. Ecol. (1994) 31, N2, 201 pp.

Kondratiev K.Ya., Losev K.S., Ananicheva M.D. Cesnokova I.V. (2004). Stability of Life on Earth: Principle Subject of scientific Research in the XXst century. Chichester: Springer-Praxis. 165 pp.

Lovelock J.E. (1982) Gaia, A new look at life on Earth N-Y: Oxford Univer. Press.

Steffen W. and Tyson P. (Eds.) (2001) Global change and Earth system: a planet under pressure. IJBP Science 4, Stockholm.

Vernadsky V.I. (1944) "Some words about noosphere" Advanced of Modern Biology" 18, N2, 44 pp.

The Potential of Conversion of Environmental Threats into Socioeconomic Opportunities by Applying Ecohydrology Paradigm[6]

Maciej Zalewski

International Centre for Ecology Polish Academy of Sciences, Tylna str. 3, 90-364 Lodz, Poland;Department of Applied Ecology, University of Lodz, Banacha str. 12/16, 90-237 Lodz, Poland, mzal@biol.uni.lodz.pl

Every strategy for success, including water resources sustainable management has to contain both elimination of threats and amplification of opportunities. The efficient strategy for sustainable development and consequently for spending of social taxes has to be focused on two elements. The first is reduction of pollutants emission by construction of sewage treatment plants and energy and mater use (Factor 4 – von Weizsaecker) as well as reduction of catastrophic floods and droughts, mostly by building dams and levees and erosion, by planting trees or reduction of slope of river valleys by terraces. The second, which is condition *sine qua non*, should be the restoration of evolution established: ecological cycles – water, nutrients, and energy flow in the ecosystems. The ecohydrology concept developed in the framework of UNESCO International Hydrological Programme, hypothesized and next empirically confirmed the possibility to use hydrology to regulate biota dynamics and vice versa: use biota to regulate hydrology. The synergic integration at the basin scale of various ecohydrological measures based on mathematical modeling provides scientific background for the enhancement of carrying capacity of the ecosystems. This in turn improves ecosystem services and in consequence enables the creation of positive socioeconomic feedback between environmental quality and society.

[6] This paper is a synthesis of author papers published in 2000-2005 in International Journal Ecohydrology & Hydrobiology

V. Burdyuzha (ed.), The Future of Life and the Future of Our Civilization, 121–131.

1. Introduction

The freshwater ecosystems are situated in depressions of the landscape. As the consequences the water quality has been to great extend dependent on human population density and its whole range of activities.

The anthropogenic impact on freshwater environments can be defined in two dimensions: technological and environmental. Emission of pollutants, which can be controlled by technologies and environmental which are primarily of physical character - modification of hydrological and biogeochemical cycles due to deforestation urbanisation canalisation.

Ecohydrology is a new paradigm which attempts to do this, developed over the lifetime of UNESCO IHP-V, 1997-2001. It has been a new concept which suggests that the sustainable development of water resources has been dependent on ability to maintain evolutionary established processes of water and nutrients circulation and energy flow at the basin scale by use two ways of regulation: biota by hydrology and hydrology by biota (Zalewski, 2001). The profound understanding of whole range processes has based on understanding of the role of aquatic and terrestrial biota in the water dynamics in the all range of scales from molecular to the basin.

The general question in formulation of Ecohydrology (EH) concept was how to regulate biological processes of freshwater ecosystem using hydrology and *vice versa* how to use biotic ecosystem properties as tool in water management.

In the face of increasing pressure on freshwater resources, there remains an urgent need for new practical tools to achieve their sustainable management. The traditional water management does not consider ecosystem properties as a potential management tool.

During the genesis of ecohydrology, it was concluded that the key questions to integrate biota and hydrology should meet the two following fundamental conditions:

1. They should be related to the dynamics of two entities in such way that the answer without consideration of one of the two components (both ways E↔H) would be impossible. In other words, this question should enable the defining of relationships between hydrological and biological processes in order to obtain comprehensive empirical data at the same spatial and temporal scales.
2. The results of the empirical analysis should test the whole range of processes (from a molecular to catchment scale), should enable their spatial/temporal integration and should be convertible to large-scale management measures in order to enable further testing of the hypotheses.

Taking into account the above conditions, the key questions for ecohydrology have been defined based on an in-depth understanding of the interplay between biological and hydrological processes and the factors that regulate and shape them. The hypotheses have been defined in the form of the following questions:

Hypothesis H1: "The regulation of hydrological parameters in an ecosystem or catchment can be applied for controlling biological processes".

Hypothesis H2: "The shaping of the biological structure of an ecosystem(s) in a catchment can be applied to regulating hydrological processes".

Hypothesis H3: "Both types of regulation integrated at a catchment scale and in a synergistic way can be applied to the sustainable development of freshwater resources, measured as the improvement of water quality and quantity (providing ecosystem services)".

It should be stressed that according to the ecohydrology concept, the overall goal defined in the above hypotheses is the sustainable management of water resources. This should be focused on the enhancement of ecosystem carrying capacity for ecosystem services and anthropogenic stress.

Such interdisciplinary, integrative approach provides the background to convert environmental threats into sustainable development – see SIL NEWS, Vol.40, cover page.

2. The Ecohydrology Principles

The concept of Ecohydrology is based on the following principles (Zalewski et al., 1997, Zalewski, 2000):

Integration of Catchment and Biota into a Platonian Superorganism

This covers such aspects as:

Scale - the mesocycle of water circulation in a basin (terrestrial/aquatic ecosystem coupling) has been the template for quantification of ecological processes such as nutrient dynamics and energy flow;

dynamics - water and temperature has been a driving force for terrestrial and freshwater ecosystems;

hierarchy of factors - the abiotic processes are dominating (hydrology), however once they become stable and predictable the biotic interactions begin to manifest themselves (Zalewski and Naiman, 1985).

Understanding the Evolutionarily-Established Resistance and Resilience of such Supreorganisms to Stress

This aspect of Ecohydrology expresses the proactive approach for sustainable management of freshwater resources. It assumes that there is not enough to protect ecosystems owing to increasing global changes because an increase of population, energy, material and human aspirations takes place. Therefore it is necessary to increase the 'absorbing capacity' (resistance and resilience) of an ecosystem against human impact.

The Use of Ecosystem Properties as Management Tools

has been expressed as Ecological Engineering (Mitsh 1993, Jørgensen 1996).

The eutrophication of inland water is a most complex consequence of various forms of human impact, which are synergistically amplified at the whole catchment of "super organism" e.g. agriculture, urbanization, recreation, point source pollution. Due to its complexity the eutrophication process was considered as a major case for formulation of ecohydrological principles and tenants. However the EH was formulated toward an universal solution. There is an increasing number of evidences confirming broad scope of aspects of water resources management e.g. application for enhancement of biodiversity (Agostinho, et al., 2001), fish production (Timchenko & Oksiyuk 2002), bioenergy, reduction of stable pollutants (e.g. heavy metal and pesticides) – Gouder de Beauregard & Mahy, 2002, socio-economic feedbacks (Zalewski, 2002) in ecosystem.

3. The Use of Hydrology for Regulation of Biotic Interactions

The first of three major key processes of the ecohydrological approach is regulation of biological cascade by hydrological manipulation is exemplified by case in Sulejow Reservoir (Fig. 1).

During eutrophication fishes depend on food in a limnetic zone, but still reproduce in littoral. So in eutrophicating reservoirs where planktonic food is not limited, the reproductive success of cyprinids, percids and centrarchids depends mostly on spawning substrata, especially on the extent to which shore vegetation is flooded (Ploskey 1985).

Following flooding of the shoreline herbal vegetation, fry survival was high, large zooplankters were drastically reduced, planktonic algal biomass increased sharply, and water quality declined. Due to overcrowding, intra- and interspecific competition among fry was high, resulting in mass shoreline migration, 30% growth retardation, and low overwinter survival. The scarsity of large zooplankters reduced growth so much in pike-perch fry (Stizostedion lucioperca L.) that they were not big enough to eat even the slow growing perch (Perca fluviatilis L.) and roach fry at the time the pike-perch would normally become piscivores in mid-July. The consequent lack of one generation of such an easily over-exploitable species might reduce its population density seriously for many years (Reid & Momot, 1985).

So, in temperate lowland reservoirs the reproductive success of dominant fish species could be regulated by their access to the shoreline ecotone by hydrology regulation on a dam. By enhancing piscivores to reduce the planktivores fish populations (Hrbacek et al., 1961; Shapiro et al., 1975), the density of efficient large filtering zooplankton (eg., cladocerans) can be increased, and the quality of stagnant water improved.

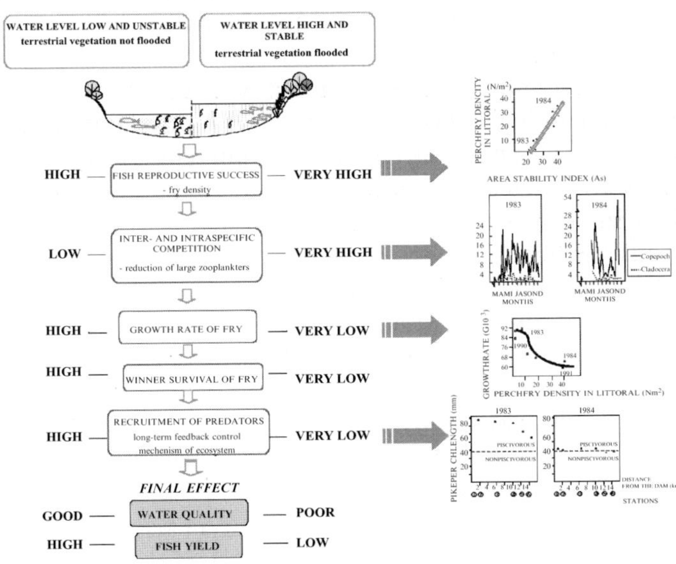

Fig. 1. The regulation of cascade of biological processes by hydrological manipulation for improvement of water quality and optimization of fish yield (for explanation see Zalewski et al., 1990; 2001).

4. Integration of Hydrological and Ecological Regulatory Measures in Basin Scale

The integrative application of all three principles at a river basin has been exemplified by Figure 2 where for reduction of eutrophication of temperate reservoir different ecological processes in the river basin has been integrated toward reduction phosphorus input and reduction of its dynamic pool. Starting from the top of the catchment, the first stage has to be enhancement of nutrient retention within the catchment by reforestation, creation of ecotone buffering zones and optimisation of agricultural practices. The buffering zones at a land water interface reduce the rate of groundwater flux due to evapotranspiration along the river valley gradient. Nutrient transformation into plant biomass in ecotone zones may further reduce the supply into the river. The wetlands in the river valley from the buffering zone: they reduce the mineral sediments, organic matter and nutrient load transported by the river during floods periods through sedimentation. Also in some artificial wetlands, nitrogen load can be reduced significantly by regulation the water level to stimulate denitrification through anaerobic processes. In shaded rivers with high nutrients load it is possible to serious extent amplify the self-purification capacity by increasing light access and maintaining filtering function - creating intermediate complexity of ecotones. If despite all above measures combined with necessary sewage treatment plants the nutrient concentrations in a reservoir are too high and potentially might be converted in to toxic algal blooms, there exist numerous methods to reduce recirculation nutrients in a reservoir by blocking its in the biomass of macrophytes, and translocation between trophic levels (e.g. biomanipulation). Since properties of a large scale system cannot be predicted from properties of its component element such complex strategy for restoring and controlling nutrients in the catarrhini landscape and freshwater ecosystem should be assessed continuously at every stage of implementation (Holling et al., 1994) and adjusted to maximize a potential synergistic effect.

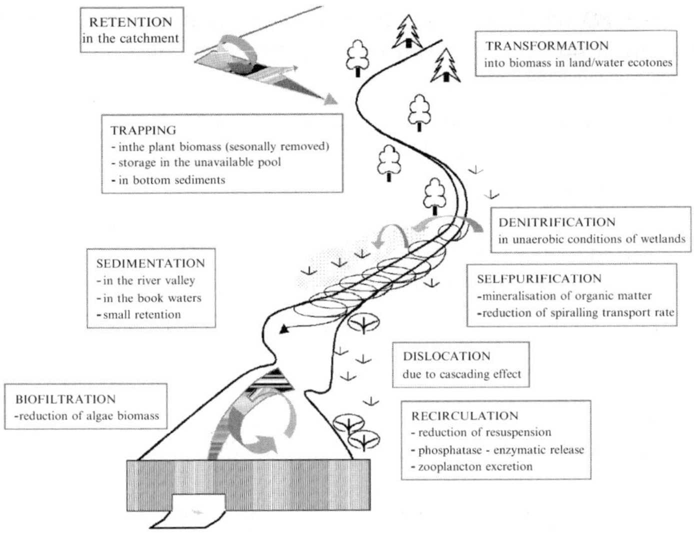

Fig. 2. The integration of three tenants of ecohydrology: Regulation (E→H (self-purification) + H→E (denitrification); Integration on synergistic way of different E and H ways of system regulations; Harmonization with existing technical solutions (use the dam as a tool of water level regulation for shaping biotic interactions toward good water quality and optimisation of fish yield (see fig. 1)

5. Conversion of Environmental Threats into Socio-Economic Opportunities for Sustainable Development

The degradation of freshwater ecosystems has been of two-dimensional character pollution and disruption of evolutionary establishing water and biogeochemical cycles in landscape. Both causes of destruction of a biotic structure of the catchments, freshwater ecosystems and water resources are declined.

Pollution can be significantly reduced or eliminated by technological progress, however, degradation of natural processes of water circulation nutrients and energy flow in the catchments scale creates a much more complex problem.

The 21st century will become an era of integrative science because understanding the complexity of our world is a key to achieving sustainable development. This is especially urgent in ecology and environmental sciences for two reasons. First, there is an urgent need for sound solutions toward declining ecosystem services and biodiversity at the global scale.

The second, further scientific progress can be made by testing existing concepts and "know-how", by implementing concepts and methods integrated at a large scale of basin landscape.

Traditional sewage treatment plants in a small town usually do not possess a sophisticated tertiary chemical treatment stage due to high costs of construction that local communities cannot afford. They reduce BOD and some nutrients but still negatively influence water quality, reducing the benefits of rivers and reservoirs and their recreational values. Extending the sewage treatment by constructing a wetland results in more efficient reduction of pollutant loads and generates additional societal benefits. Improvement of water quality increases the appeal of water resources for tourism, which contributes to the inflow of capital to a region (Fig. 3).

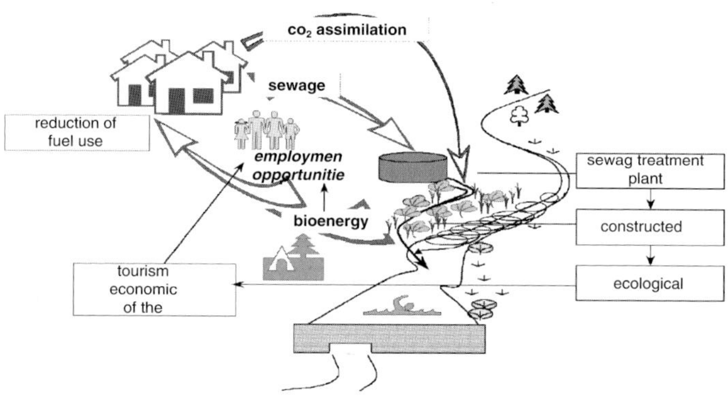

Fig. 3. The development of the ecohydrology concept for improving water quality, ecosystem services and the creation of positive socioeconomic feedbacks.

Moreover, the establishment of multispecies willow plantations using local species that can tolerate the resulting high ground water level, maintain river valley landscape biodiversity and provide an alternative source of energy (bioenergy) that can help to reduce CO_2 emissions from burning fossil fuels. The resultant ash can be used to fertilize forest plantations. Thus, pollutants are converted into bioenergy. Producing bioenergy and timber also generates new employment opportunities and revenue flows while reducing capital outflows for fossil fuel use. Bioenergy can be use for conversion of non-degradable plastic wastes by low energy technology in to paraffin's linking water and waste management. The use of ecological knowledge, therefore, results not only in a good quality environment but also can help to elevate the economic status and level of sustainable development in local communities.

Such implementation case as a UNESCO/UNEP demo site has been recently under development at the town of Przedborz on the Pilica River, a western tributary of the Vistula River, above Sulejow Reservoir in Poland (Fig. 3).

Following this as a result of several years of cooperation by an interdisciplinary team of scientists involved in International Hydrological Programme UNESCO, the key tenants of ecohydrology were formulated:

1. REGULATION of hydrology by shaping biota and, vice versa, regulation of biota (eg. elimination of toxic algal blooms) by altering hydrology (e.g. Zalewski et al., 1990, Zalewski et al., 2001) – Fig. 1.
2. INTEGRATION – at the basin scale various types of regulations (E↔H) should be integrated towards achieving synergy to stabilize and improve the quality of freshwater resources at a basin scale (Zalewski, 2000).
3. HARMONIZATION of ecohydrological measures with necessary hydrotechnical solutions (e.g. dams, irrigation systems, sewage treatment plants, levees at urbanized areas etc.) - Hellegers & Witte, 2002; Timchenko & Oksiyuk , 2002)

The empirical testing of three tenants of ecohydrology at different river basins are urgently needed toward their implementation for sustainable development.

References

Agostinho A.A., Gomes L.C., Zalewski M. (2001). The importance of floodplains for the dynamics of fish communities of the upper river Parana. In: M. Zalewski, F. Schiemer, J. Thorpe (eds.) 2001. International Journal of Ecohydrology and Hydrobiology. Vol. 1 (1-2). Special issue on: Catchment processes land/water ecotones and fish communities. Warsaw 2001.

Gouder de Beauregard A-CH. & Mahy G. (2002). Phytoremediation of heavy metals: the role of macrophytes in a stormwater basin. . International Journal of Ecohydrology &Hydrobiology, Vol. 2(1-4). Proceedings of the Final Conference of the First Phase of the IHP-V Project 2.3/2.4 on Ecohydrology "The Application of Ecohydrology to Water Resources Development and Management" Venice, Italy, 16-18 Sep. 2001.

Hellegers P.J.G.J., Witte J-P.M. (2002) Towards a simple integrated model for the re-wetting of nature reserves. International Journal of Ecohydrology &Hydrobiology, Vol. 2(1-4). Proceedings of the Final Conference of the First Phase of the IHP-V Project 2.3/2.4 on Ecohydrology "The Application of

Ecohydrology to Water Resources Development and Management" Venice, Italy, 16-18 Sep. 2001.

Holling, C.S., Gunderson, L.H. & Walters, C.J. 1994. The structure and dynamics of the Everglades system: guidelines for ecosystem restoration. In: S. Davis, J. Odgen (Eds.) The Everglades: The Ecosystem and its Restoration. St. Lucie Press, Delray Beach.

Hrbacek J., Dvorakova M., Korinek V. & Prochazkova L. 1961. Demonstration of the effect of the fish stock on the species composition of zooplankton and the intensity of metabolism of the whole plankton association. Verhandlungen der Internationale Vereinigung fur Theoretische und Angewandte Limnologie 14, 192-195.

Jorgensen, S.E. 1996. The application of ecosystem theory in limnology. Verh. Int. Verein Limnol. 26, 181-192.

Mitsch,. W.J. 1993. Ecological Engineering - a co-operative role with planetary life-support system. Environ. Sci. Technol. 27, 438-445.

Ploskey G.R. 1985. Impacts of terrestrial vegetation and preimpoundment clearing on reservoir ecology and fisheries in the United States and Canada. Food and Agriculture Organization of the United States, Fisheries Technical Paper 258, 1-35.

Reid D.M. & Momot W.T. 1985. Evaluation of pulse fishing for the walleye, Stizostedion vitreum vitreum, in Henderson Lake, Ontario. J. Fish. Biol. 27 Suppl. A, 235-251.

Shapiro, J., Lamarra, V. & Lynch, M. 1975. Biomanipulation: an ecosystem approach to lake restoration. pp: 85-96. In: P. L. Brezonik, J. & L. Fox (Ed.). Proceedings of a Symposium on Water Quality Management Through Biological Control, University of Florida.

Timchenko V., Oksiyuk O. 2002. Ecosystem condition and water quality control at impounded sections of rivers by the regulated hydrological regime. International Journal of Ecohydrology &Hydrobiology, Vol. 2(1-4). Proceedings of the Final Conference of the First Phase of the IHP-V Project 2.3/2.4 on Ecohydrology "The Application of Ecohydrology to Water Resources Development and Management" Venice, Italy, 16-18 Sep. 2001.

Zalewski, M., Naiman, R.J. 1985. The regulation of riverine fish communities by a continuum of abioti-biotic factors. In: Alabaster, J.S. [Ed.] *Habitat Modification and Freshwater Fisheries*. 3-9. FAO/UN/Butterworths Scientific, London.

ZalewskI, M., Brewinska-Zaras, B., Frankiewicz, P., Kalinowski, S.1990. The potential for biomanipulation using fry communities in a lowland reservoir: concordance between water quality and optimal recruitment. Hydrobiologia, 200/201: 549-556.

Zalewski, M., Janauer, G. A., Jolankai, G. 1997. Ecohydrology. A new paradigm for the sustainable use of aquatic resources. UNESCO IHP Technical Document in Hydrology **7**. IHP - V Projects 2.3/2.4, UNESCO Paris, 58 pp.

Zalewski M. 2000. Ecohydrology-the scientific background to use ecosystem properties as management tools toward sustainability of water resources. Guest Editorial, Ecological Engineering 16:1-8.

Zalewski M. (ED.). 2002. Guidelines for the Integrated Management of the Watershed – Phytotechnology and Ecohydrology. United Nations Environment Programme, Division of Technology, Industry and Economics. International Environmental Technology Centre. Freshwater Management Series No. 5, 188pp

Zalewski, M. 2002. Ecohydrology – the use of ecological and hydrological processes for sustainable management of water resources. Hydrological sciences Journal. 47, 825-834

Zalewski M., R. Robarts. 2003. Ecohydrology – a new Paradigm for Integrated Water Resources Management. SIL News 40, Sep. 2003:1-5

Advances in Space Meteorology Modeling and Predicting - the Key Factor of Life Evolution

Mauro Messerotti

INAF-Trieste Astronomical Observatory, Loc. Basovizza n. 302, 34012 Trieste, Italy and Department of Physics, University of Trieste, Via A.Valerion. 1, 34133 Trieste, Italy, messerotti@ts.astro.it

Both the emergence and the evolution of life on a planet is favored by the existence of suitable environmental conditions, which are in turn determined by the interplay of factors intrinsic to the planet itself and factors driven by energetic inputs originated in the interplanetary space and even in the galactic and extragalactic environments. The observation, modeling and prediction of the perturbation phenomena associated with the known variety of astrophysical sources, ranging from the central stars to Gamma Ray Bursts, are the goal of Space Meteorology. In this framework, a review on the known phenomenology relevant to life evolution will be given by emphasizing the most advanced prediction techniques available to date and stressing the need of a significant improvement in the light of terrestrial life preservation, as the space environment seems to force the ecospace conditions much more effectively than any anthropogenic factors.

Keywords: Space Meteorology; Space Climatology; Space Weather; Bioastronomy.

1. Introduction

The chemistry and the energetics of Lithosphere, Hydrosphere and Atmosphere of a planet, such as the Earth, determine the presence of suitable biogenic materials and possibly the emergence of life. In particular, physicochemical processes are driven by the interplay of local ones as well as by radiation and particle inputs of solar, interplanetary, galactic and extra-

V. Burdyuzha (ed.), The Future of Life and the Future of Our Civilization, 133–143.

galactic origin (Space Weather and Space Climate), which directly and indirectly have been influencing the terrestrial weather and climate by triggering substantial to minor modifications in the Earth's atmosphere. The outer energy inputs can have played a role in the evolution of life and certainly can play a direct or indirect role in life preservation. Furthermore, the evolution towards more complex life forms up to the human being and the evolution of the human being into a technological civilization introduced new forcing agents of anthropogenic origin in the environmental scenario. The respective weights and the interplay of the natural and the artificial perturbing agents are still not quantitatively defined and are not understood to a satisfactory level of detail yet. Whatever the final answer, it must be stressed that both the anthropogenic and the natural agents play a concurrent role in favoring global changes on the Earth, which can affect the future survival of our civilization. With regard to that, in section 2 we will comment on the basic requirements for life emergence and evolution. The relevance of Space Meteorology to Bioastronomy is outlined in section 3. The impacts of space conditions for life emergence and preservation are considered in section 4 and the advanced modeling of solar activity in section 5. The conclusions are drawn in section 6.

2. Requirements for Life Evolution

Both the emergence and the evolution of life require a set of physical and chemical conditions to occur and to be maintained (Popa, 2004a; Killops and Killops, 2005). In this framework, the availability of an adequate level of energy inputs is fundamental (Popa, 2004b). In fact, chemical bonds and components as well as life structures contain a certain amount of energy. The temporal stability depends on basic materials and external conditions. In turn, life: - needs certain energy input; - keeps energy degradation high; and - is degraded by energy excess. As various models for life emergence exist and the evolution process from chemical to organic is still poorly understood (see e.g. Killops and Killops, 2005), no quantitative thresholds for the radiation levels can be given for life catalysis or suppression in such first stages, but many inferences can be derived for life under conditions of ionizing radiation (e.g. Baumstark-Khan and Facius, 2001). Evolution of life from simple to complex and to varied forms has been also dependent on energy inputs, which has been determining genetic mutations. In this context, Galactic Cosmic Rays (GCR) and Solar Cosmic Rays (SCR), high energy particles originated in our Galaxy and on the Sun respectively, have played a role (Belisheva et al., 2002).

Similarly, high energy electromagnetic radiation from the Sun, such as the ultra-violet (UV) one, represents an effective energy input which affects the evolution of biological systems (e.g. Cockell, 2000) via synergization or inhibition.

3. Space Meteorology and Life

Life emergence and evolution has been affected by external and internal factors, which are related to coupling physical environments at different spatial scales, i.e. the local terrestrial one, the planet Earth, the Solar System, our Galaxy as well as, perhaps, other galaxies. In fact, all such environments are source of energy inputs via electromagnetic and particle radiations. In particular, Space Meteorology (SpM) is aimed at studying the physical state of the outer space and of the ecospace, which can be considered, in a conservative sense, as the region of space which is interested in biological and human activities or as the whole Universe observable by the human being, in an extended sense. The physical state on a short to medium term is defined as Space Weather (SpW), whereas on a medium to long term is defined as Space Climate (SpC), in accordance with the terrestrial analogs (Messerotti and Lundstedt, 2004). Space Weather and Space Climate can act as catalysts or inhibitors for life at its early stages by means of the various outer space perturbations (e.g. Messerotti, 2003, 2004) and at evolved stages, for example by affecting biological systems via altered magnetic fields (e.g. Ghione et al., 2003).

4. Impacts of Space Conditions on Life

The interrelationships between Space Weather, Space Climate and life emergence, evolution and preservation can be summarized by means of a concept map such as the one reported in Fig. 1 (Messerotti, 2003; Messerotti and Lundstedt, 2004), which was generated with the software developed by the Institute for Human and Machine Cognition (IHMC, Florida, USA; http://cmap.ihmc.us).

The concept map clearly points out how the action of Space Climate affects the emergence of life by biasing the formation of the solar system and the formation of one or more habitable planets, as well as the emergence and evolution of life up to its propagation elsewhere. In fact, a set of physicochemical conditions must be minimally maintained to preserve the habitability on a long time scale.

The action of Space Weather, which occurs on shorter time scales, similarly affects the same stages relevant to life other than the solar system formation. SpW is instead quite important for the human and animal physiology on the mother planet and anywhere, and affects the human and animal activities in different ways.

This latter aspect is particularly relevant for the future of our civilization as detailed in the following.

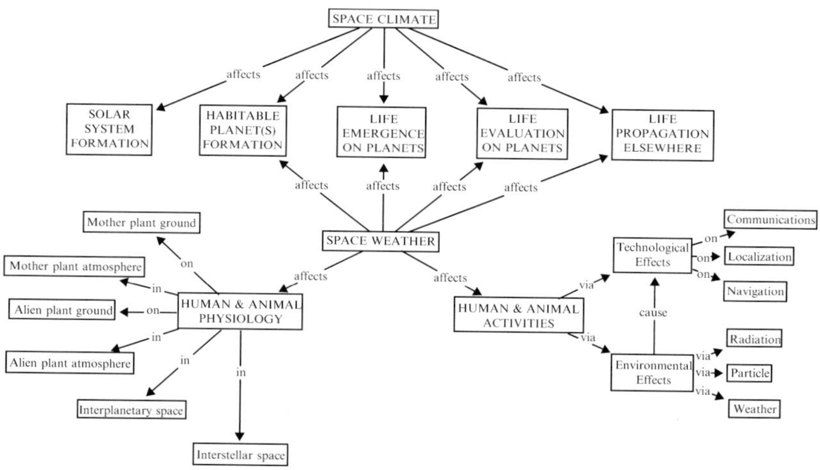

Fig. 1. Interrelationships and impacts of Space Weather and Space Climate.

In fact, Space Weather and Space Climate are characterized by a series of environmental effects, such as high energy electromagnetic radiation outbursts and particle storms, which originate in our Solar System, in our Galaxy and even farther away. Moreover, remnants of the Solar System formation, like asteroids and comets, can occasionally collide with planets and, dependently on their size, can cause significant changes in the planetary climate up to triggering biological catastrophes. The activity of the Sun is characterized by solar flares: highly intense and localized magnetic fields of opposite polarity associated with sunspots annihilate and impulsively release the stored energy. This results in local heating of the solar plasma to tens of millions ^{o}K, electron and proton acceleration (SCR; Belisheva et al., 2002), outbursts of X, UV and radio radiation, capable to transfer energy to the magnetosphere by perturbing the geomagnetic field, to enhance the ionization and the density of the ionosphere, to deplete the ozone layer and to feed the upper atmosphere with a large amount of energy. Another solar driver of SpW are Coronal Mass Ejections (CME), plasmoids ejected from the solar corona following e.g. a prominence eruption or the evolution of a flare. Halo CMEs and Earth-directed ones are geoeffective, as in such cases the plasmoid reaches the magnetosphere and can transfer energy to it: charged particles can penetrate the magnetic shielding when the polarity of the interplanetary magnetic field is opposite to that of the geomagnetic one and reconnection is favored. Outer sources, such as supernovae-associated outbursts, massive star collapse to a black hole or neutron star merging into a black hole, originate Cosmic Rays (GCR), Extreme Energy Cosmic Rays (EECR), Gamma Ray Bursts (GRB) and Ultra-High energy GRBs (UHGRB) (e.g. Gialis and Pelletier, 2005). SCRs occur on a sporadic basis in-phase with solar activity but GCRs constitute a continuous, quasi-isotropic flow of high-energy charged particles, which is modulated by the solar wind in anti-phase with the solar activity cycle. When impacting the Earth's atmosphere, they generate a shower of secondary particles upon interacting with the atmospheric atoms and molecules. All above cited SpW and SpC effects have been operating with different effectiveness for life during the Sun's and planets' evolution.

4.1 The Young Sun and the Early Solar Weather

The Sun-In-Time Program has been devoted to study the radiation and magnetic evolution of the Sun along the Main Sequence as inferred via multi-wavelength observations of solar analogs in the age range from 0.1 to 9.0 Gyr (e.g. Guinan and Ribas, 2004). The outcomes of such research indicate that the early solar weather was determined by a young Sun that was: - extremely active with more frequent high-energy flares (2-5 per day); - originating massive winds (500-1000 times the present time); - producing high-energy emissions (1000 times the present time). The following trends with increasing age were derived: - spin-down of the rotation period (two orders of magnitude); - decrease of Far UV surface flux (one order of magnitude); - decrease of relative fluxes in the high energy wavelength bands. The intensities of fluxes that can be estimated for the young Sun at an epoch 4 Gyr BP (Before Present) with respect to the present time values are: - 11 times in the X band; - 6 times in the XUV (extreme ultraviolet) band; - 4 times in the FUV (Far ultra-violet) band; - 3 times in the UV band; - 0.7 times in the visible band. According to this scenario, when the radiation output was 70% of the today Sun, it was emitting a significantly higher output at shorter wavelengths. Consequently the effects of the young Sun on the Earth paleoatmosphere can be summarized as: - an increased energy deposition; - an enhanced photodissociation; - an increased exosphere heating, expansion and photoionization; - an effective atmosphere erosion.

4.2 Evolution of the Atmosphere and Life on Earth

The evolution of the Earth's paleoatmosphere (Killops and Killops, 2005), as driven by the early solar weather and, in particular, by the evolution of the UV radiation flux (Cockell, 2000), was critical to life emergence. In turn, later it was driven by the evolution of life itself. In fact, the evolution of the atmosphere from anoxic to oxygenous was driven by the interplay between the UV radiation level from the Sun, which was driving the photo-reaction processes, capable to favor the evolution of life from chemical to organic, and the later action of organic life, capable to directly or indirectly modulate such processes.

In this context, the understanding of the early Sun state is a key factor in modeling the evolution of the paleoatmosphere, as the adoption of the Standard Solar Model leads to an early Sun characterized by lower brightness and activity with respect to the present time, whereas recent solar models (e.g. Sackmann and Boothroyd, 2003) describe the early Sun with higher brightness and activity consistently with the observational estimations outlined in section 4.1.

4.3 Possible SpW Drivers of Earth's Climate Changes

Global changes in the mean temperature of the atmosphere can severely affect the biological systems.

The Sun plays an important role in affecting the climate via its radiation inputs and their evolution in time. Such a role is presently estimated of the order of 25 to 30% in the global change models, but much has still to be understood via diachronic, the high resolution observations, as stressed by the International Panel on Climate Change.

For more information the reader is referred to the paper on the science of climate change by Weaver (2003) and to the paper on celestial climate drivers by Veizer (2005), which present different perspectives. The long-term solar variability and climate change was extensively studied e.g. in Muscheler et al. (2004) and the modeling of the effect of solar variability on climate e.g. in Schlesinger and Andronova (2004).

In the following we just mention some observed aspects, which could be (and could have been) relevant to the climate change but whose related processes and effects has still to be fully understood and quantified: - the change in the total solar irradiance (0.09 % in a solar cycle), which contributes to the heating of the atmosphere; - the change in length of the solar cycle (e.g. 9-12 years in 150 years and from 11.8 to 9.8 years in the period 1885-1930). The higher frequency of activity peaks modifies the duty cycle of energy deposition; - the change in UV radiation flux (the Extreme UV flux increases 8 times in a half cycle), which affects the temperature, chemistry and dynamics of the atmosphere; - the direct influence by high energy solar protons (SCR), which favor the ozone depletion; - the indirect influence by GCRs modulated by solar activity (lower flux during maximum; 15% decrease in 100 yr due to doubling of solar magnetic flux), which were proposed as drivers of the global cloud coverage by Svensmark and Friis-Christensen (1997), based on the ionization favoring the cloud formation, and the lower temperatures possibly associated with a larger cloud coverage.

4.4 Space Weather Effects and Our Civilization

As a natural evolution of intelligent life forms, our civilization has reached a technological level of ever increasing complexity. Everyday life on the Earth as well as space activities heavily rely on technological devices and on very sophisticated technologies, which, in turn, are more sensitive to SpW effects. For example, SpW-originated modifications of the ionosphere can affect radio communications, in particular making less reliable the navigation based on the Global Positioning System (GPS), causing black-outs in the propagation of radio waves at certain wavelengths or even jamming the communications via mobile phones. Energetic particle storms from the Sun can cause malfunctions in satellites or, in the worst cases, can electrically destroy them. Furthermore, they are potentially harmful for the astronauts in space and for the crews of airplanes flying at higher altitudes. Lastly, the perturbation of the geoelectric and geomagnetic systems can induce high intensity electric currents in long wires and power plant transformers, causing blackouts and serious damages. The successful prediction of such SpW effects well in advance is the only way to allow for a viable minimization of the most dangerous ones.

5. Advanced Solar Weather Predictions

Many scientific models have been proposed for the various solar drivers (Lathuillere et al., 2002). On the other hand, very few (if any) of them can be incorporated in a prediction system due to their numerical complexity and hence computational demand or to the limitations in the considered phenomenological domain. As most critical SpW effects depend on solar weather, it is of fundamental importance to study effective approaches to overcome the present limitations (Messerotti and Lundstedt, 2004). In order to improve the forecasts of the impact of solar activity on the terrestrial environment on time scales longer than days, improved understanding and forecasts of the solar activity are needed. Promising results of a new approach of modeling and forecasting solar activity are based on the Lund Solar Activity Model (LSAM) (Fig. 2), which is a hybrid, physics-based neural network (NN) model. Time series of solar activity indicators, such as sunspot number, group sunspot number, F10.7 (radio flux index at 10.7 cm), E10.7 (UV flux proxy index), solar magnetic mean field, Mount Wilson plate and sunspot index, analyzed by new wavelet methods, which point out deterministic and stochastic features (Lundstedt et al., 2005), are used to produce a pre-processor neural network. The model uses NNs to discover new numerical laws and incorporate the known solar activity theory. On one hand, this facilitates a reliable forecasting, and, on the other hand, it allows the merging of all the relevant knowledge in a post-processor NN. Until the level of knowledge of solar activity will be significantly improved, such a hybrid approach is the most effective predictive technique.

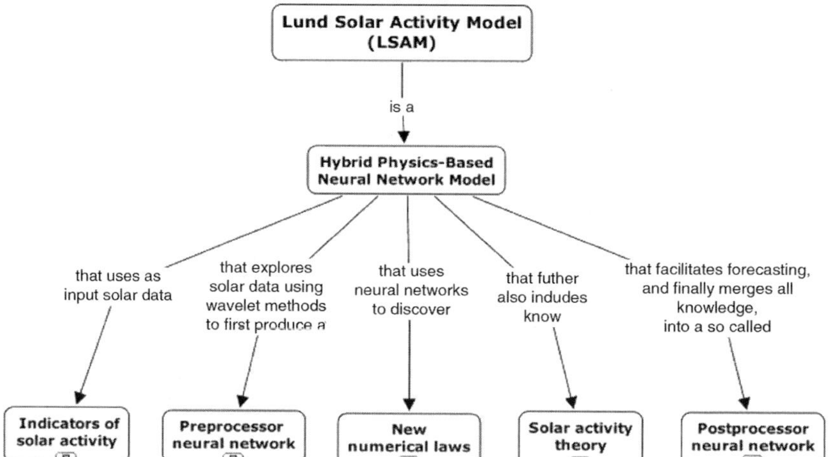

Fig. 2. Concept map of the Lund Solar Activity Model.

6. Conclusions

Life emergence and evolution on any planets are heavily biased by the environmental conditions of the outer space as well as by forcing agents of anthropogenic origin. Hence no catastrophe originated by inner and outer or anthropogenic drivers can be a priori ruled out. Anyway, no sufficient knowledge of the phenomenology exists for establishing predictive techniques for such catastrophes that can affect to various extents the preservation of life on our planet. In fact the time series of relevant observations even derived via refined indirect methods for the past are still fragmentary and their interpretation suffers by the uncertain working hypotheses. On the other hand, the global models, even if reasonably indicative of future trends, are still unsatisfactory due to the poor knowledge of some concurrent factors, such as the role of outer drivers, which might be presently un-underestimated. According to our opinion, the unpredictability of natural catastrophes is a major source of concern for life evolution to date. Technological civilizations are more and more dependent on advanced technological systems, and more advanced technologies are more sensitive to Space Weather effects. The to-date knowledge indicates the Sun as the primary driver of outer perturbations, via its direct electromagnetic and particle emissions as well as via the modulation of e.g. Cosmic Rays. Even if this is still a debated aspect, the Sun may have a first-order influence on Earth weather and climate, perhaps more relevant than any concurrent anthropogenic forcing. Notwithstanding, the human being must be anyway seriously committed in limiting its perturbations to the biosphere, as, even if it will be scientifically proved that the action of space drivers prevails, the human action on the environment is by no means negligible and can only worsen the global situation. As a final remark, we stress that other outer drivers, like GRBs and EECRs, cannot be completely disregarded as potentially dangerous for life evolution, but their action seems having synergized life via genetic mutations instead of suppressing it.

If we restrict our attention to the future of our civilization, we can conclude that, assuming it will evolve towards more and more advanced technologies for everyday life (but also for extended space explorations), which we can guess, according to the today knowledge, will be highly sensitive to ecospace perturbations, the advancement in Space Meteorology will be a must for reliably predicting both short-term Space Weather effects and long-term Space Climate effects, and for reliably predicting global changes.

Acknowledgements

This work has been supported by ESF/COST Action 724 "Developing the scientific basis for monitoring, modeling and predicting Space Weather". The SOC is gratefully acknowledged for the kind invitation, the Editor for his full collaboration during the preparation of the manuscript, and J. Chela-Flores (ICTP, Trieste) for many fruitful discussions.

References

Baumstark-Khan, C., and Facius, R. (2001) Life under Conditions of Ionizing Radiation, in: Astrobiology, The Quest for the Conditions of Life, G. Horneck and C. Baumstark-Khan (eds.), Springer, Berlin, pp. 261-284.

Belisheva, N. K., Semenov, V. S., Tolstyh, Yu. V., and Biernat, H. K. (2002) solar flares, generation of cosmic rays, and their influence on biological systems, ESA SP-518, pp. 429-430.

Cockell, C. S. (2000) The ultraviolet history of the terrestrial planets - implications for biological evolution, Planetary and Space Science 48: 203-214.

Ghione, S., Del Seppia, C., Mezzasalma, L., and Messerotti, M. (2003) Possible Relevance of Space Weather Effects to Medicine: Influences of Altered Magnetic Fields on Biological and Clinical Phenomena, in: ESA Space Weather Workshop: Developing a European Space Weather Service Network (3-5.11.2003, ESA/ESTEC, NL); online proc.: http://www.estec.esa.nl/wmwww/wma/spweather/workshops/spw_w5/proceedings.html.

Gialis, D., and Pelletier, G. (2005) High-Energy Emission and Cosmic Rays from Gamma-Ray Bursts, Ap. J. 627 (2): 868-876.

Guinan, E. F., and Ribas, I. (2004) Evolution of the Solar Magnetic Activity over Time and Effects on Planetary Atmospheres, in Stars as Suns: Activity, Evolution and Planets, IAU Symp. 219, A. K. Dupree and A. O. Benz (eds.), Atron. Soc. of the Pacific, USA, pp. 423-430.

Killops, S., and Killops, V. (2005) Introduction to Organic Geochemistry, 2nd ed., Blackwell Publishing Ltd, USA.

Lathuillere, C., Menvielle, M., Lilensten, J., Amari, T., and Radicella, S.M. (2002) From the Sun's atmosphere to the Earth's atmosphere: an overview of scientific models available for space weather developments, Ann. Geophys. 20: 1081-1104.

Lundstedt, H., Liszka, L., and Lundin, R. (2005) Solar activity explored with new wavelet methods, Annales Geophysicae 23 (4): 1505-1511.

Messerotti, M. (2003) Solar and Stellar Space Weather and Space Climate: Relevant vant Issues in the Birth and Evolution of Life? in: ESA Space Weather Workshop: Developing a European Space Weather Service Network (3-5.11.2003, ESA/ESTEC, NL); online proc.: http://www.estc.esa.nl/wmwww/wma/spweather/workshops/spw_w5/proceedings.html.

Messerotti, M. (2004) Space weather and space climate: life inhibitors or catalysts?, in: Life in the Universe From the Miller Experiment to the Search for Life on Other Worlds. Seckbach, J.; Chela-Flores, J. Owen, T. Raulin, F. (eds., series: Cellular Origin, Life in Experince Habits and Astrobiology, pp. 177-180.

Messerotti, M., and Lundstedt, H. (2004) What is next in Solar Weather Monitoring, Modeling and Forecasting? in First European Space Weather Week (29.11-3.12.2004,ESA/ESTEC,NL);onlineproc.:http://www.esa-spaceweather.net/spweather/workshops/esww/proceedings.html.

Muscheler, R., Beer, J., and Kubik, W. (2004) Long-Term Solar Variability and Climate Change Based on Radionuclide Data from Ice Cores, in: Solar Variability and Its Effects on Climate, J. M. Pap and P. Fox (eds.), American Geophysical Union, USA, pp. 221-235.

Popa, R. (2004a) The Early History of Bioenergy, Between Necessity and Probability: Searching for the Definition and Origin of Life, Springer, Berlin, pp. 15-34.

Popa, R. (2004b) Energy and Life, in Between Necessity and Probability: Searching for the Definition and Origin of Life, Springer, Berlin, pp. 165-166.

Sackmann, I. J., and Boothroyd, A. I. (2003) A Bright Young Sun Consistent with Helioseismology and Warm Temperatures on Ancient Earth and Mars, Ap. J. 583 (2): pp. 1024-1039.

Schlesinger, M. E., and Andronova, N. G. (2004) Has the Sun Changed Climate? Modeling the Effect of Solar Variability on Climate, in: Solar Variability and Its Effects on Climate, J. M. Pap and P. Fox (eds.), American Geophysical Union, USA, pp. 261-282.

Svensmark, H., and Friis-Christensen, E. (1997) Variation of cosmic ray flux and global cloud coverage - a missing link in solar-climate relationships, J. Atm. Terr. Phys. 59 (11): 1225-1232.

Veizer, J. (2005) Celestial Climate Driver: A Perspective from Four Billion Years of the Carbon Cycle, Geoscience Canada 32 (1): 13-28.

Weaver, A. J. (2003) The Science of Climate Change, Geoscience Canada 30 (3): 91-109.

Ocean Circulations and Climate Dynamics

Mojib Latif

Leibniz Institute of Marine Sciences at Kiel University (IFM-GEOMAR), Dusternbrooker Weg 20, 24105 Kiel, Germany, mlatif@ifm-geomar.de

The research work of the research unit is directed toward the dynamics of the ocean circulation, in order to obtain a better understanding of the role of the ocean in climate change and with respect to environmental problems. The tools used are numerical models on regional, basin-wide and global scales, with which the complexity of oceanic variability can be studied on time scales from weeks to millenia and beyond. These models are developed and analyzed in close co-operation with the observational groups at IFM-GEOMAR.

V. Burdyuzha (ed.), The Future of Life and the Future of Our Civilization, 145.

How We Are Far From Bifurcation Point of the Global Warming?

Oleg Ivaschenko

Sovetskaya str.104, apart.4, Chistenkoe, Simferopol, Ukraine, info@poteplenie.ru

Today the problem of global warming is discussed as one of anthropogenic influence on climate owing to emission of greenhouse gases during economical activity. However, a strong influence may give violation of the temperature regime of carbon circulation. The anthropogenic factor may be an initial push to including other process of mass separation of greenhouse gas from their natural reservoir where they are in conservation state. This will come to further increase of global warming process. That is the anthropogenic factor acts as a trigger mechanism of serious changes of climate.

Questions of separation CO_2 from oceans are discussed many years. The problem of CH_4 separation from oceanic methane hydrates is not discussed practically. But CH_4 is the greenhouse gas and oceans contain in 3000 times more methane than atmosphere. The reservation of methane hydrates may be subjected to sufficient destruction for warming of oceanic depths on some degrees only. This comes to the increase of greenhouse effect.

Similar effects were in past history of the Earth. 55 millions years ago during some millennia took place disintegration of 1200 Gton methane hydrates. In the result in the end of Paleocene water temperature of World Ocean increased sharply on 8°C. But in the present time reservation methane hydrates is 10 times more than before. It is the unique situation.

Our estimates have shown that in the end of this century an additional effect of warming may be up to 15-20°C although these were preliminary calculations in which did not take into account weakening of temperature grow for strong increasing of methane concentration in atmosphere. They were carried out together with I.K. Larin from Institute of Energetic Problem of Chemical Physics (Russian Academy of Sciences).

V. Burdyuzha (ed.), The Future of Life and the Future of Our Civilization, 147.

Mankind Can Strive Against the Global Warming

Michel E. Gertsenstein, Boris N. Shvilkin

Physical Department of Moscow University, Leninskie gori, 119992 Moscow, Russia, shvilkin@polly.fis.msu.ru

Most likely the cause of the global warming can be carbon dioxide venting into the atmosphere as a result of organic fuel burning though it isn't proved rigorously. Now it is clear that mankind is in charge of the beginning of the global warming and must prevent it. The point of bifurcation- the point of the loss of stability can't be crossed on any account. So it is necessary to find the sources of energy, which don't use organic fuel. The nuclear power represents such a source in production quantities. But the nuclear power also has disadvantages, connected with the problem of a utilization of nuclear waste and hazard of possible accidents. The methods of elimination of these disadvantages are discussed in this article.

1. Does the global warming exist or it's a myth-a 'scarecrow' for people? What are the causes of the global warming?
2. What time do people have to eliminate the global warming?
3. How can we strive against the global warming?

In Y.L.Hotuntsev's review (this issue, p. 5) on the collector of the Symposium reports 'The Future of the Universe and the Future of Our Civilization' (Singapore, 'World Scientific', May, 2000) it is stated that a crisis, connected with the loss of climate stability and the global warming, can break out on the Earth.

The nature is seriously damaged even under rigorous observance of the modern technology. Carbon dioxide blowout at a fuel burning is an example of it. Fuel gases can contain other substances, which can be absorbed, but it is impossible to avoid a blowout at the fuel burning. Carbon dioxide produces to the greenhouse effect-the global warming. As Prigozhin noted (1) that it can lead to the instability of the climate after coming over the bifurcation point, then the process will become irreversible and uncontrolla-

V. Burdyuzha (ed.), The Future of Life and the Future of Our Civilization, 149–153.

ble. The Earth can gradually transform into something like Venus – hot opaque dense atmosphere absorbing solar radiation in visible spectrum. Only infrared radiation can reach the surface of the planet, photosynthesis is impossible and all living things die. To avoid this perspective it is necessary to pass to new technology, especially in power engineering, when using it the blowout of carbon dioxide into atmosphere must be strongly reduced. This is a problem of science and engineering, economy and politics; it can't be restricted only by improvement of laws. Politicians all over the world try to avoid discussing troublesome questions, affirming that there is no problem. So some years ago the president of the USA John Bush didn't ratify the Kyoto protocol. His opinion was based on letters by 17000 'scientists' proclaimed that the global warming doesn't exist (2). According this protocol advanced countries engaged to contribute to under-developed countries on the pollution of the atmosphere. The president thought that the money wouldn't be used properly by the countries-receivers, it would disappear, and he was partly right. Were these letters by 'scientists written by request or not? It is difficult to answer, but it isn't very important. However, this method of 'approval of the scientific truth' can hardly be considered. The fact of using such 'methods' is the evidence of the problem of the global warming. Floods have damaged many countries for the last two years that witness to the essential changes in climate. In my opinion, terrible fire in California is the result of the global warming.

Now we are proceeding to the main question: what to do to prevent the global warming? How can the mankind go through the global energy and economy crisis?

In the nearest future fossil fuel can't be burnt because of the greenhouse effect. What to do then? Alternative power engineering is only being discussed it doesn't exist in production quantities. Let's remember, that one of the causes of the war in Iraq is the struggle for resources. But the problem is not only in resources; it's in impossibility of burning it in considerable quantities! But how can the prohibition be brought about? It is a political problem. Carbon dioxide emerges into the atmosphere at fires of forests and ignition of peat bogs. People use wood to prepare food including bonfires, which is the least effective method. Mostly it can be observed in underdeveloped countries. These processes can't be controlled. So we don't share the confidence of optimists in positive results of the Kyoto protocol ratification.

What is the alternative? Ecologists speak about necessity of population reduction. But church speaks about necessity of consumption reduction! However, the consumption reduction will inevitably lead to the necessity of population reduction. The question about reduction will arise repeatedly, if the process starts. Who will live and who will

die? Which country should be reduced first of all? That will be decided by the atomic war! How should the reduction be performed? What is Church's attitude to it? Unfortunately, both the Church and too emotional ecologists didn't analyze all the results of their appeals. An atomic warfare, it is a logical conclusion from the statement of the nuclear power vainness. We are faced with the dilemma: nuclear power today or atomic warfare in the nearest future! So we have the right to choose, but the choice is simple – nuclear power, which is the safest and ecologically pure in the operating mode.

We live in Russia, so we formulate the question: can Russia manage without nuclear power today? The answer is no. It can be proved. Winter in Russia is cold and people freeze in their houses, especially in the regions without nuclear power. The Russian government has shown its disability to put in order ordinary power engineering, even without any ban on the fuel burning. Therefore the conclusion about destroying nuclear power isn't logical. We'd like to believe, that appeals of politicians to ban nuclear power on TV show are spoken not for their self-interests but because of misunderstanding.

The residue from coal incineration is radioactive, there are radioactive substances in coal and their content depends of the coalfield. Air is necessary for incineration, hence incineration products escape into the atmosphere, but in a nuclear reactor all products are inside behind the protecting wall. Besides, radioactive acid rains are produced and forests are damaged. People suffer from smog, containing carcinogens.

The first step is the acknowledgement of necessity of nuclear power by politicians. This step is inevitable – it is difficult to argue with arithmetic! However, most politicians and too emotional ecologists haven't done it. The nuclear power is necessary, but what must it be like? Nuclear power has its open problems and they have to be discussed. First of all, it is the problem of **nuclear waste**, which is hazardous for people. There are some methods:

1. Realization of reactions with minimum of nuclear waste. These are pure 'deuterium explosions', described in detail in the book (3). This work was done in the town of Snezhinsk. The suggested device is based on using of impulse nuclear fusion. An explosion power is about 25-50 kilotons of trinitrotoluol equivalent the same as a nuclear bomb has. There are variants with less power. Pure deuterium explosions are brought about in practice and the physics of the process is obvious. The energy at deuterium explosions is mainly produced at the incineration of deuterium; reaction products are less dangerous than the deuterium fission products. The ignition of the reaction is realized by heavy elements fission, but their mass in

a fuse is on dozens of times less then in the standard nuclear bomb. The main problem is outside nuclear physics. This problem is an engineering one. It is necessary to work out a technology of construction, operative monitoring and repairing a blasting chamber, where the deuterium explosions take place. The blasting chamber must withstand a great number of explosions. It is about 50 explosions in 24 hours and 18 000 a year. At a working life equal 30 years – more than 0.5 million explosions. Is it possible to provide? Maybe, but it is difficult to guarantee. We need non-destructing radio electronic systems controlling the state of a blasting chamber in breaks between explosions. An analysis of the sound by ear after a knock to detect defects in metals has been used for a long time. An automatic analysis with the help of digital devices must be done too. It is only an engineering problem, which is beyond nuclear physics, and it must be discussed. The second problem is repairing walls of a blasting chamber at wear or appearing defects. The repair must be done without the presence of people because of the high level of radioactivity. There must be used robots. A repair problem should be discussed after discussing technology of constructing of the blasting chamber; there are some problems connected with the big size of the blasting chamber. At the first stage it useful to **test cheap models of small sizes and only then work with big sizes**.

2. It is necessary to change radically the construction of the fission reactor and technology of recycling of fuel so that the necessity of transportation of dangerous nuclear waste must be excluded. Nuclear debris have the highest level of radioactivity and emission of heat, therefore their transportation is the most dangerous. Nuclear waste must stay inside the reactor behind its walls. The energy, containing in debris, is almost two orders less than the energy emerging at fission. So the wall of the working blasting chamber will withstand this additional radiation. Now the nuclear fuel is pressed in ceramic tablets, which then are dissolved in big volumes of acid. This process requires taking out the worked out fuel from the reactor and its transportation to the chemical factory. It is necessary to work with liquid or gas fuel to eliminate transportations. Then it will be a hope to separate in the working place. That is only an idea, but it requires lot of work to fulfill it.

3. To find a way out from the ecology, fuel and energy crisis it is necessary to obtain much energy to organize recycling and utilization of the afterproducts. At the mining of minerals a lot of rocks goes to dump terraces that occupy large areas. At recycling there won't almost be these dumps. **Therefore the opportunity of fast increasing of nuclear engineering power is of great importance. Besides, new fuel-heavy elements are**

produced in some reaction. We can use the working reactions, but change the construction of the reactor and the technology of fuel processing to eliminate the necessity of transportation of debris.

We would like to believe that the mankind will use rational ways to prevent the global warming.

References

1. G. Nicolis, I.Prigozhin. Poznanie slozhnogo. (The knowledge/perception of complicated.) M., Mir.1990
2. L. Kropp. The greenhouse effect: who is in charge and what to do. America and we. 2001, 53
3. G. A.Ivanov, N.P. Voloshin, A.S. Ganeev, F.P. Krupin, B.V. Litvinov, S.U. Kuzminih, A.S. Svaluhin, L.I. Shibarshov. " The blasting deuterium engineering." Snezhinsk. 1997.

Can Advanced Civilization Preserve Biodiversity In Marine Systems?

Menachem Goren

Department of Zoology, The George S. Wise Faculty of Life Sciences, Tel Aviv University, 69978 Tel Aviv, Israel, gorenm@post.tau.ac.il

Biodiversity is defined in *The Stanford Encyclopedia of Philosophy* as the variety of all forms of life, from genes to species, through to the broad scale of ecosystems (Faith D.P. 2003).

Should We Preserve Marine Biodiversity?

Prior to any discussion on the problems related to the preservation of biodiversity in marine systems we have to refer to a question often asked. Why should we make so much effort to preserve marine biodiversity? I think that this question can be answered on two levels: the moral-ethical and the practical-economic. On the moral level, we can justify the preservation of marine biodiversity by claiming that *Homo modernicus* should not deprive future generations of the privilege of enjoying the same biodiversity that we do. And for me this is a sufficient reason.

The practical level, however, is the better understand one for most people. Marine biota is important for life on our globe as they perform almost 50% of the global photosynthesis (Field et al., 1998) and produce much of the oxygen consumed by the biosphere. The sea is also an important source of food for mankind, with the annual consumption of marine food (e.g. fish, mollusks, crustaceans, etc.) being ca. 80-90 million tons (FAO 2002). An increasing number of medicines of marine origin are discovered and used (Gudbjarnason, 1999, Mahler, et al., 2003). In addition marine biota purified sea water from anthropogenic pollutants and contributes significantly to the tourism industry.

V. Burdyuzha (ed.), The Future of Life and the Future of Our Civilization, 155–164.

The massive anthropogenic impact on the environment is relatively new. For hundreds of thousands of years our ancestors (various species of Homo and Australopithecus) had a negligible impact on their environment, but about 10-12 thousand years ago they changed their economic structure, turning from nomadic groups of hunter/gatherers to people who settled in villages, grew plants and domesticated animals (Cury & Cayré, 2001). This was the beginning of massive changes to the environment caused by mankind. The impact has been tremendous, many habitats have been destroyed, biodiversity in many regions has declined sharply, many species of animals and plants have been, deliberately or accidentally, transferred from one region to another, and recently we have even managed to affect the climate of our world. The marine environment has also suffered, but until the 1960s most cases of damage could be defined as local or regional. However, due to the use of advanced technology, within 50 years we have almost closed the gap and today marine habitats suffer to the same extant as the terrestrial ecosystems. "Homo modernicus", characterized by high technological abilities, an endless demand for marine bio-products and other products, and little concern for the environment, affects marine biodiversity in many direct and indirect ways. The most destructive activities are:

1. Fishery; 2. Habitat destruction; 3. Pollution; 4. Climate change and 5. Invasion of biota.

In the present talk I shall present an overview of three of these activities that are less known to the public and decision makers, and try to make an educated guess about the possibility of co-existence of advanced civilization and marine biota, which is measured by biodiversity indices.

Fishery

The impact of fishery is expressed in several ways:

Over-fishing. The average annual global crop, since mi 1980s, is ca 80-90 million metric tons (FAO 2002). However, this alleged stability is achieved due to the advanced technology and exploitation of new fishing grounds. In fact many traditional fishing regions show clear symptoms of over-fishing. The cod fishery industry in North America and of most of the north – east Atlantic has collapsed (Kurlansky, 1999), anchovy stocks off South America have declined sharply, the catch in rich fishing areas such as off Namibia (Nichols 2003) is much lower than a few years ago, the catch per unit effort has declined in 13 major fishing regions by 80-90% (Worm & Myers 2003), and above all, an analysis of the global catch clearly shows that we are fishing down the food web (Pauly, et al., 1998;). This means that the upper level in the food web, which includes predators such as the blue marlin (*Makaira nigricans*) and the sailfish (*Istiophorus platypterus*) are becoming exhausted. Their proportion in the catch has declined whereas the proportion of fishes belonging to lower trophic levels (and of lower economic value) has increased. As Worm and Myers (2003) show, even in cases where it seemed at first that there was a kind of compensation in which one predator (the codfish) was replaced by another predator (the flatfish), after a decade or so the population of the replacing predator too has declined sharply. If this trend will continue, as it seems to being, we can expect a decline in the high and mid trophic levels in the oceans and the future fish industry will rely on fishes belonging to the lower trophic level only namely: herbivores and planktivores. This is a fundamental change in marine biodiversity.

Is there any solution to the problem of over-fishing? The answer is conditional. In cases where the damage is severe, such as with cod fishery off the north-eastern coast of America, it seems that the fish community has not recovered even after two decades of controlled fishing (Kurlansky, 1999). Regarding to other destroyed fishing grounds, there is insufficient information. However, in cases where principles of responsible fishery, based on scientific knowledge, were enforced, the annual catch has remained stable. Such are the cases of the Atlantic cod, the haddock and other fishes in Iceland (Icelandic Ministry of Fishery, 2005), and Norway (Norwegian Ministry of Fisheries. 2003).

Trawling. Another aspect of the impact created by mankind on marine biota relates to the fishing methods. The most traumatic method is bottom trawling. Kura et al., (2000) suggest that the total world trawling grounds are approximately 20 million km^2, nearly two and half times the size of Brazil. The huge nets dragging on the bottom, for 4-6 hours each haul, pick up everything on their route. The FAO (Food and Agriculture Organization of the United Nation) estimates the by-catch as "In total, between 18 and 40 million tons of fish are thought to be discarded annually, representing roughly 20 per cent of the total marine harvest" (Pascoe, 1997). Very few animals survive the trawl. The no target catch includes numerous noncommercial invertebrates as well as under-sized fishes. Most of them die even if returned to the sea. One of many examples of marine community collapse is that of the Israeli Mediterranean coast. In the early seventies a catch of an average trawl haul of 4 hours included hundreds sea urchins and sea feathers. In recent years sea urchins have become very rare and only single broken sea feathers are found.

Another aspect of the by-catch issue is the flourishing of the opportunist species that replace other species.

Mismanaged Mariculture. The increasing demand for marine product has led to increased production of cultured fish, crustaceans and mollusks. The annual production of marine fish and invertebrates is ca. 17 million tones (FAO 2005). This mariculture is based on cages in sheltered marine habitats and a variety of artificial reefs, and on ponds along the seashores, especially in the tropics.

Fish cages have become an important alternative source for marine fishes such as salmon, bass, sea bream and flounder. The cages are installed in sheltered habitats such as bays and fjords where they are protected from the energy of the waves. The consequences of large scale farming in closed areas, characterized by limited water circulation, are eutrophication, and a sharp increase in the number of parasites and pollutants such as the remains of medicines and food (Athanassopoulou et al., 2004; Machias et al., 2004; Dimech et al., 2002; Douglas-Helders et al., 2003; Alongi et al., 2003; Mazzola and Sara, 2001.) In tropical regions, this may affect the health of coral reefs, as is suspected to have happened in Eilat (northern Gulf of Aqaba). In Eilat, where the annual fish production is 2500 tons the fish fed on about 5000 tons of food that had caused a nutrient enrichment equivalent to the total sewage of the city of Eilat (which was diverted to a treatment facility about 10 years ago)(Zalul, 2005). This is suspected as the cause of destruction of the coral reefs in Eilat (Loya et al., 2004).

The seawater ponds, especially the shrimp ponds, in the tropics replace mangrove forests. The mangroves, which cover about 40 million acres in the coastlines of the tropics all over the world (Blueplanetbiomes 2004), are rich ecosystems. They filter sediment and nutrients of terrestrial water flowing into the sea and serve as wave breakers that protect the seashores. The reckless destruction of many mangrove areas such as in the Honduras, Costa Rica and South-east Asia, caused environmental problems that could have been prevented if the ponds had been properly planned.

The solution to the problem of mismanaged mariculture is relatively simple: responsible planning of fish and shrimp farms with respect to the local environmental conditions, as well as the development of technologies of fish farming in the open sea, where the circulation of water minimizes the damage.

Ghost Nets

Ghost nets are nylon gill nets that were abandoned or lost. These nets, which can be kilometers long, are not degradable and catch finfish, crustaceans, ink fish and occasionally also marine mammal and sea birds. The trapped animals become bait for others, and this deadly cycle can last for years. There is no reliable information on the extent of this phenomenon, but there is an estimate that the Japanese fishermen alone lose 17 km of net each night during the North Pacific drift net fishing season (Dolphin Action and Protection Group, 2005).

The solution to this problem too is simple: first to reduce the use of drift nets, as many countries have done in their territorial waters, and also to enforce the tagging of each net. This will force the fishermen to treat their nets properly and to remove torn nets from the sea.

Global Warming

The next anthropogenic negative effect is that of climate change - global warming. Thousands of papers have been published on this subject (which I suspect also contributed to global warming); however, very little attention has been given to the possible impact of these changes on the structure of marine communities, especially the coral reefs. These communities are highly sensitive to temperature rise. In a report published by Al Gore, Vice President of the USA who headed a task force that studied the status of the coral reefs throughout the world, it was noted that: "Assessments to late 2000 are that 27% of the world's reefs have been effectively lost, with the largest single cause being the massive climate-related coral bleaching event of 1998" (Gore, 2000). The ocean warming also affects the marine ecosystem through an increase in precipitation (which increases nutrient flow from land to ocean) and through stronger storms (NOAA 2000). All this affects the coral reefs and consequently the fish communities. The relation between coral reef structure and fish community composition has been clearly demonstrated by many scientists (Pomerance et al., 1990; Strander et al., 2000; Brokovich et al., in press, and others). The various effects together have a synergistic effect on the coral reefs and fish community, as summarized in figure 1. It is believed that this web of changes will result in the following: 1. An increase in the number and intensity of bleaching phenomena; 2. A lower recovery rate of damaged reefs; 3. An increase in the proportion of dead corals; 4. An increase in algal cover; 5. A negligible increase in reef area due to warmer water and increasing sea level. All these expected changes will lead to lower structural and biotic complexity and lower family diversity (= fewer species in specious families, Goren, 1993). The consequences of these changes will be: Fewer species obligatorily associated with corals; fewer small species; more herbivores and substrate crushers and fewer coral feeders such as butterfly fish.

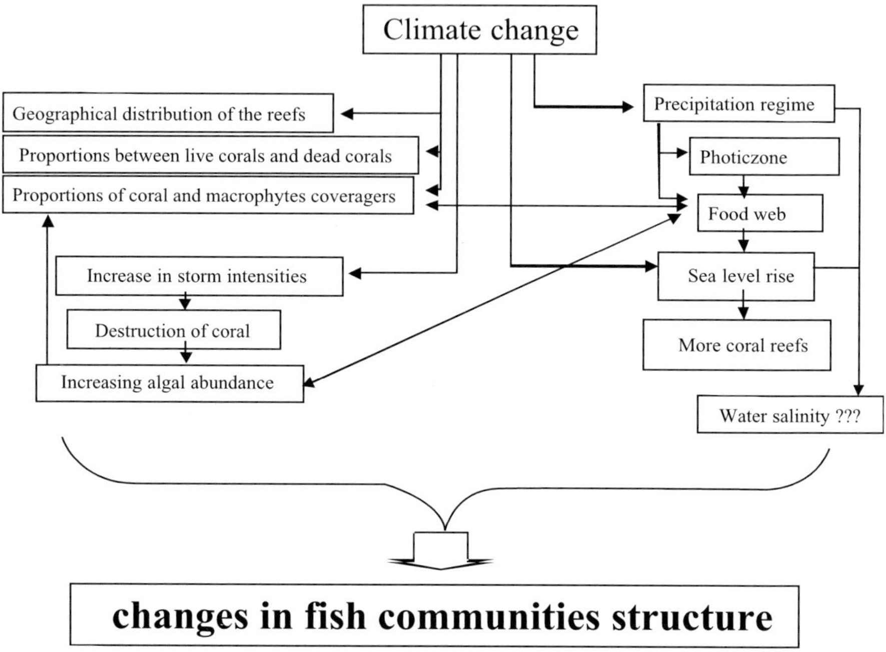

Fig. 1. A model of the effects of climate changes on coral reefs.

It seems that such changes, like global warming, are inevitable in the near future. We can only monitor the changes and try to reduce other negative anthropogenic effects that synergetic with the direct effects of climate change.

Invasion

The invasion of marine biota can cause fundamental changes to the structure and dynamics of many marine ecosystems. In many cases the local biota are not able to compete with the invaders and the alien species take over. One of the well-known cases is the appearance of *Caulerpa taxifolia* in the Mediterranean. *C. taxifolia* is a marine green alga accidentally introduced into the Mediterranean Sea (probably in Monaco). The algae is now spreading over most of the Mediterranean due to its successful spreading strategy (Smith and Walters 1999), and it has caused severe changes to the habitats. In some places these have changes reached the scale of a catastrophe (ECOSIM, 2005). *C. taxifolia* is replacing in many cases other algal species that were used by fish and some invertebrates as a nursery ground, and thus it also causes indirect damage to the ecosystem.

Another example of a harmful invader is the jellyfish *Rhopilema nomadica,* which appeared in the eastern Mediterranean in the late seventies and since then has dominated the sea shores during July and August. This seasonal massive bloom of *R. nomadica* causes a great deal of damage as it interferes with fishermen, blocks the cooling systems of power stations and stings people.

Perhaps the best example of invasive biota, which can be used as a model for marine invasion, is that of Red Sea species that penetrated the eastern Mediterranean via the Suez Canal.The opening of the Suez Canal, connecting the Red Sea with the Mediterranean, allowed hundreds of Indo-Pacific species, among them ca. 60 species of fish, to settle along the east Mediterranean coast and to establish flourishing populations.

In order to understand the success of the invaders in the eastern Mediterranean, we have to look at the unique ecological condition of this region. The surface water temperature in the Levant basin is relatively high (17°C to 31°C) and similar to that of the northern Golf of Suez. The original fauna in this region arrived after the last glacial period (about 12-15 thousand years ago) from temperate regions and thus is not well adapted to the local conditions. Some of the Red Sea invaders, which are of tropical origin, out-compete the local fauna. The results of this invasion are: that the invaders now form ca. 50% of the fish crop in the shallow water (<100 m.); ca 50% of the fish biomass in shallow rocky habitats are of Red Sea origin; and ca 50% - 90% fish in shallow sandy habitats are of Red Sea origin (Goren and Galil, 2001; Golani, 2002; Galil, 2003; Galil, 2004; Goren and Galil, in press). In addition, the penetration of two herbivore rabbit fishes has dramatically altered the food web in this region (Lundberg et al., 2004).

Can We Preserve Marine Biodiversity?

After this brief glance at the ways in which *Homo modernicus* affects marine biodiversity, the questions that arise are: Can all these changes be reversed? Can we rehabilitate marine biodiversity? And beyond all this, can advanced civilization preserve marine biodiversity and make sustainable use of marine biotic resources? The answers are not simple. Some of the changes are irreversible. Nonetheless, for the moment, most marine ecosytems can still be restored and biodiversity can be preserved. However, this can only be achieved if a global emergency action plan, based on the best available scientific knowledge, will be prepared and implemented. Thus, the answer to the key question “Can Advanced Civilization Preserve Biodiversity in Marine Systems?” is yes, and no.Yes, because Advanced Civilization can preserve marine biodiversity if it wishes to do so; an no, because unfortunately, I fail to see anyone or any authority undergoing the the task of preparing such a plan and coordinating the numerous actions needed to preserve marine biodiversity in time.

Acknowledgment

I thank Ms. N Paz for editing the manuscript.

References

Alongi, D. M., Chong, V. C., Dixon, P., Sasekumar, A. & Tirendi, F. (2003) The influence of fish cage aquaculture on pelagic carbon flow and water chemistry in tidally dominated mangrove estuaries of peninsular Malaysia. Marine Environmental Research. 55(4): 313-333.

Athanassopoulou, F, Groman, D., Prapas, Th. & Sabatakou, O. (2004) Pathological and epidemiological observations on rickettsiosis in cultured sea bass (Dicentrarchus labrax L.) from Greece Source. Journal of Applied Ichthyology. 20(6): 525-529.

Blueplanetbiomes,(2004) http://www.blueplanetbiomes.org/mangrove_forests.htm.

Brokovich, E., Baranes, A. & Goren M. In press. Habitat structure determines coral reef fish assemblages at the northern tip of the Red Sea. Ecological Indicators.

Cury, P. & Cayré, P. (2001) Hunting became a secondary activity 2000 years ago: marine fishing did the same in 2021. Fish and Fisheries, 2:162-169.

Dimech, M., Borg, J. A. & Schembri, P. J. (2002) Changes in the structure of a Posidonia oceanica meadow and in the diversity of associated decapod, mollusc and echinoderm assemblages, resulting from inputs of waste from a marine fish farm (Malta, Central Mediterranean. Bulletin of Marine Science. 71(3): 1309-1321.

Dolphin Action and Protection Group. (2005) http://sacoast.uwc.ac.za/education/resources/ envirofacts/gillnets.htm.

Douglas-Helders, G. M., O'Brien, D. P., McCorkell, B. E., Zilberg, D., Gross, A., Carson, J. & Nowak, B. F. (2003) Temporal and spatial distribution of paramoebae in the water column: A pilot study. Journal of Fish Diseases. 26(4): 231-240.

ECOSIM (Ecosystems simulation) (2005) http://www.isima.fr/ecosim/ct.html.

Faith, D. P., (2003) "Biodiversity", The Stanford Encyclopedia of Philosophy (Summer Edition), Edward N. Zalta (ed.), http://plato.stanford.edu/archives/sum2003/entries/biodiversity/.

FAO, (2002). The state of world fisheries and aquaculture. FAO Publishing Management Service, Rome, Italy.

FAO, (2005). World review of fisheries and aquaculture. FAO Publishing Management Service, Rome, Italy.

Field, C. B., Behrenfeld, M. J., Randerson, J. T. & Falkowski, P. (1998) Primary production of the biosphere: integrating terrestrial and oceanic components. Science 281, 237-240.

Galil, B. S. (2003) Control and eradication of invasive aquatic invertebrates. In: Encyclopedia of Life Support Systems. EOLSS Oxford, UK.

Galil, B. S. (2004) Exotic species in the Mediterranean Sea and pathways of invasion. In: Davenport, J. and Davenport, J. L. (eds.), The Effects of Human Transport on Ecosystems: Cars and Planes, Boats and Trains, Royal Irish Academy:1-14.

Golani, D., Orsi-Relini, L., Massuti, E. & Quignard, J.-P., (2002) CIESM Atlas of Exotic Species in the Mediterranean. Vol. 1. Fishes. F. Briand, (Ed). CIESM Publishers, Monaco. 256 pp.

Gore, Al. (2000) "The struggle to conserve coral reefs is now at a critical stage" http://www.aims.gov.au/pages/research/coral-bleaching/scr2000/scr-00gcrmn-report.html.

Goren, M. (1993) Statistical aspects of the Red Sea ichthyofauna. Israel Journal of Zoology 39: 293-298.

Goren, M. & Galil, B. S. (2001) Fish biodiversity in the vermetid reef of Shiqmona (Israel) Marine Ecology, 22(4):369-378.

Goren, M. & Galil, B. S. In Press. Impacts of Invading Fish on Levantine Inland and Marine Ecosystems. Journal of Applied Ichthyology.

Gudbjarnason, S. (1999) Bioactive marine natural products. Rit Fiskideildar. 16. 107-110.

Icelandic Ministry of Fishery, (2005) http://www.fisheries.is/.

Kura, Y., Burke L., McAllister D. & Kassem K. (2000) The Impact of Global Trawling: Mapping our Footprint on the Seafloor: Adapted from World Resources 2000-2001 and PAGE: Coastal Ecosystems.

Kurlansky, M. (1997) Cod – A biography of the fish that changed the world. Printed by Nørhaven Paperback, Viborg, Denmark. 294 pp.

Loya, Y., Lubinevsky, H., Rosenfeld, M. & Kramarsky-Winter, E. (2004). Nutrient enrichment caused by in situ fish farms at Eilat, Red Sea is detrimental to coral reproduction. Marine Pollution Bulletin. 49(4): 344-353.

Lundberg B., Ogorek, R., Galil, B. S. & Goren, M., (2004) Dietary choices of siganid fish at Shiqmona reef, Israel. Israel Jour. Zool. 50:39-53.

Machias, A., Karakassis, I., Labropoulou, M., Somarakis, S., Papadopoulou, K. N. & Papaconstantinou, C. (2004) Changes in wild fish assemblages after the establishment of a fish farming zone in an oligotrophic marine ecosystem. Estuarine Coastal & Shelf Science. 60(4): 771-779.

Mahler, S. M., Chin, D.Y. & Van Dyk, D. (2003) The application of emerging technologies in genomics and proteomics to drug development. Journal of Pharmacy Practice & Research. 33(1): 7-11.

Mazzola, A. & Sara, G. (2001) The effect of fish farming organic waste on food availability for bivalve molluscs (Gaeta Gulf, Central Tyrrhenian, MED): Stable carbon isotopic analysis. Aquaculture. 192(2-4):361-379.

Myers, R. A. & Worm, B.. (2003) Rapid worldwide depletion of predatory fish communities. Nature. 423: 280-283.

Nichols, P. (2003) A developing country puts a halt to foreign overfishing. http://usinfo.state.gov/journals/ites/0103/ijee/nichols.htm.

NOAA. (2000) The potential consequences of climate variability and change on coastal areas and marine resources noaa's coastal ocean program. Boesch D, F., Field J.C. and Scavia, D. eds. Decision Analysis Series # 2. 181 pp.

Norwegian Ministry of Fisheries. (2003) Facts about the Norwegian fisheries industry, http://odin.dep.no/filarkiv/212064/Vite_ENG_2003.pdf.

Norwegian Ministry of Foreign Affairs, (2005) http://odin.dep.no/odin/engelsk/norway/environment/032091-120004/index-dok000-b-n-a.html.

Ostrander, G. K., Armstrong, K. M., Knobbe, E. T., Gerace, D. & Scullyi, P. E. (2000) Rapid transition in the structure of a coral reef community: The effects of coral bleaching and physical disturbance. Proc. Natl. Acad. Sci. USA, 97 (10): 5297–53021.

Pascoe, S. (1997) Bycatch management and the economics of discarding. In FAO Fisheries Technical Paper No.370 http://www.fao.org/fi/publ/abstract/t370f.asp.

Pauly, D. & Christensen, V. (1995) Primary production required to sustain global fisheries. Nature 374: 255–257.

Pauly, D., Christensen, V., Gu'nette ,S ,.Pitcher ,T .J ,.Rashid Sumaila U., Walters C. J., Watson R. & Zeller D. (2002) Towards sustainability in world fisheries. Nature 418: 689–695.

Pauly, D., Christensen, V., Dalsgaard, J., Froese, R. & Torres, F. Jr. (1998) Fishing down marine food webs. Science 279: 860–863.

Pomerance, R. (1999) Coral Bleaching, Coral Mortality, and Global Climate Change Released by the Bureau of Oceans and International Environmental and Scientific Affairs U.S. Department of State. 16 pp.

Roberts, C. M. (2002) Deep impact: the rising toll of fishing in the deep sea. Trends Ecol. Evol. 242: 242–245.

Smith, C. M. & Walters L. J. (1999) Fragmentation as a strategy for Caulerpa species: Fates of fragments and implications for management of an invasive weed. Marine Ecology. 20(3-4): 307-319.

Tegner, M. J. & Dayton, P. K. (1999) Ecosystem effects of fishing. Trends Ecol. Evol. 14: 261–262.

Worm, B. & Myers, R. A. (2003) Meta-analysis of cod–shrimp interactions reveals top–down control in oceanic food webs. Ecology 84: 162–173.

Zalul, (2005) http://www.zalul.org.il/InCat.asp?cid=3&topcid=2.

Can We Personally Influence the Future with Our Present Resources?

C. Gros, K. Hamacher, W. Meyer

Department of Physics, Frankfurt University, 60438 Frankfurt am Main, Germany, gros05@itp.uni-frankfurt.de
Center for Theoretical Biological Physics, University of California, San Diego, La Jolla, CA 92093, USA, hamacher@matisse.ucsd.edu
Center for Evaluation (CEval), University of the Saarland, D-66123 Saarbrücken, Germany

Influencing the future of mankind for the better, supporting sustainable development and realizing ambitious plans as the colonization of outer space? But how? In this paper we will differentiate two distinct approaches: (I) immediate action/projects - most NGOs follow this avenue and (II) long-term growth within an evolvable and improvable organization to accumulate both financial resources and methodical, scientific and procedural knowledge to support the goals.

We explain how the initiative 'Future 25'[7] tries to develop a foundation whose goal is to maintain a platform for the second approach. The long-term perspective requires special organizational prerequisites to support a stable structure and by that the desired long-term growth. The direct member participation results as a possibility in contrast to most foundations that are governed by a small group of people. Our foundation will be heavily based on the utilization of the Internet to realize what we will introduce as maximum participation. We show how current sociological research prompts for concepts such as monitoring and evaluation to support the stability of the organization and facilitate new developments and changes within the foundation.

We discuss organizational, sociological, legal, technological and financial issues.

[7] http://www.future25.org

V. Burdyuzha (ed.), The Future of Life and the Future of Our Civilization, 165–178.

Avenues to the Future

We don't have presently, and this is one thing we are definitely sure about, the capabilities of the psycho-historians of Isaac Asimov's Foundation trilogy to predict our future. Uncertainty about consequences is the fate of all our decisions and we suppose that our own personal action can not significantly change the way things go. Although we are struggling for the capacities to foresee our future since the beginning of our history, we are still lacking this capability and it seems to be improbable that humanity will ever acquire it.

However, our interests are not only limited to the nearby future of our personal lives and of the respective societies we are living in. We are also curious what the long-term future will harbor for mankind and if we, as ordinary individuals living today, can do something about it. Confronted with horrible scenarios on world wide climate changes and other catastrophes that probably will be the outcome of our deeds and those of our predecessors, responsibility for the future and sustainable action with respect for our descendants seems to be more important than ever. So this is the rational of this paper: what can we do to influence the long-term future or, more precisely, what must be done to ensure (positive) long-term consequences for our action?

Thinking a little while on the subject, we may come-up with two, fundamentally different, avenues of approach:

(I) First Avenue to the Future: Act Now

We try to change the present and our immediate future in such a way, that the consequences of our actions ripple down in time and history, influencing in a desired way events yet to happen in the far future. In a more defensive form, this is the approach of sustainable development, defined as a development 'that meets the needs of the present without compromising the ability of future generations to meet their own needs' (World Commission 1987: 8) in the famous report "Our Common Future" of the UN World Commission on Environment and Development. This approach is very intuitive: Change the present for the better now, or at least trying to do so, hoping for a positive outlook. Sustainable action, in this sense, means to make decisions that produce long-lasting positive outcomes and being able to avoid negative side-effects continuously. Unfortunately, we are facing two difficulties with this approach.

The first obstacle is our above mentioned in-capability to actually predict the long-term future. We are not only missing data and knowledge about the complex structures and relationships of the world we are going to live in future times. Moreover, we have to predict unpredictable events and have to implement solutions for problems we do not even have a clue whether they will arise or not and, if ever, at which time they will occur. The longer the time period in discussion, the more probable we are facing a lack of information on the problems that have to be solved. Furthermore, not only our knowledge decreases but also the difficulties for further generations to ascribe measured impacts to our action will increase. Even Asimov's Foundation gets in some trouble on this point: while Hari Seldon (the Founder) could not precisely calculate all future developments, the discussion within the Foundation society raise, whether he had been able to foresee the actual difficulties or not.

The second problem lies in the limited resources we dispose of, as simple individuals or as a small group of engaged citizens (the smaller the group the weaker is in general the assumable impact of their action). One may object that there are many individuals who did actually change the course of history alone for the better or the worse and that it is nowadays possible, in principle, for ordinary citizens to achieve positions, like the American presidency, which allows to exert personally and directly an enormous political and economical influence. However, even if we assume this being correct in an open society, it is exceedingly difficult to achieve such an influential position without the devotion of a lifetime. Even then, when devoting all ones energies and time to achieve an influential position, any single person has not more than a tiny chance to actually achieve this goal. It may even be incompatible to devote energies for the advancement to such a position and to work for long-term changes simultaneously.

However, history teaches us that even people who did not have any political or economical power are able to influence the course of our development significantly. Sometimes individual action can start an epidemic spreading across huge regions and time periods (for some popular examples see: Gladwell 2000). Unfortunately, the impact of such kind of social diffusion processes remains unpredictable especially as long as there is no continuous monitoring and routing of the process towards the requested direction. We need sustainable impacts not only on the first dimension of sustainability (implementing sustaining solutions) but we also need social institutions for sustaining production of solutions (fourth dimension of sustainability, see figure 1). Even Isaac Asimov's Founder Hari Seldon implemented a second foundation for steering the development.

(II) Second Avenue to the Future: Act Then

As a consequence, it seems not to be sufficient if we are acting now to influence the course of future events directly even if our actions have some sustainable positive impacts for our descendants. Future events have to be influenced precisely at the time when they will happen (or when they can be prevented or supported) possibly long after we ceased our own individual activities on earth. As far as we are not able to predict the future, we will need to motivate other people now and then in the future, to carry out, and to carry on, this endeavor. Furthermore, we have to hand our ideas how to steer the future (e.g. the concept of sustainable development) over from generation to generation, or give basic guidelines how to evolve these concepts to a changing socio-cultural environment.

Table 1. Four dimensions of Sustainable Impacts.[8]
(Basing on Meyer (2002) and Stockmann (1997), modified.)

		Is there any potential for innovations implemented?	
		No	Yes
Are there any new Social institutions implemented within the	No	Sustaining Solutions (Dimension I)	On-going solution development (Dimension II)
	Yes	Sustaining Institutions (Dimension III)	On-going institutional development (Dimension IV)

While our life-time is strongly limited to a short period, our ideas may be able to last for longer. A successful historical example from the occident is Jesus Christ, whose speeches and deeds in a very small and politically marginalized country spread throughout the whole world and survived more than two thousand years, impressing millions of people in the course of history. Similar examples of sustaining ideas can be found in other religions as well as in arts, politics and sciences.

These processes had been guided for the better *and* the worse by social institutions throughout the centuries. Durable social organizations had been implemented to protect the idea against challenges, to support the diffusion of the idea, to adopt it to new social developments and to decide between different interpretations of the idea. Therefore, social institutionalization of an idea is at least as important as the idea itself for its survival.

[8] Examples: (Dimension I) Books presenting new ideas and proposals. (Dimension II) Scientific research published in journals. (Dimension III) Social institutions with a fixed program (e.g. a company). (Dimension IV) Social institutions capable of autonomous development (e.g. the economy).

If we want to influence the future then, we have to use our present resources for building up sustainable social institutions. The objective of these institutions have to be the empowerment of future generations to overcome the troubles we are not able to foresee (or to battle successfully) and to give them the capacities and resources for doing that successfully. We will point out in this article that a suitable organized non-profit organization, growing continuously over time, could do the job. Our main task is to ensure this organization for developing an ever growing amount of resources (in form of both money and knowledge).

This approach then solves immediately the predictability problem. We do not need to be psycho-historians in order to discern possible important issues for the future development of humanity, let's say, for the next few centuries. Our descendents will then be there to ascertain the specific issues to be dealt with, hopefully with vast resources at their disposal. This second approach also solves, at least partially, the problem of the limited amounts of resources available to us individually, since long-term growth will lead eventually to massive financial resources, even when starting with an initially modest endowment. Regarding the concept of sustainability, the capital stock will be left untouched (and is probably growing if more and more people invest in the fund) while interest proceeds can be used for activities. Depending on growth and interest rates, the organizations ability to finance human actions will increase over the centuries.

Having this idea to influence the future in mind, some very important questions have to be answered:

- What are the objectives of this organization? Which future tasks should be handled in which way?
- What kind of social measures can be used to ensure that the resources of the organization will be durably used for the goals attended by us? How can we prevent individual abuse for personal benefits?
- How can we permanently control the impact processes produced by the activities of this organization and guarantee that the development is directed towards the intended goals?

The Objectives: 'Big Questions' on the Future of Mankind

The first question to be answered by building up a sustaining organization is: what are the objectives of such an organization? For the purpose of this article we consider now the fundamental motivations of the nascent initiative 'Future 25'. Considering the 'big picture', the future of live and mankind on earth and in the universe, we may formulate some basic, partially related, issues:

- Will large-scale ecological habitats survive human activities on earth? In the long-run?
- Will humanity ever support large-scale space exploration and support the development of life on other planets?

Nobody can, of course, give definite answers to above questions, or, in this respect, to any other 'big question'. In fact, this is one of the reasons that they are called 'big questions'. Some may even argue that those questions are not important because they are too big and an answer is almost impossible to find from our recent perspective. Others may criticize these questions from a moral point of view and refuse them as aspiring goals or dispute them on an ethical basis. We may, however, consider general socio-economical and structural conditions needed to resolve above issues positively without discussing whether these questions are really the right topics for a sustaining organization or not. Here, we concentrate instead on the question of why non-profit organizations are needed to handle the issues above.

If we leave these tasks to governments and governmental agencies then the answer to both big questions is probably no. The primary concern of democratic governments is to satisfy the needs of their people here and now. Only secondary considerations, if at all, will lead governments to decisions having a positive and lasting impact in distant future. Governments are accountable only to their living electorate, not to their descendents. To keep the electorate in a spirit to demand sustainability over long periods would still need social structures to hold up these ideals and to achieve a majority for these goals. Non-democratic governments are even not feeling responsible for their people at all.

While the political system seems to be unsuitable to contribute to the big questions of humanity, some may set some hope in the market forces of the economical system. Unfortunately, market failures limit the regulating power of market forces: the function of market have to be guaranteed by legal systems that assure fair trade and correct price fixing. To realize sustainable development one has to assure the precise recognition of distant future costs which have to be taken into account and weighted with present benefits. What, if all external cost would be 'internalized', would be the real oil price? Moreover, striving for individual benefits and not protecting common goods are the driving forces of the market system.

Most recently, non-profit oriented civil-society organizations understand themselves as gate-keepers of common goods. Their ability in mobilizing people and organizing effective collective actions is an important threat for those who want to abuse common goods for their own profit. While politicians may be dependent on commercial interests (e.g. for financing their election campaign), the voluntary participants of civil-society organizations are more difficult to control. Unfortunately, civil-society organizations itself are dependent on the deliberate decision of their members to engage for common issues. In general, people are concerned by actual problems and not by the distant future.

Nevertheless, there are many people who are interested in future topics. The avenue to the future, when applied to problems afflicting us today, is shared by a wide range of non-governmental associations and non-profit foundations. All contributing in their ways to the progress of our civil society mostly is pursuing the first avenue to the future. We believe that the civil society would benefit qualitatively if one or more non-profit associations, dedicated to the second approach to the future, would complement, in a kind of symbiosis of different approaches, the estimable array of private associations and foundations.

The Organization: Challenges and Principles

The establishment of a globally active, non-profit organization, dedicated to long-term external and internal growth for the benefit of our descendents presents a formidable challenge. This organization would have to grow financially and membership wise over very long periods. The annual growth rates would not need to be large, internal financial growth rates of 3-4 percent are sufficient and attainable. This organization would need an internal structure allowing to hand down over many generations the ideas of the founders, our ideas for the future of humanity. Basic economics tells us, that the financial resources of this foundation would grow, within a few centuries to such massive size, that our descendents would then have the means, by using them wisely, to actually change, in certain aspects, the course of history, for example by promoting the lasting and successful expansion of mankind to outer space.

This organization would not be initially inactive. Immediately it would start to finance regularly projects as any other non-profit foundation does it nowadays. But at a reduced rate, in order to retain enough money for internal growth. Considering the respective national laws for non-profit organizations we find that this strategy is possible for a German non-profit foundation, but not for an American trust, as we will discuss further below.

When successful, this organization would contribute in quite a few different ways to the advancement of civil society, besides financing large-

and larger-scale projects with the passing of time. It would need to evolve into a test-bed for applied sociology and political sciences.

Legal Aspects

National laws regulate the financial activities of non-profit associations. A charity in the US has to spend on the average 5 per cent of the fair value of its endowment every year, limiting strongly its growth potential. A non-profit foundation in Germany has to spend on projects the majority of the annual yields, like dividends and interests, the actual endowment is protected and cannot be used for charitable or other means. With a suitable investment strategy, one possibility is to invest the endowment predominantly in stocks, a long-term growth of the endowment can be achieved.

German law permits two kinds of tax-deducible donations to a non-profit foundation: the normal donation to be used for the charitable activity and donations towards the endowment, increasing it in size. This venue of external financial growth is especially important in the starting phase of a nascent non-profit foundation. It is therefore natural for a growth-oriented foundation to identify co-founders, i.e. everybody donating a certain amount towards the endowment, with members.

The Organizational Structure

As of today, we find two dominant generic types of organizational structures for non-governmental bodies: with and without internal democracy. Organizations without internal democracy typically have self-electing governing organs, for instance a board of directors: When one board-member leaves, the remaining ones co-opt a new member. According to national laws and basic-democratic ideals the vast majority of civil-society organizations are democratically ordered with members that elect their board for a defined period. In bigger organizations the election process is divided in several different levels and the members of the lower level elect delegates for the higher level. Nevertheless, strong democratic principles are established in nearly all cases.

As far as clubs and other registered associations are acting principally for the benefit of their members, the internal democracy is strongly limited to members and membership is voluntary giving each member a voice (the right to get his or her interests represented) and an exit (the freedom to leave the organization if he or she wants to) option (see Hirschman 1970). Therefore, the main problem of each civil-society organization is to stabilize membership and to continuously mobilize them for common objectives.

Stable Internal Democracy

Self-electing organs constitute very efficient governing organs for organizations with a well specified aim and purpose. Most charities, to give an example, spend the available money in very regular ways every years within recurring programs. Major decisions, like the discontinuation of an existing program, are only rarely up to decision by the governing organ. The most important work is carried out on a daily basis by the administrative staff.

An association dedicated to the long-term perspective cannot have, on the other hand, a fixed a priori program. Important decisions will be needed along its way into the future. A self-election organ is not suited to carry out this mission. Its human resources are, by definition, limited and constant and independent control of its actions are difficult to achieve. The association dedicated to the second approach to the future therefore needs a properly thought-out democratic structure.

To find a suitable internal democratic structure for a globally active organization with a long-time horizon might appear, on the first sight, a trivial task, since so many democratically organized associations already do exists. To guarantee from the outset the stability of an association over very long time-spans is, however, a challenge and we can only discuss now some of the most important and difficult points.

Member Participation

Private associations and non-governmental organizations enter frequently, especially with growing size, a stage with a dramatic reduction in the effectiveness of the internal democratic decision-making and participation processes. Quasi oligarchic informal internal networks may form and effectively exclude the other members from the decision-making process. Or the internal communications channels may be insufficient to keep up with growing demand (Meyer 2004). A tendency for larger organization to develop indirect systems of democratic participation, stemming from limited pre-Internet communication channels, with delegates speaking for local subunits or internal interest groups, also harbors the risk of reducing the effectiveness of democratic communication. On the other hand, the request for effective communication increases by the number of people involved in decision processes and professional communication measures and management is definitely needed. Therefore, a foundation for the future has to implement professional communicational management and to optimize them with respect to the communicational development within the organization.

This permanent development process draws support, beside other, from current economical, sociological and psychological research, confirming the positive influence of participation possibilities on the satisfaction of the members of an organization. This is an all-important basis prerequisite for a non-profit association, which is dependent on the voluntary participation of active and motivated members. The term 'procedural utility' has been coined in economics to describe the satisfaction somebody receives from undertaking an action (procedure), irrespectively from the outcome. Studies have shown that procedural utility is increased not only by the actual participation processes, but already from the conceived possibility to participate, even if the actor chooses not to make use of this possibility (Frey and Stutzer, 2002).

The rise of the Internet has opened new communication and participation channels for non-profit organization yet unused in many traditional associations. A key example in this respect is the direct involvement of all members of a non-profit foundation in selecting the projects to be financed yearly by the proceedings of the endowment. We are not aware presently, of any foundation allowing for such a direct participation of its members. The vast majority of non-profit foundations do not accept members at all; they are controlled by self-electing execution organs installed by the original founders. And the small fraction of foundation allowing real members does mostly not yet exploit fully the power of Internet participation.

Monitoring and Evaluation

It is unrealistic to assume that it is possible to conceive out from the start already the optimal structure for an association, optimal for stability, internal and external long-term growth. What is possible though, is to formulate basic guiding principles. The foundation 'Future 25' would try to continuously improve its own internal organization, communication channels and participation possibilities. It would therefore present a platform for experiments in applied sociology and political sciences. Proper evaluation principles, a functional continuously used impact monitoring system, and regular critical documentation of the effectiveness and efficiency of its own structure are therefore mandatory and might be beneficial for other non-profit organization aiming to improve their own organizational structures. Yet, most civil-society organizations are very poor in these aspects and they do not, in general, have any kind of monitoring and evaluation system implemented.

Effective and continuous internal evaluation, combined with a thorough open internal and external information policy, should help to detect appropriate warning signals. Comprehensive free availability of information would imply, to give an example, the publication of all financial details and of all meeting-protocols on the home-page of the foundation. It should also allow the free and effective discussion of the evaluation results and such stimulate counter-actions. Future 25 will therefore serve, if successful, as a test-bed for various possibilities of internal evaluation with the aim of optimizing both the evaluation-tools themselves and the growth potential of the association.

Principles for Long-Term Stability and Growth

Following the discussion above, we can now formulate three basic principles helping to support a stable and evolving internal democracy:

1. Optimal participation possibilities for all members.
2. A comprehensive and fully open flow of information.
3. Continuous internal and external evaluation.

We believe that these three guiding-principles provide a basis for long-term stability and long-term growth, and that other requirements follow from here. The need for an internal balance-of-powers, to give an example, results from (1) and from the external requirement that, by law, any registered foundation needs to nominate legal representatives, e.g. in the form of a board of directors.

The Role of the Internet

The utilization of the Internet for direct communication between all members, sub-groups and administrative contacts is a cornerstone of our plans for three reasons:

- The Internet is nowadays a world-wide communication tool, faster and more reliable for communication between different regions of the world than any other system and therefore most suited for a global organization.
- The costs are reduced in comparison to paper-based mailing and face-to-face meetings involving substantial traveling.
- The Internet is up to now our only way to achieve the goal of direct member participation and the possibility of systematic organizational evolution.

While the first two points are obvious the third needs some further elaboration. There are several aspects where an Internet-based communications platform is useful.[9] As discussed above, we estimate direct participation and the facilitation of procedural utility to be crucial for establishing a growing and internally stable organization. One of several approaches to discuss intra-organizational 'hot topics' is to propose new ideas for further development is the Open-Space-Concept (Owen 1997). The concept itself has been used up to now only for face-to-face meetings in real places, in our case people will come together virtually. In an open-space meeting there are no speakers or round-tables as in traditional conferences, people establish an agenda with priorities on their own. It would be impractical to have several dozens up to several thousands of Future 25-members invited to some place. In addition to the large organizational and travelling costs we would build up barriers for direct participation, since many members would not be able to participate in-persona due to other constraints (business, family, etc.). Virtual open-space meetings would also allow for longer-lasting discussions, increasing the possible success rate, i.e. that proposals resulting from these discussions will be eventually implemented. Therefore this approach makes it also feasible to discuss topics of highest importance such as consequences from evaluations and monitoring activities in length with an ending call for voting; giving every member the chance to review the exchange in the electronic archive of such a meeting.

The concept of an Internet-based open-space conference discussed above is an example of structured opinion formation. Effective intra-organizational communication needs also informal channels. Every member of Future 25 would be empowered to invoke a discussion forum on his or her own for some topic he wants to discuss, creating a marketplace for ideas. It is important that important discussion forums have a resonance beyond the people actually participating in it. A forum in which a topic is discussed that many participants feel to be important for Future 25 needs to receive attention. For this purpose we have developed the concept of 'results-oriented discussion-forums'[10].

Direct communication by a message system or by 'normal' e-mails from member-to-member, will lead to personal networks of members. These networks of personal contacts, trust and common interests are best described by the small-world-phenomena (Watts 1998, Albert 2002) that gives raise to the hope that the open participation opportunity leads to per-

[9] Please see also for the concept of 'Structured opinion formation via Internet': http://www.future25.org/structure.html#opinionFormation

[10] Please see: http://www.future25.org/structure.html#discussionForums

sonal contacts accelerated by the high ‘connectivity’ and short ‘distances’ between people in small-world-networks, accelerating in the end the advance of the foundation on its own. To take psychological effects into account, Future 25 will provide every member the storage to save a small ‘homepage’ to give everybody the chance to look at a photograph or get informed about one's respective hobbies and such.

Technological Prerequisites

The necessary infrastructure in terms of computers and Internet-connections is small, affordable and the technology is already developed. Beside regular newsletters and an official web-site which should not only inform members about current topics but also non-members about the philosophy, activities of Future 25 and about participation opportunities we propose the use of the Internet for internal democracy and participation activities.

For voting and other communications that is sensitive, we plan to use an established, well known and by several official bodies approved cryptographic system a-la GnuPG (GPG). It will not only enable the private exchange between people but also the authentication in voting procedures or discussions. A public key for the main administrative account will be certified by a central instance and used to sign further cryptographic keys for administration that then will certify member keys.

The Foundation Future 25 is forward looking: Awareness of the future implies consciousness of the past and Future 25 will therefore maintain a complete archive system. A Document Management System would further increase the effectiveness of this archive. It will allow to substantiate the past development and to document the outcome of organizational reforms and experiments.

Outlook

The nascent initiative Future 25 is, in its very nature, open for participation, like the open-source movement. We extend a friendly invitation for collaborations.

References

Frey, Bruno S. and Stutzer, Alois (2002) 'Beyond Outcomes: Measuring Procedural Utility'. Berkeley Olin Program in Law and Economics, Working Paper Series. Paper 63. http://repositories.cdlib.org/blewp/art63 .

Gladwell, Malcom (2000) 'The Tipping Point. How Little Things Can Make a Big Difference', Little Brown & Co.

GPG: http://www.gnupg.de/ .

Hirschman, Albert O. (1970) 'Exit, Voice and Loyality. Responses to Decline in Firms, Organizations, and States', Cambridge: Harvard University Press.

Meyer, Wolfgang (2002) 'Sociological Theory and Evaluation Research. An Application and its Usability for Evaluating Sustainable Development', Saarbrücken: CEval-Working-Paper No. 6, http://www.ceval.de.

Meyer, Wolfgang (2004) 'Regulation, Responsibility and Representation . Inter-organizational co-ordination and its relationship to intra-organisatonial communication'.

Owen, Harrison (1997) 'Open Space Technology: A User's Guide', Berrett-Koehler.

Stockmann, Reinhard (1997) 'The Sustainability of Development Cooperation', Baden-Baden: Nomos.

Watts, Duncan J. and Steven H. Strogatz (1998) 'Collective dynamics of small-world networks', Nature, 393, 440-442.

World Commission on Environment and Development (1987) 'Our Common Future', New York: Oxford University Press.

World Energy Development Prospects

Anatoly Dmitrievsky

Oil and Gas Research Institute, Russian Academy of Sciences, Gubkina 3 str., 119991 Moscow, Russia, a.dmitrievsky@ogri.ru

Forecasting future energy sector development, profiling rational scopes of energy supply, early formulation of structural, organizational and engineering solutions – all of these have to adequately meet the challenges facing society in the period.

In the coming 20–30 years, these challenges appear to be as follows:

- demographic growth in the developing countries: their ratio in the world population would be over 90%;
- global deterioration of the environment (the greenhouse effect, ozone holes, and radioactive wastes);
- aggravation of the resource problem;
- raising the efficiency of energy conversion and searching for new energy sources.

Therefore, the challenge of building up sustainable development of the energy sector should take account of these diversely vectored tendencies and meet the demands of energy, economic, and environmental efficiency.

Within the entire foreseeable future, the global population will basically grow due to the developing countries. One of the mankind's most important strategic targets is to supply them with energy and to close the gap between those countries and the developing countries.

Considering long-run prospects of energy development, one should bear in mind the following features of the world energy pattern:

- the current traditional centers of world energy consumption, i.e. the industrially developed countries, would stay being major consumers of

V. Burdyuzha (ed.), The Future of Life and the Future of Our Civilization, 179–186.

either flat or slightly dropping levels of energy, which is primarily due to their energy-saving policy and evolution of their economic structure;

- there will be new centers of fast-growing energy consumption; above all, these will be South-East Asian and Oceanic countries that are intensely drawn into industrialization and post-industrial development, as well as Latin American countries, Latin America also pursuing the trend;
- China and India would join the major energy-consuming countries within the next decade;
- the current spacing between energy production and energy consumption would stay; it is likely, however, that it will be somewhat smoothed down due to broad development of shelf energy resources, internationalization of energy companies, progress in nuclear power generation, and a reduced role of oil;
- the role of gas in the world energy supply, considering gas resources and favorable environmental features, is potentially very high; however, the potential would be implemented in competition with other kinds of fuel and energy: within a short-term period – primarily in competition with oil and coal, and within a longer period – in competition with coal and nuclear energy generated by safe nuclear reactors, and later with renewable energy sources and nuclear energy.

The competition would be focused on the consumer's choice of a preferable energy carrier, the choice being based on its consumer properties and price. In fact, while gas consumer properties are substantially advantageous as compared with those of many other energy carriers, the position of gas in terms of price competition would depend on solutions for a number of large-scale engineering and economic problems. These are primarily as follows:

- to cut down gas production and transportation costs;
- to advance technologies in traditional gas uses, first and foremost in power generation;
- to develop the promising incipient and novel areas of gas use, such as use of compressed or liquefied gas for transport, methanol production, etc.

Fundamentally new and cost-efficient technologies will have to be built for production of non-traditional gas resources for the future gas industry of 2030–2050 (from coal beds, tight reservoirs, gas-hydrate pools, etc.).

In many countries, R&D is carried out along all of these lines, and Russia is no exception in this respect.

Performed both by international organizations and by national teams, numerous research studies reveal, while energy saving remains being important and significant, that further escalation of energy consumption and production is unavoidable (at least for the nearest 25–30 years).

The rising significance of natural gas in the world energy balance is a long-standing tendency. At the same time, there is good reason to believe that in future, in any event within the next 2–3 decades, the tendency of the past years will stay like that – on the global scope.

To date, there is a profound inconsistency, and even inadequacy between both the level and structure of world energy production on the one hand and the mankind's energy demand on the other. Out of almost 6 billion population of the Earth, 1 billion people live in the industrially developed countries, and they are supplied with qualified energy, while 2 billion lack electric power or commercial energy resources; the efficiency of their 'open hearth' heating devices is miserably low, and the population is busy wiping out the forests. The other 3 billion people provide some energy resources for themselves with difficulty and at high cost. The population is likely to grow to 8 billion by 2020–2030, and to 10 billion by 2050. Most of the population growth would belong to the developing countries. The problem of their energy supply is one of the most significant challenges for Society.

A joint report published by the Vienna Institute of Systems Analysis, IAEA, OPEC and UN (Industrial Department) outlines the world scope and structure of energy production until the year 2050 as follows:

Total	25900-30000 tons of fuel equiv.	(100%)
in that number:		
oil	7720-8300 tons of fuel equiv.	(30-28%)
natural gas	7150-8720 tons of fuel equiv.	(28-29%)
solid fuels	5720-7150 tons of fuel equiv.	(22-24%)
water and nuclear power	5290-5860 tons of fuel equiv.	(20%)

Therefore, overall energy consumption and energy production will more than double. Energy production will be carried out mostly employing resources of organic origin, and the gas industry will underlie the advance, being the only industry increasing its share in the energy balance.

In the energy balance of practically every world region, natural gas is rising in its significance. That is a prevalent tendency, which is supported and intensified by a number of energy-related, environmental, and economic factors pertaining both the current situation and the foreseeable future.

The World Energy Council is working on the prospects for world energy development until the year 2050 and further on. The work has not yet been completed, but preliminary data are as follows: the average production forecast scenario yields consumption of 23 billion tons of fuel equiv. in 2020 and 28.6 billion tons in 2050. By the end of the 21st century, the world population will have been close to 12 billion, while primary energy resource consumption (in the average forecast scenario) will have been approximately 40 billion tons of fuel equiv.

A report made by the well-known consulting company Mitre at the Tokyo Congress of the World Energy Council in 1996, identified the world demand for primary energy resources in 2030 at 17 billion tons of fuel equiv., in that number for natural gas at approximately 4–4.5 billion tons of fuel equiv. As to the adverse environmental impact, the annual emission of CO_2, which amounts to 6 billion tons of carbon, will rise to 12 billion tons per year.

According to calculations, the European demand for energy resources is estimated for 2030 at 4 billion tons of fuel equiv., in that number at about 800 billion m^3 of gas.

Natural Gas in the World Power Production

Natural gas currently occupies a specific position in the world power production: it belongs to a group of the most broadly used energy carriers (alongside with oil and solid fuel, appreciably ahead of nuclear energy or renewable energy sourccs, including water power) and the most promising energy resources.

Every principal vector of the energy resource selection criterion, i.e. maximum energy-related, economic, and environmental efficiency, point at natural gas in many instances.

There are a number of factors constraining the scope of employing natural gas. As a rule, these factors include considerably rigid and highly capital-intensive infrastructure for natural gas delivery. For implementation of new projects, it requires a certain level of confirmed volumes of solvent demand. In case of pipeline transportation, it also requires appropriate solution of transit issues, as well as stability of the political and economic situation in the areas involved into the project for the entire long term of project payback. For the economies of the countries and regions, important are also factors of the hazard of energy dependence on one or a small number of energy sources, while the rigid infrastructure and capital intensity of the gas sector more often than not bring about a situation like that. There are also natural delimiters of the "affected zone" associated with certain sources of natural gas deliveries depending on the expected levels of gas prices at the appropriate markets. Any change in their macroeconomic situation and employment of new technologies may have an impact on the relevant quantitative estimates and final inferences.

During the past two centuries, the mankind has been developing due to change in technological structures, each of which is associated with a dominating energy resource. Last century, coal was such a resource, and later it was oil. The new, so-called post-industrial technological structure that is coming up is based on the rising role of natural gas in primary energy sources, as well as that of electric power in the end consumption of energy.

During the recent 30 years, the share of natural gas in the primary energy resources rose from 19% to 23%, while the share of oil dropped from 49% to 39%, and that of coal from 30% to 26%.

During the recent 30 years, world energy consumption rose by 60%, while natural gas consumption increased 2.2 times. It is precisely by the rate of its spread that gas would come to the leading position in the first decades of the 21st century. According to the International Energy Agency, the share of gas in the fuel-energy balance will have risen by 2020 to 27%, that of oil will further drop to 38%, that of coal – to 24%, and that of nuclear power – from 7 to 5%.

The gas industry has reached the worldwide scale indeed, erasing geographical differences and therefore intensifying the advantages pertinent to gas. Discussing the future of gas, one should not miss the paramount issue of environmental protection. Being environmentally the cleanest of all kinds of fuels used to date, gas is undoubtedly one of the central elements in our common struggle against harmful emissions, the ones that bring about the notorious 'greenhouse effect.'

The benefits of gas in terms of environmental protection do not elicit any doubts. As an environmentally clean fuel, it yields a low level of hazardous emissions, in particular, of sulfur dioxide; as a fossil hydrocarbon fuel, it produces a low level of carbon dioxide, lower than that of coal or oil. In consequence, replacement of other fuels by gas may improve the greenhouse effect situation fast and efficiently.

Assessing gas from the viewpoint of demand and supply, attention should be focused on the competitiveness of gas as compared with other energy carriers.

A number of factors are advantageous for natural gas:

- the overall tendency of growing energy consumption related to the progress of the world economy;
- emergence of ever more improved technologies of gas use in the most diverse industries and economic sectors;
- growing importance of gas being environmentally clean as compared with other kinds of fossil fuel.

Presently, the World Energy Council jointly with the Institute of Systems Analysis (Vienna) are working out a long-term forecast for the world energy development until 2050. Evidently, essential for such forecasts is identification of certain tendencies from the viewpoint of the current position rather than detailed quantitative estimates inasmuch as these depend on too may uncertainty factors. At the next mid-century, the forecasts reveal appreciable dispersion of potential levels of natural gas production – from 5 to 8 billion tons of fuel equiv. It is essential that even the minimum level assumes further advance of the gas industry. At the same time, moderate 'pragmatic' scenarios assume a rise in the share of natural gas in the balance to 25%.

Raw Materials Resource Base for the Gas Industry

Present-day world proved reserves of natural gas have reached 150 TCM (proved gas reserves are enough on the average for 70 years), which is better than the oil situation, with due regard to the current and expected production level.

The International Gas Union (IGU) estimates the natural gas resources at approximately 400 TCM. It is clear, however, that making them proved would call for a large amount of exploration. Besides, it is noteworthy that the international classification of reserves takes into account the economic efficiency of their production at the current level of prices. This parameter may vary with a change, in particular, with a drop in prices, as it is occurring at the moment. A geological estimate of natural gas resources in the former USSR, which incompletely considered the economic efficiency of resource production in the former USSR, amounted to 275 TCM, including 236 TCM in Russia only. An estimate provided by the IGU for the former USSR territory includes only about 157 TCM of natural gas; nevertheless, that is approximately 40% of the world resources. The share of promising resources in Russia is at its maximum, which matches its enormous potential despite the high level of gas production at the moment. At the same time, the share of Russia in the world natural gas resources is over 45%.

Energy of the Future

One can easily assert that the energy sector is a basic component for the progress of civilization. Fire and heat facilitated survival of the primitive people, and their striving for more efficient use of the energy capabilities assured their transition from cave-dwelling to modern civilization benefits and advantages.

Naturally, some questions do occur: how long can the mankind use these benefits? For how many years would organic fuel be sufficient? What kind of energy would replace oil and gas? Because of unsettled energy and environmental issues, will the mankind go back to caves?

Let us try and answer these questions. During the past two centuries, the mankind has been developing due to change in technological structures, each of which is associated with a dominating energy resource. In the 19th and early 20th centuries, coal was such a resource, and starting from the second half of the 20th century it was oil. The technological structure that is coming up will be more and more oriented towards natural gas.

The share of oil and coal in production of primary energy resources have dropped substantially. In the mid-20th century, coal and oil prevailed in energy production. Currently, the share of coal is hardly over one fifth, and that of oil is a little over one third in production of primary energy resources. While in the second half of the 20th century coal was being busily replaced by oil, during the past 30 years 10% of the oil share went over to gas (the share of oil dropped from 45% to 35%).

Now what is going to happen then? The mankind can count on the use of organic fuel for a sufficiently long term. While prioritizing the intensity of employing an energy source, one should bear in mind that oil reserves are enough for only 40 years. It is worthwhile to recall a well-known Dmitry Mendeleev's statement on oil being extremely valuable as a chemical raw material: ("To fuel oil for heat is the same as to burn ovens with bank notes").

Gas will be sufficient for the mankind for at least 60 years ahead (demonstrated reserves), or maybe for 90 years (estimated resources). According to the World Gas Union, gas resources in out planet amount to 420.0 TCM, in that number explored reserves reach 160.0 TCM.

On top of the trillion cubic meters, our civilization may count firmly upon the so-called non-traditional gas resources. These include: gas in tight reservoirs, coaled methane, gas dissolved in formation water, gas-hydrate pools, and deep gas.

Why is it that we can count firmly upon that? Because the first two sources have already become operational. There are technologies making it possible to produce gas from tight reservoirs, which gave a chance to the USA to increase potential gas resources by several times. According to BGR (1998), gas resources in tight reservoirs are estimated at 114.0 TCM. There is intense production of coalbed methane as well. The overall coalbed methane resources amount to 233.0 TCM.

Gas resources associated with gas-hydrate pools are over gas resources in traditional gas fields. Gas-hydrate resources of gas in the seas and oceans are more than 20,000.0 TCM. Currently Russia, the USA, Japan, India, and other countries carry out intense research assessing gas-hydrate resources in their territorial waters and elaborating efficient gas production technologies.

Large-scale studies are necessary to assess the potential of using gas resources associated with formation water and work out efficient technologies for their development. The overall resources of gas dissolved in aquifer water of the world sedimentary basins reach 24,000–30,000 TCM .

As drilling equipment and technology evolve, the mankind will have access to the enormous deep gas resources. In the meanwhile, an inexhaustible energy source is associated with hydrogen.

Hydrogen is the principal element in the Universe. Large amounts of hydrogen were lost in the Earth's evolution. Nonetheless, according to numerous researchers, hydrogen is one of the main elements in the products of Earth degassing. However, great hopes are associated with production of hydrogen in natural gas conversion or water electrolysis. The concept of hydrogen energy production assumes employment of nuclear energy for production of hydrogen (e.g. in decomposition of water into hydrogen and oxygen) and further use of hydrogen as fuel.

The benefits of hydrogen energy are self-evident. First, the raw material is water, i.e. a practically unlimited source, and, second, that is an environmentally clean process.

Nevertheless, much has to be done for production of hydrogen from water (due to electrochemical, thermochemical, plasmochemical, or other processes) to become acceptable in its both energy and economic parameters.

There are no doubts that successful and efficient scientific, engineering, and technological solutions would be found. Not to trust that is not to trust the grandeur of human intellect.

Energy in the Universe and its Availability to Mankind

Josip Kleczek

Astronomical Institute Czech Ac. Sci., 25165 Ondřejov, Czech Republic

The **importance of energy** should also be mentioned at our Symposium on "Future of Life and our Civilization". Thanks to energy we have reached a high living standard. A fast increase of the world population, the continuing growth of the industrialized world, the natural desire among developing nations to get also a higher standard of living – all that induce an inevitable increase of energy consumption. The contemporary consumption of energy represents about 13 TW (TW is terawatt, or 10^{12} Watt).

Burning fossil fuels is the main source of the contemporary energy consumption. They contain solar energy in biomass stored by photosynthesis during many millions years. The ancient biomass was deposited underground and soiled (e.g. by sulphur). Fossil fuels represent a valuable material for the chemical industry. Their amount is limited and their prices are rising. The products of fossil fuel burning are harmful to health, are damaging the biosphere and are changing the global climate. People are perished (in mines and in wars) and life is destroyed on a large scale (e.g. by spilling oil from great tankers). And from a physical and astronomical point of view, burning fossil fuels is the least effective way to get energy from matter.

Energy is the capacity of matter and radiation to do work (*en* means in Greek "in" and ergos means "work"). It is everywhere – in matter (Einstein: mc^2), in radiation (Planck: hf) and in space (dark energy). The dark energy is supposed to be the most important energy in the whole Universe. However, its nature and even its existence are still highly hypothetical.

Solar Energy represents by far the most important energy source for the Earth, its biosphere and for mankind. There is no doubt, that the energy from the Sun will become the principal energy resource for the future generations in the post-fossil-fuel era. It is pure, of high quality and practically eternal (because the amount of hydrogen in the Sun is sufficient for seven billion years). It is always free for all and everywhere.

V. Burdyuzha (ed.), The Future of Life and the Future of Our Civilization, 187–200.

The flood of solar radiation falling on the Earth – 180 000 TW – is enormous. How can to explain the fact that all terrestrials together need only 13 TW, whereas they neglect the generous energy gift of the Sun?
Why do we neglect the solar 180 000 TW and harm at the same time the biosphere? Why do we destroy our cosmic home by using fossil fuels? Why do we want to pass on a poisoned dump to the next generations instead of the beautiful BLUE PLANET? Why are people killed and why is nature destroyed for the 13 TW? Why do we pay more and more for our thoughtlessness? Are the classical philosophers right in defining the human being as intelligent ("Homo creatura rationalis est in qua anima et corpus coniuncti sunt")? Or was Einstein right (?) when he said: "Ich kenne zwei Unendlichkeiten: das Weltall und die menschliche Dummheit. Aber bei dem ersten bin ich mir nicht ganz sicher"

1. Energy of Matter

Our home – the planet Earth – and our organisms are only a tiny part of the Cosmos. All changes in the Cosmos, on the Earth, and in our organisms are governed by the energy laws valid in all Universe. Energy has a fundamental role in the structure and evolution of the Universe, in life of the biosphere in general and in our organism in particular.

Each matter consists of many elementary particles: protons p, neutrons n and electrons e. Each elementary particle is a droplet of energy. If a particle is isolated and at rest, its energy is called rest energy m_oc^2.

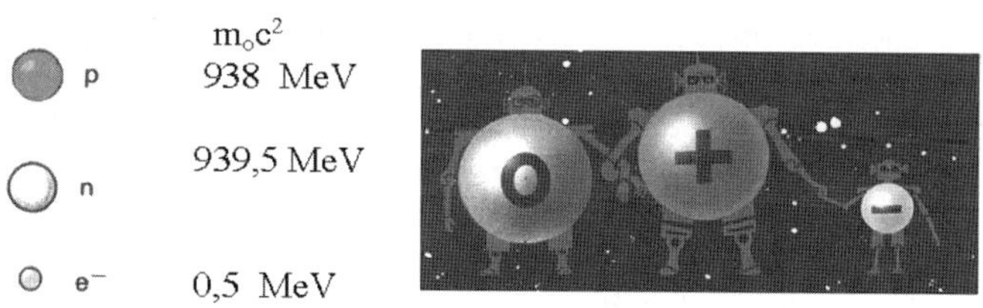

Proton is the hero of the Universe. In the core of our Sun it gives away 7 MeV from its rest energy (938 MeV). In this way 560 tons of protons release together $3{,}8 \times 10^{26}$ watt, i.e. the solar luminostiy.

The rest energy may be increased or decreased. The energy used by mankind is drawn from the rest energy of matter (= from agglomeration of elementary particles). One could say that our energy is "squeezed out" from matter. The squeezing out of energy is performed by the fundamental forces (interactions) acting between elementary particles: electric, nuclear and gravitation). In burning fossil fuels the electric force is active, which is the least effective way to get energy.

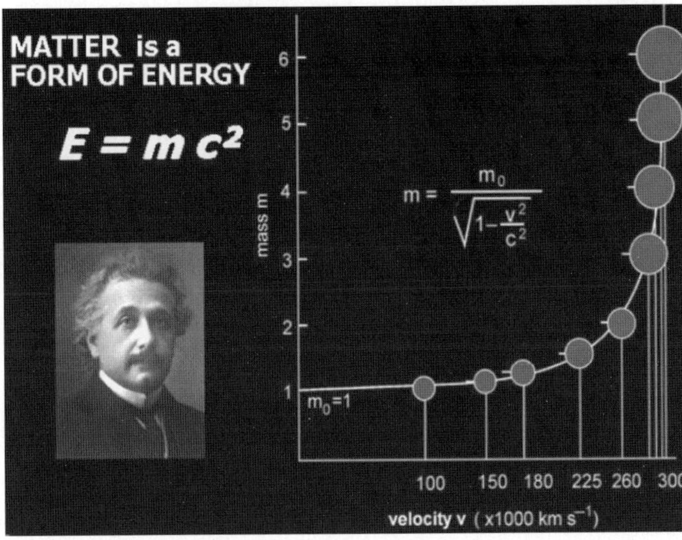

Rest energy m_0c^2 of an elementary particle (or of an object) may be increased by acceleration. This increase is due to increase of it mass as the velocity of light *c* is everywhere and always the same. The Einstein's expression shows that motion of material object cannot exceed the velocity of light.

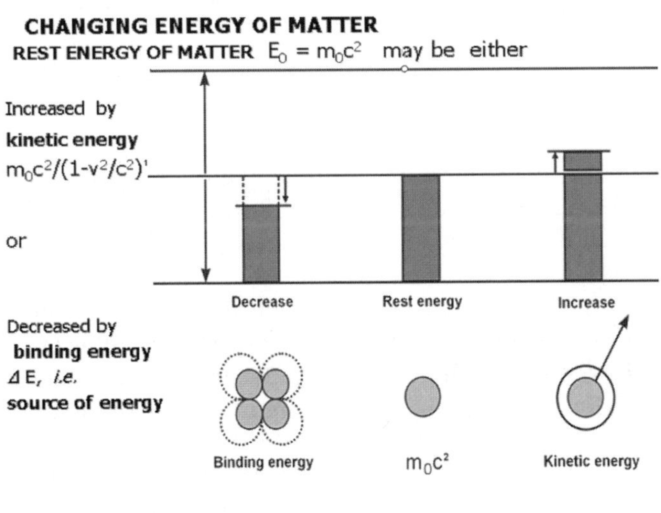

A small part of rest energy m_0c^2 may be squeezed out from matter (i.e. from a system of elementary particles). The squeezing out is realized by one of the fundamental forces. On the other hand, the decrease of rest energy represents binding energy of the particle system. E.g. by fusion of four protons into one alpha particle in the core of the Sun each proton releases 7 MeV. This is its binding energy in the alpha particle.

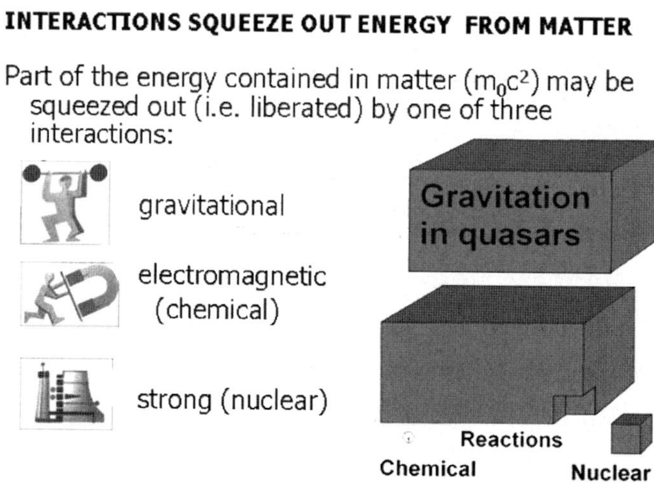

The complete cube represents m_oc^2. Energy can be squeezed out from m_oc^2 by three fundamental forces, *viz.*

a) gravitational (e.g. in quasars),
b) electromagnetic (e.g. burning of fuels),
c) strong (e.g. in stars, nuclear power plants).

The yield is worst (10^{-9} mc^2) in burning. It is about 10^{-3} mc^2 for strong interactions and up to 10^{-1} mc^2 for gravitation in the Universe. On the Earth the gravitational energy is much smaller (e.g. water in dams).

2. Energy of Radiation

Photons are particles (quanta) of electromagnetic radiation. They are little wads of oscillating electric and magnetic force. Each photon is a droplet of energy which depends upon frequency of its oscillation f and is determined by *Planck expression* $E = h\,f$.

The letter h in the equation is the universal Planck constant $h = 6{,}6 \times 10^{-34}\,J\,s$. Our eyes perceive the photons with energies between 2 eV and 4 eV as light. Photons with higher energy than 4 eV are called ultraviolet radiation, with lower energies than 2 eV are infrared radiation.

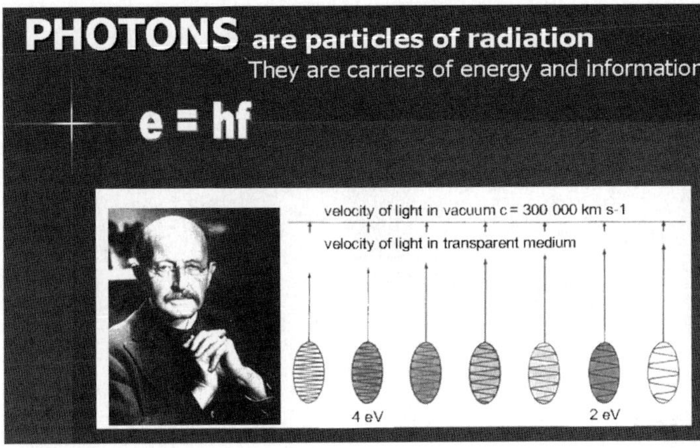

3. Energy in the Universe

History of the Cosmos (i.e. of ordered Universe) is History of energy. It began 13,7 billion years ago by Big Bang, when energy was born with radiation, matter, time and space. From chaotic extremely hot and primitive material (called "quark plasma") all the present systems have been created (atoms, planets, stars, galaxies et al.). Here, in this coin of the Universe – after 13,7 billions of years, on one of the planets (= the Earth) accompanying a star (= the Sun) which is quite a common star among the 150 billion stars of the Galaxy (= Milky Way) – intelligent animals were created.

The law of energy conservation means that the energy can not be destroyed or created from nothing. This implies that any form of energy can be traced backwards in time till the Big Bang when all energy was created. Big Bang means not only the beginning of space and time 13,7 billion years ago, but also of matter and of radiation. But matter and radiation are forms of energy (see.1 and 2).

The horizontal line in our Figure represents the time axis. It is not linear.

Three-dimensional space expansion is represented by one dimension r only, which is the distance of any two points in the Universe.

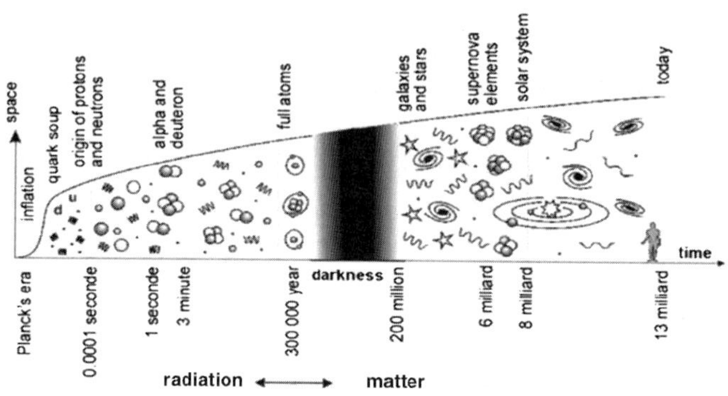

By increase of r with expansion any volume increases as r^3. The number of particles and the number of photons in the volume decrease equally, i.e. as r^{-3}.

The energy of particles m_0c^2 is not influenced by expansion, so that energy density of matter ρ_m changes as r^{-3}.

On the contrary, the energy density of radiation ρ_r decreases faster with expansion than the density of matter. The wavelength of radiation λ expands as r. That means that frequency of photons $f = c/\lambda$ decreases with time as r^{-1} so that the energy density of photons ρ_r decreases by expansion as r^{-4}.

The number density of particles n_m has always been (and still remains) much smaller than the number density of photons. Immediately after Big Bang the photons had a very high energy (gamma photons). As a result, during the 300 000 years after Big Bang:

$\rho_r > \rho_m$ (radiation era or photon era)

and later until present

$\rho_r < \rho_m$ (matter era or particle era)

After Big Bang the temperature was extremely high, but it is decreased because of expansion. When the Universe was 300 000 years old, its temperature dropped to some thousand degrees (K) and all free electrons recombined with protons. This period of the Universe history is called cosmological recombination. Since the time recombination the cosmic space has been transparent. Today, the astronomers can therefore observe very ancient events in the history of the Universe.

4. Energy in Life of Stars

The binding energy of a proton in atomic nuclei is released by nuclear force from proton mass energy $m_0c^2 = 938$ MeV. Protons are fused into nuclei of heavier atoms in the core of stars. The curve in our graph is called *valley of stability*.

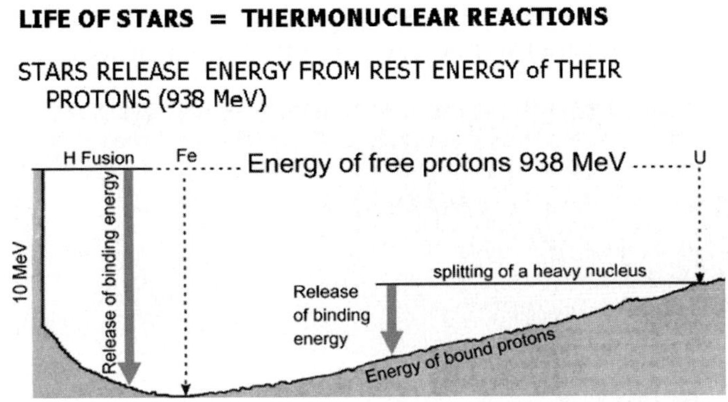

The number of protons (= atomic number) is on the horizontal axis. The depth of the valley (on vertical axis) corresponds to the binding energy of a proton in nuclei. As it may be seen nuclear fission of heavy nuclei (e.g. of U-235 in nuclear power-plants) or nuclear fusion of light nuclei (e.g. protons in the Sun) releases energy. Produced nuclei are in both cases lower in the valley of stability. The energy difference is radiated away.

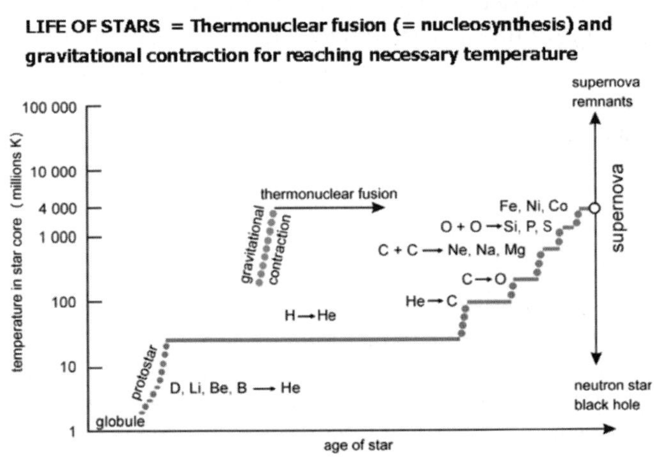

The life of a star is determined by gravitational and nuclear energy. The vertical line represents the central temperature of the star. The horizontal line is the time axis. The vertical axis is the temperature in the central core of a star. With age the temperature of the core increases. A globule (= stellar embryo) is very cold (about 10 K, which is –263°C). Its gravitational potential energy is transformed into heat (by contraction). Gravitational contraction and heating are marked by oblique segments of the curve. Thermonuclear fusion produces nuclei of different chemical elements. E.g. fusion of 3 alpha particles at the temperature 100 million degrees results into carbon nucleus (Salpeter's reaction). All carbon atoms in the Universe (including our organisms) were created by this 3-alpha process.

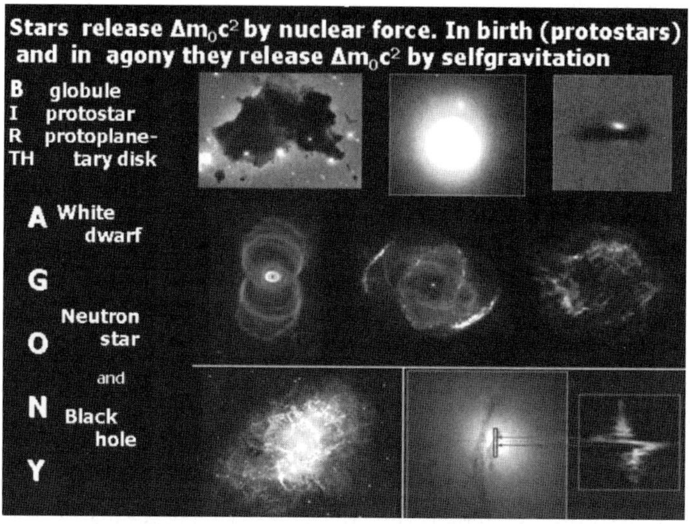

A dark globule (stellar embryo) is an enormous reserve of gravitational and nuclear energy. Gravitational energy is due to the globule size (a few light years) and the mutual attraction of any pair of its particles (by Newton Law of gravitation). Hydrogen is the most abundant element in the globule and as well as in the whole Universe. It represents the best thermonuclear fuel for the stars. The life of a star consists of getting rid of its enormous reserve of gravitational and nuclear energies.

During birth of a star gravitation acts as "midwife". Self-gravitation compresses a globule (a cool cloud of dust and gas in interstellar space). By compression the globule is heated and becomes a radiating protostar. By further compression the temperature reaches 7 million degrees (K) in the central part. At such a temperature, protons (nuclei of hydrogen atoms) fuse to alpha particles (nuclei of helium atoms). Nuclear force replaces gravitation in releasing energy and the protostar becomes a grown-up star. The life of a star consists of a sequence of thermonuclear reactions. Each star we see with our naked eye is a thermonuclear reactor.

When the nuclear fuel is exhausted, self-gravitation is a source of radiation. Gravitation assists also in the final agony as a "gravedigger" of stars. See penultimate graph.

5. The Sun is a Perfect Fusion Reactor

The Sun is the best known star. It is the nearest star, observed continuously in all detail. Its physiology and anatomy is studied by solar physics. It is one of one hundred fifty billions of stars in the Milky Way Galaxy, which itself is one of hundred of billions of galaxies in the known Universe.

The Sun is an enormous ball of very hot gas (= plasma). A jet plane would need nearly half a year to fly around it. Its mass (2×10^{30} kg) is 333 000 masses of the Earth. Nearly all the solar mass is invisible, hidden under the visible surface called the photosphere. The visible part of the Sun is called the solar atmosphere. The hidden part is called the solar interior. The atmosphere is very extended and rarified. The interior of the Sun contains ten billion (10^{10}) times more matter than the solar atmosphere.

The core is the "power plant" releasing $3{,}8 \times 10^{26}$ J s^{-1} ($3{,}8 \times 10^{26}$ W). The energy is released by fusion of hydrogen.

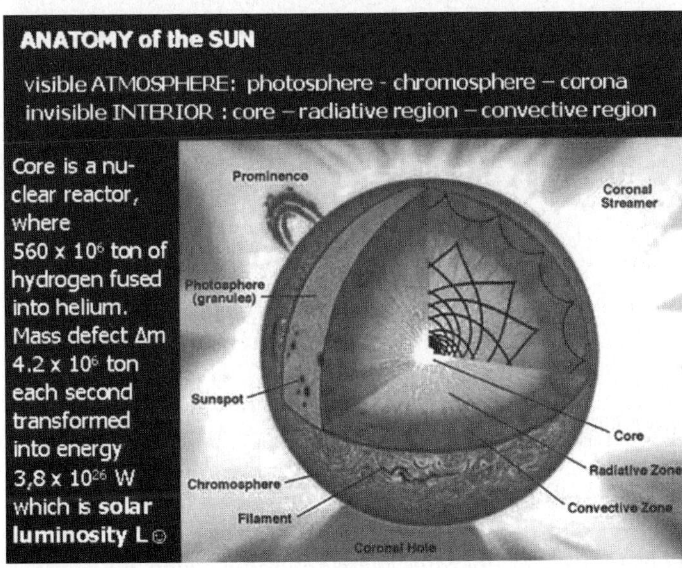

Each second 560 million tons of hydrogen is transformed into helium and at the same time mass of 4,3 million tons is transformed into radiation. The core has a temperature of 15 million degrees (K) and the released radiation is in hard X-ray form.

The X-ray photons slowly propagate from the core upward to upper cooler layers. They are absorbed and reemitted. Sometimes one photon is absorbed and two photons with less energy are reemitted. On the long journey from the very hot core to the visible and cooler surface

(= photosphere), one X-ray photon is being transformed gradually into (approximately) 2 – 3 thousands of light photons. From the photosphere the light photons escape into the surrounding cosmic space. The journey of the photons from the hot core to the visible surface is very erratic and lasts about one million years. (It may be compared to a walk of a drunk man through a forest in a dark night.)

6. Energy of the Earth

The Earth is our cosmic home, moving around the Sun like a lonely spaceship with 6 billion human beings on board. It is a blue fragile Beauty with its own energy sources which are limited and soon will be depleted.

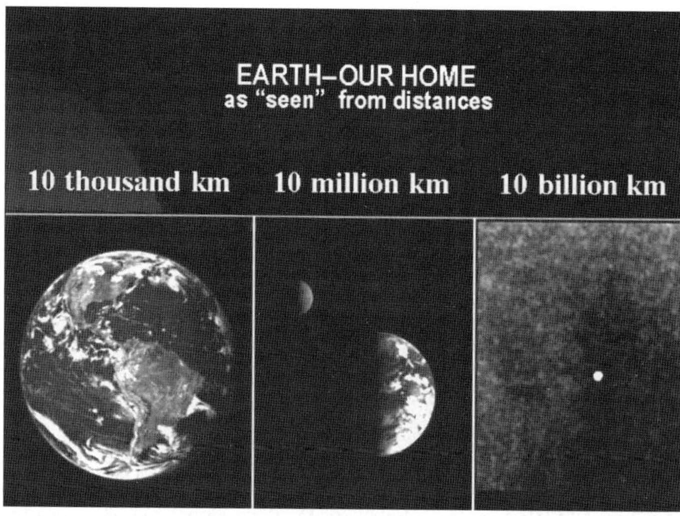

The Earth has only two different types of energy resources:

a) Kinetic (revolution, rotation) and nuclear (geothermal and heavy hydrogen). These very old energies were inherited when the Earth was born from the protoplanetary disk 4,5 billion years ago.

b) An enormous contemporary flood of solar energy is falling upon the Earth. It is nourishing the life on Earth and will do so for the 7 billion years ahead.

Energy given by the Sun to the Earth each second is 180 000 TW, which is 14 thousand times more than 6 billion humans need (i.e.13 TW). It is an immense gift of the Sun to all, clean, of high quality, everywhere given free to everybody and inexhaustible. Reserve of the solar fuel – the hydrogen - in the core of the Sun will suffice for the next 7 billions years (7×10^9 year). Inexpensive technologies know how to transform solar radiation in useful forms of energy (chemical, heat, electricity, mechanical energy.)

The solar radiation falling on the Earth is being transformed into indirect forms of solar energy, *viz.* wind, currents of water, heat of oceans and continents, waves on water surface and by photosynthesis in biomass. Fossil fuels (oil, coal and earth gas) contain solar energy accumulated by photosynthesis millions of years ago.

7. Energy and Life

Our body is a minute part in the architecture of the infinite Universe and a tiny link in its evolution. But the human beings are enormous by their immaterial soul, because they understand the infinite Universe, predict its future and know how to use its energies.

Photosynthesis (in green chloroplasts) deposits solar energy into water (H_2O) and carbon dioxide (CO_2) and releases oxygen (O_2) into the atmosphere. Stored energy is in biomass (in particular in food). We receive the energy stored in food. By oxidization of food in mitochondriae (marked in red) the energy is released. Breathing is a reversed process to photosynthesis.

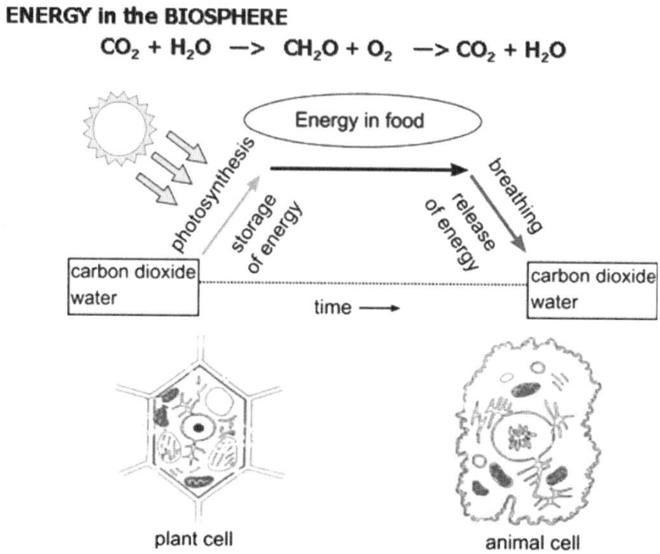

8. Solar Energy in Our Service

Solar radiation cannot be accumulated as such. Instead it can be transformed into a convenient form of matter energy. The term "materialization" of solar radiation could be used, because it means (according to Einstein relation) increase of mass.

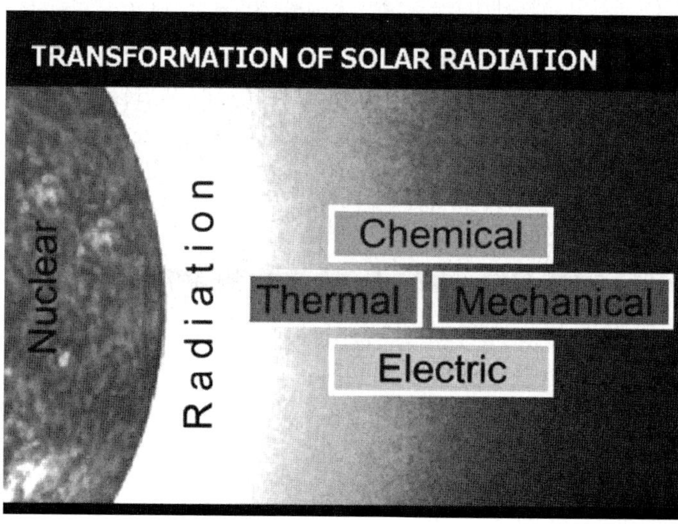

A) **Chemical** - e.g. hydrolysis decomposition of water into oxygen and hydrogen. Oxygen with hydrogen used in fuel cells to produce electric current. However, the best known chemically stored solar energy as biomass energy – solar radiation deposited by photosynthesis. Biomass (organic matter) can be used to provide heat, make fuels, chemicals and other products, and generate electricity.

B) Absorption transforms solar radiation into **heat**. Solar collectors for heating water or air for heating buildings may be seen in many houses. In focus of concentrating collectors water evaporates and the vapor is used in classical powerplants. Nature itself absorbs huge quantities of solar radiation to heat land, hydrosphere and atmosphere. Without this absorption the mean temperature of the Earth would be only minus 260°C. The heat drives winds, water cycle and is used by thermal pumps and oceanic thermal energy converters (e.g. OTEC).

C) In photovoltaic (= solar) cells, the incident solar radiation is transformed directly into **electricity.** The cells are still expensive if compared with normal powerplants. But in some cases the solar cells are irreplaceable, e.g. on lonely places, satellites, spacecraft and International Space Station which needs 110 kW. The International Space Station receives the 110 kW from solar cells on panels. Many powerplants use kinetic energy of water or of wind biomass energy – which is transformed solar energy.

D) Solar radiation is transformed into **mechanical energy** (i.e. in kinetical or potential energy).

Indirectly the transformation occurs by biofuels, in heat machines, electric cars or airplanes (see Helios photo of NASA). Our cars and airplanes are also driven by the solar energy – but by the ancient one.

On a huge scale, the solar radiation is being transformed into mechanical energy by nature. The kinetic energy of winds and of streaming water, as well as the energy of water circulation is indirect solar energy.

9. Conclusions

It is obvious that the direct solar radiation in the form of photons (180 000 TW) or the indirect solar energy (deposited in wind, water motions, heat and biomass) represent the only energy solution for the Future of Life and for our Civilization. One cannot make an economical comparison with the ancient solar energy used in form of fossil fuels. To their rising price the "price" of many human lives, of a clean healthy environment for our descendents and of a peaceful Blue planet should also be added. Then our discussion will be right.

Deuterium Explosion Power

N.P. Voloshin, A.S. Ganiev, G.A. Ivanov, F.P. Krupin, S.Yu. Kuzminykh, B.V. Litvinov, L.I. Shibarshov, A.I. Svalukhin,

Russian Federal Nuclear Center, Institute of Technical Physics (VNIITF), Snezhinsk, Russia.

Oil and natural gas are crucial for the industrial development of modern civilization. They supply three fourths of the world energy. Some people hope oil will run for ever, others hope for a wonderfull discovery, informed people don't.

Is it possible to make up the energy shortfall after natural oil resources are exhausted? When does the time come for sounding alarm and taking all possible measures to escape the disaster if account for large-energy inertia? Or should we hope for a new, timely discovered resource?

All possible energy sources (according to mechanics and thermodynamics laws) come from celestial motions (only tidal currents seem usable) or temperature difference: natural or artificial (chemical and nuclear fuels). No other energy sources exist.

Natural temperature difference sources available are geothermal sources and solar radiation.

Tidal currents and geothermal sources are very weak.

Solar radiation has derivatives:

biomass,

wind,

water (river and sea currents).

Biomass + water are estimated to satisfy no more than 10% of energy demand.

For **wind and solar radiation**, the percentage is yet smaller.

The *low density of energy flux* is their fundamental shortcoming which results in:

V. Burdyuzha (ed.), The Future of Life and the Future of Our Civilization, 201–204.

- a low energy output - to - energy input ratio (input is all energy spent for exploitation, power plant construction, its materials production and mining, transportation, etc.),
- a high cost of one kilowatt-hour of mean installed power, and
- substantial environmental effects if the production scale is large.

Non-continuity is another shortcoming. *Thus wind and solar radiation are supplementary power sources only.*

Chemical Fuel

A lot of chemical compounds can be found in the entrails of the earth but only few of them can be used as fuel.

Different estimates suggest that oil and natural gas reserves will run out in 20-50 years. The reserve of coal is sufficient for several hundreds of years but it is much more expensive and “dirty”.

Are there other (yet unknown) chemical fuels available in abudancy?

Nuclear Fuel

mainly U-235;

Secondary nuclear fuel:

mainly Pu and U-233.

In fact, nuclear power plants are safest and pollution-free sources of energy (!). *But the total NPP capacity is limited* by the reserve of U-235. NPP deployment plans do not provide for substituting exhausted oil.

Fuel Exhaustion Times

are 30-100 years for most profitable fuels (oil, natural gas, uranium-235), and 200-300 years for less profitable ones (coal, uranium, and thorium).

None of the energy sources mentioned can substitute oil in the future. They are either exhaustible (fuel) or cannot be widely deployed (renewable).

But there is another energy source which can! It is deuterium. *It is cheaper than oil (in energy equivalent) and abundant in nature. In thermonuclear fusion, 1 kg of deuterium (extracted from 30 tons of water) releases as much energy as 10 thousand tons of oil.*

Is it possible to burn deuterium through **controlled fusion**?

ITER project plan for deuterium-tritium mix includes:

Phase 1: Achieve break-even in 2015,

Phase 2: Demonstrate a thermonuclear power plant in 2030.Tritium is absent in nature. As an energy source, it is much more expensive than oil.

Large amounts of tritium required for large energy produce bring about a lot of techical, safety, and security problems.

The reaction rate of deuterium fusion is 2 orders lower. Deuterium fusion seems achievable in the very far future.

Deuterium Explosion Power (DEP)

The break-even deuterium problem has been resolved long ago: industrial deuterium charges with a high fusion-to-fission energy ratio exist (fusion factor FF = fusion energy/ fission energy).

DEP containment: blasts in "deuterium explosion chambers" (DEC) – underground steel-coated cavities.

For DEP, material input and electricity costs are lower than for water power stations of the same electric power.

Scientists from Russia and other countries at various times proposed producing energy by fission/fusion explosions but not for DEP.

Deuterium fusion: eventually, $3D \rightarrow n + p + He^4$. The additional energy is released in neutron capture in the coolant ($Na^{23} \rightarrow Na^{24}$) and the radioactive decay of Na^{24} with a half-life of about 24 hours.

The deuterium fusion "primer" is the low energy released in the fission chains of fissile materials (FM): Pu-239, U-235, U-233.

For FF = 100:

of the total energy produced, only 1% comes from FM (uranium and thorium contained in seawater can be used?),

200 neutrons per 1 FM nucleus are produced; 1-2 of them is needed to regenerate FM and a few may be needed to produce secondary nuclear fuel.

The **DEC** is an underground cavity coated with steel and filled with a noble gas. Blasts are assumed to be conducted at intervals in several hours.

A liquid-sodium curtain protects the coating from neutrons and fireball emissions, and absorbs the shock. Heated sodium then serves as a primary coolant. Energy is transferred to electric generators, as it is done at the Beloyarsk NPP; if necessary, artificial fuels are produced. Heat (for example, steam 3MJ/kg) and electricity can be transmitted to long distances.

According to calculations to construct one DEC it is necessary to make building and earth-moving works for the same or even smaller volumes than at the construction of the hydroelectric power plant of the same electrical output power.

Safety

Deuterium explosion safety requirements will be stricter than those for mining and peaceful nuclear explosions.

Radiation safety requirements are formulated following from experience gained in numerous underground nuclear tests. The probability of nuclear releases is extremely small due to containment; nuclear releases in dangerous amounts are ruled out. The amount of radioactive substances in the DEC is 3-4 orders of magnitude smaller that in the reactor core:

the amount of radioactive substances per unit energy is 100 times lower, and

the time between decontamination procedures will be short. Seismicity is small becouse of explosion decoupling in large cavities; shocks are not felt at distances of several kilometers.

Security

Some measures to make nuclear energy charges inaccessible for terrorists:

Nuclear energy charges should be produced in underground facilities near the DEC, with protected transportation.

Materials and charges used must be least attractive for proliferation.

No storage (use immediately after production).

Very high radioactivity of material used

Some DEP Essentials:

Production and long-distance transmission of electricity, heat and artificial fuel (hydrogen, hydrocarbons, magnesium), Generation of "heat oases", Desalination of large amounts of seawater for irrigation and other purposes,

Nuclear power growth due to the use secondary nuclear fuels produced in unlimited amounts in DECs.

The unique expertise of VNIITF scientists holds out a hope that deuterium blast energy is feasible. It is assumed that prior to designing full-scale DECs, experimental explosions of yields between 0.1 and 1 kt will be conducted in small-size DECs. An experimental DEC can be constructed and operated if international community is agreeable to nuclear explosions in peaceful purposes. Nuclear experiments in the DEC can be started after an international treaty on peaceful nuclear explosions is concluded, as provided for by the Comprehensive Test Ban Treaty. **The sooner the treaty is concluded, the sooner we can start DEP development and deployment.** It is better to make secure, «to spread straw», and start to develop DEP than to hope for a wonder only.

Accelerating Changes in our Epoch and the Role of Time-Horizons

Kay Hamacher

Center for Theoretical Biological Physics, University of California San Diego, La Jolla CA 92093-0374, USA, hamacher@ctbp.ucsd.edu

Technology shapes modern society at a scale unknown to our ancestors. During the last decades many observers noticed that improvements in technological capabilities seem to occur at an increasing rate. The emerging applications of nanoscience, biotechnology, information technology and cognitive science ("NBIC") are prominent examples - the NBIC-convergence an universally accepted fact. Progress especially in the field of life sciences, biophysics, and biology itself promise even more accelerated change/progress than the one shaped by IT so far. We show examples of these observations, discuss the modeling and simulation of those developments and discuss the effects in terms of technology and economics that prompt for more interdisciplinary research in this direction

Keywords: Price dynamics; accelerated technological change; cobweb model.

1. Introduction

People living some 200 years ago found their century most exciting and promising. The inventions such as steam powered machinery and the subsequent economical, sociological and cultural change must have been overwhelming. The people thought about their predecessors as living in a medieval age. Since then new technologies arise more and more rapidly leaving us not only thinking about 1805B.C. as another medieval age but also now speculating about computation by means of single elementary particles [1], building space stations on Mars [2] or evolving artificial life & conscience [3]. This feeling about a fast-pace progress can also be put

V. Burdyuzha (ed.), The Future of Life and the Future of Our Civilization, 205–215.

into a more fact-based framework by inspecting several numbers. In the following we give some examples, describe one approach to quantify progress and finally introduce a model on our own from the field of econophysics.

1.1 Some First Observations

The rate by which new technologies are not only developed but also introduced/placed successfully in the marketplace is increasing. For example to reach 10 million customers in Germany manufacturers of communication technologies needed [4]: Telephone - 40 years, Fax - 20 years, Cell phone - 10 years, Internet - 4 years. Other established technology based products are deconstructed and replaced in the market place at a rate unknown before [5]. But not only technology based products are in general on an accelerated scheme, also some other developments show the accelerated change in our epoch. Figure 1 shows for example the increased financial effort in education as the basis for further technological progress as a function of time in comparison to the Gross Domestic Product (GDP) in the US. While the rate at which the ratio grows is more or less linearly one has also to take into account that the US GDP has increased dramatically (nearly exponential) during that period thus boosting the expenses for education to new heights more or less every single year. While at the present it is unclear how educational levels translate into economic growth or technological change it is justified to assume a positive correlation between those observables.

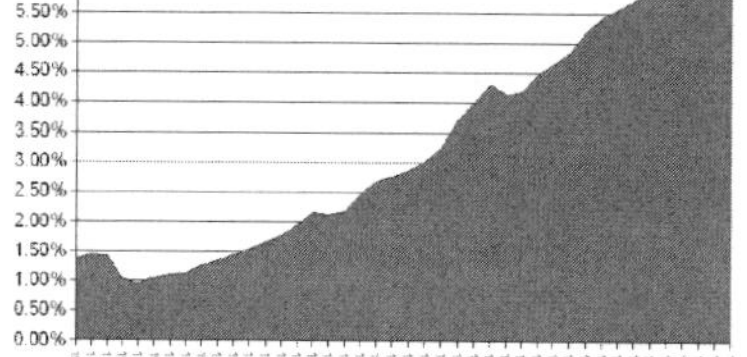

Fig. 1. How much does the US spend on education with respect to its GDP? Note the (nearly exception less) monotonic increase and the fact that the GDP itself increases, too. [data from various .gov-sites].

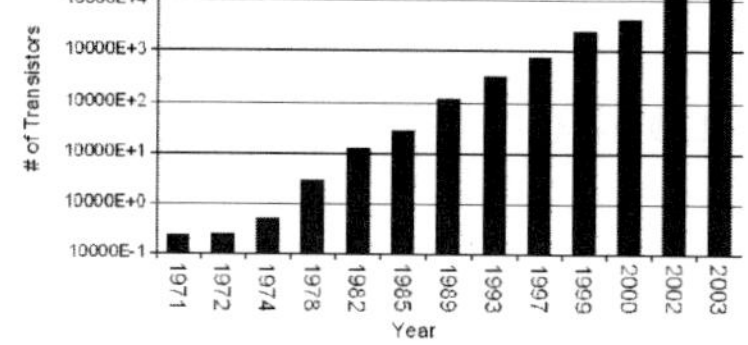

Fig. 2. Moore's Law [data from www.intel.com].

Figure 2 shows the well-known Law of exponential increase in computational power/number of transistors on a CPU – named after the first business leader who mentioned it: Moore's Law. Although there are actually several versions of Moore's Law to take side effects into account and although it was argued that the strong exponential increase is not a strict as it was thought to be [6] we have to acknowledge a close-to-exponential increase in the computational performance anyway.

1.2 Implications and Emerging Questions

These findings prompt for some questions about the development to come.

- First of all we have to be sure about the observables that are reasonable, that provide answers for important questions. What exactly are suitable measures to indicate an accelerated change? Is it about economic figures? And most important: How to measure technology? Should we do so in more technical terms like computational capabilities or in economic terms such as utility of computation?
- If a satisfying measurement system is found, do we really observe acceleration in the change? Or does it manifest itself only in figures that are not important?
- If we accelerate any changes in our society, our technology - does this necessarily imply the existence of time horizons? What does happen beyond this horizon that indicates a phase transition (from the statistical physicists point of view)? Either we encounter a singularity (see below) in the advancement of civilization or there will be an upper bound to acceleration as the capacity of society and technical facilities reaches some saturation level. In the case of a singularity: Can we derive any knowledge when some technological Big Bang will occur?

2. Modeling (Accelerated) Changes

2.1 General Remarks

In 1987 the Nobel price laureate in economics Robert Solow stated his concern [7] about an observation, which became known as the Solow Paradox: "You can see the computer age everywhere but in the productivity statistics."

Since then production theory tried to overcome this judgment with various approaches but it is still occupying a large fraction of researchers in this field to develop a suitable approach how to account for 'knowledge-related' productivity, wealth creation, value measurement and so on. For details see Triplett's discussion [8].

The quote itself is not necessarily meant to deny the productivity effects of IT but more a concern about the lack of a suitable measure. This is a most important issue as due to public awareness and media coverage the productivity numbers of national economies tend to influence policy makers all over the world – and the supra-national organizations such as the WTO or the IMF enforcing global trading policies. An accurate way to measure the influence of technology is therefore a prerequisite to use the results of productivity studies on the economic well doing and decisions that effect technological developments. In modeling we have always to keep in mind that direct comparison to empirical data is therefore complicated.

2.2 The Kurzweil Ansatz

Ray Kurzweil [9] introduced an approach to model accelerated change on a very coarse-grained scale – 'macroeconomics for technology' we would like to call it. He solely focused on computational power as a measure for progress – implicitly assuming that computation alone is feasible of bringing progress. Suppose there is a quantity V that somehow measures the computational speed available by the most advanced technology. Consider further the sum of all of our knowledge W. Both quantities are functions of time t.

Kurzweil assumes now that the speed of the best computational technology that can be built is proportional to the knowledge at a particular time: if you are capable of dealing with"twice as much" technology you can build machines faster a factor of two in processing speed:

$$V(t) = c_1 W(t) \tag{1}$$

2.2.1 Two Models

Now his first model just states that the increase in knowledge is proportional to the computational effort we undertook in a time-period:

$$dW = V(t)\, dt \qquad \Rightarrow \qquad W(t) = c_2 \int_0^t dt' \, V(t') \tag{2}$$

W is therefore assumed to be cumulative. Inserting Eq. (1) in Eq. (2) immediately brings us to $W(t) = \alpha_W \exp(\beta \cdot t)$ and from that $V(t) = \alpha_V \exp(\beta \cdot t)$. So we obtain an exponential increase in computational power over time – which he relates then to Moore's law.

Model no. 2 tries to take the growing number N of resources capable of computation into account. But then the increase in knowledge is $dW = N(t) \cdot V(t)\, dt$. The computational resources N are also growing at an exponential rate $N(t) = \alpha_N \exp(\beta_N t)$ and again $V(t) = c_1 W(t)$. Then

$$W(t) = c_2 \int_0^t dt' \, V(t') \cdot N(t') \text{ can be led to } W(t) = \alpha_W \exp(\beta \cdot \exp(\gamma t))$$

and, therefore, we obtain a very fast increase in computational capabilities $V(t) = \alpha_V \exp(\beta \cdot \exp(\gamma t))$.

These models come actually in several flavors. A published discussion actually shows that the authors tried to come up with a model that has actually a mathematical singularity [9]. But even without the mathematical form the increase is very fast. The term singularity is then used for the point of time when machines are able to sustain and improve themselves. From there on the acceleration is self-sustained as synergies across all technologies can be exploited. The predictability cease to exist at this point as we can make no assumption about what this 'new' world of machines will come up with – imposing a time horizon on our capability to make forecasts.

2.2.2 Some Remarks on Kurzweil's Ansatz

Here we want to give a non-comprehensive list of criticism on the above-mentioned model:

- Coordination problem: Both model I and model II assume that the ratio between some measure of computational effort and knowledge is (nearly) linear. This is however a very optimistic assumption as the increase in the number of computational units makes it more likely that computations are repeated due to a lack of central coordination. If we however introduce some coordinating instance we face the problem that the growing number of processing units also leads to an exponential increase in the complexity of coordination – thus loosing most of the gained technological progress in bookkeeping.
- Moore's law: The ansatz itself relies heavily on the validity of Moore's Law. There are however different flavors (like number of transistors per

square-inch, per dollar, per....). These gradual changes of the 'Law' over time was critically discussed by Tuomi[6] and might lead to the invalidity of at least model II.

- General sociological remarks:
  The raise of a singularity can only occur if technology can improve itself without human interference. If humanity however does interfere due to social, ethical or philosophical (even selfish, e.g. like Veenhoven [10] suggested) considerations than we have to rethink the model or even discard it.
  Competition is not taken into account. While perfect competition can lead to an efficient market in comparison to a monopoly or an oligopoly the competition (between human beings, companies, countries) itself can waste resources that are assumed to work for a common goal in Kurzweil's ansatz.
- Resource allocation: While one variant of Moore's Law explicitly connects computational power and costs (that is Dollars) it is not clear how resources to design, built and maintain computational devices should be allocated. A market mechanism might lead to a very different response – if e.g. the marginal utility of computation is not perceived by (human) customers as valuable as the gain from some other investment.

2.3 Getting More Detailed

We want not to be as bold as Kurzweil, but instead investigate the mechanisms at a more detailed level – in the analogy of above - a 'microeconomic approach' to modeling the accelerated change. This is however not a comprehensive description of all effects.

2.3.I Appropriated Measures

While it would be an intuitive approach to measure the impact of technology on our lives by values that are displaying some 'technological power' we do not follow this path here. First of all one has to admit that due to technological change the relevance of a certain number changes over time, in an accelerated world even more so. Take for example people some 500 years ago their living was influenced by e.g. the speed of a horse or the speed of a boat but definitely not by the speed of computation. As a second argument for a non-technological measure of the impact of further technological progress one has to bear in mind that all technology has to be created from scarce resources. The allocation of those is however an economical process. From now we assume that a market economy will be the prevailing way to solve this distribution problem.

2.3.3 Economical Modeling

We reduce our investigation on economical matters and thus broadening what has become popular as econophysics over the last years, see e.g. Feigenbaum's review [11]. Science as a production factor has become of increasing interest for economist [12]. Technological changes influence several aspects of the market:

- new products and services emerge
- production costs can be reduced
- prices can decrease due to the emerging competition or increase as a new technology might be perceived as more useful (more demand) and is protected e.g. by patents or comes in combination with another technology:
- substitution effects change the way goods are produced/consumed, e.g. a new technology replaces at least partially existing products thus decrease the demand for the substituted 'old' products
- positive correlation: one product might increase the demand for another, e.g. MP3-technology and CD-media to burn your MP3-library

We do not want to speculate on new technologies and the characteristics of those (see above for some thoughts about measures). We want however model and simulate the influence of abstract technologies on the economy. We are not concerned with problems of global trade (like taxes etc) and externalities (such as environmental pollution) here but assume just one global market place without any friction[11].

2.3.3 Classical Price Determination

One of the most well studied models is the cobweb-model. Here we assume market equilibrium at a particular time t (that is: everybody who wants to sell at a determined price sells, everyone who wants to buy at that price is able to buy → market clearing price). The market for just one good is described by a demand function and a supply function relating price to quantity of the good traded. This model already shows rich dynamics, even chaotic behavior [13-16] that was found to be present not only in the model but also in real-world data [17]. The onset of chaos manifests itself in positive Lyapunov-exponents of the time-series, implying a non-vanishing Kolmogorov-Entropy K. In the case of $K \neq 0$ we are confronted

[11] These effects will be investigated in a forthcoming study.

with a system that already has a finite time-horizon of predictability [18] as any predictions become inaccurate on a time scale $\propto K^{-1}$. The original cobweb-model explains just the price formation for one good. It doesn't account for product substitutions. The original studies focus on periodic, synchronous price determination (like seasons). Therefore the cobweb-model is most applicable for commodities like agricultural goods. These findings prompt for a careful modeling and analysis of accelerated change as time horizons might occur even in technological constant markets.

2.3.4 Aspects of Accelerated Economics

Here we want to list some aspects that have to incorporate into a model for an 'economy' undergoing accelerated technological change.

Limited history: Accelerated change means that all market participants – might they be consumers or producers – can only take limited information for granted. Prices, demands & supplies change also at an increasing schedule with improved technological capabilities. Price and demand/supply histories become less and less accurate.

Substitutions: Technologies are capable of replacing each other, like Email and fax machines, gopher and WWW services, or cars and trains. There is however a limited substitution capacity, e.g. without electronic signature you cannot transmit a legal binding document by Email while you might be able to do so by fax (depends on your government). If prices for technology A are higher than the amount we want to pay for it, we can substitute a competing technology B for it – at least a certain amount thereof.

Convergence and Synergies: Substitutions might even lead to phenomena called convergence nowadays. An example is the progressing integration of entertainment electronics into telecommunication devices (like games, cameras, MP3-player, radios etc). The progress in one area (like miniaturization) might help to improve another technology, too. If those synergies lead to a product providing every benefit of the former single technologies than we can replace those by that newly emerged one. The transition might however be characterized by a substitution behavior of buyers because the 'old' technologies might be cheaper and as well suited as the new one (see Substitutions above). A threshold exists below the maintenance of production units does not pay and a substituted technology gets discarded from the market (like telegraph services) completely.

2.3.5 Modeling such a market

At market clearing every item to be sold will find its buyer, so the price that is achievable is determined by the demand $d_t(p_t)$ where t is the time. The supply is determined as $s_t(p_t^e)$where p_t^e is the expected price at time t. At market equilibrium we have: $d_t(p_t) = s_t(p_t^e)$. This already leads to a price dynamics, which is quiet rich as can be seen in figure 3.

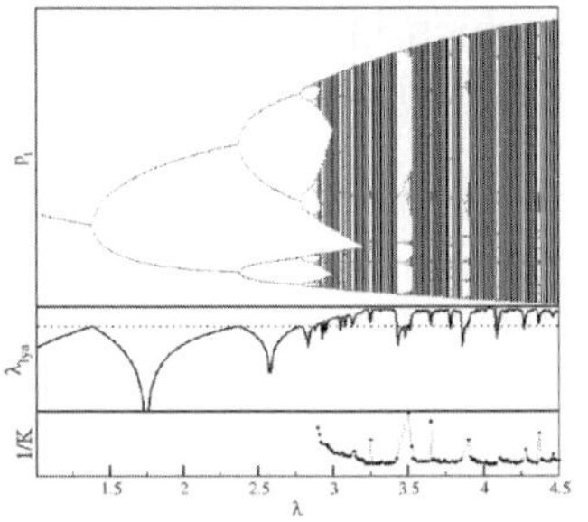

Fig. 3. Rich Dynamics of prices, their Lyapunov-exponents and the resulting Kolmogorov-Entropy. A non-vanishing entropy implies a time horizon for the price predictability.

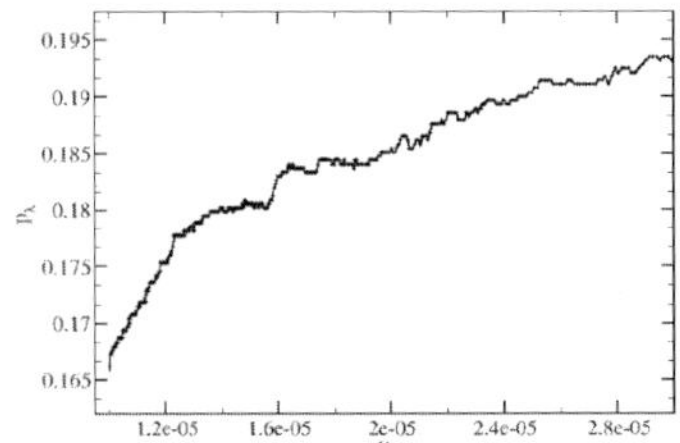

Fig. 4. Probability to encounter a time horizon in the price dynamics with respect to rate of technological change.

The details of the model and the subsequent analysis can found elsewhere [19]. If now introduce a rate of change μ to mimic substitution as well as decrease in the relevance of historical knowledge about the market, we can observe various effects that most of all lead an increase in the non-trivial behavior.

Figure 4 shows the result for a particular parameter set. The dynamics gets richer the more change is facilitated in the production of goods. We thus conclude that the accelerated change of technological production capabilities induces more chaotic price dynamics, thus more often time horizons.

3. Summary & Conclusions

In this contribution we described some developments that modern society faces. Technological advances pose a challenge for existing technologies, the economy and the society. We showed three indications of an acceleration of change and discussed possible impacts and accompanying questions We mentioned the ongoing controversy about suitable measures and described one coarse-grained model that produces a technological singularity (Kurzweil). We continued with some remarks on the limited applicability of this model and introduced a more detailed model in the spirit of econophysics (cobweb and the extension presented here). We described some of the findings that are characteristic for this model and put the results into the context of real-world developments. We hope to have encouraged further interdisciplinary research, which in our point of view is the only feasible approach to grasp future development in society and the economy.

Acknowledgements

The author is supported by a Liebig-Fellowship of the Fonds der chemischen Industrie. He is grateful to the organizers of the Symposium for the invitation.

Notes

 For details on this and future work please visit the author's website http://www.kay-hamacher.de

References

1. J. Stolze and D. Suter, (2004) *Quantum Computing: A Short Course from Theory to Experiment* (John Wiley & Sons, New York).
2. NASA, http://weboflife.nasa.gov/currentResearch/currentResearchGeneralArchi-ives/inspiration.htm.
3. C. Gros, (2005) Autonomous dynamics in a dense associative network for thought processes, (submitted).
4. R. Bargsten et al., (2004) Digitale Trends, Erwartungen, Realität und Perspektiven, Wirtschaftsrat der CDU e.V. Landesverband Hamburg.
5. P. Evans, T. S. Wurster, (2000) *Web Att@ck*, (Hanser Verlag, Hamburg).
6. Ilkka Tuomi, (2002) FirstMonday **7**, 11.
7. R. Solow, (1987) New York Review of Books.
8. J. Triplett, (1990) Canadian Journal of Economics 32, 309-334.
9. http://www.kurzweilai.net.

10. R. Veenhoven, (1998) Social Indicators Research 20, 333-354.
11. J. Feigenbaum, (2003) Rep. Prog. Phys. 66, 1611-1649.
12. Paula E. Stephan, (1996) Journal of Economic Literature 34, 1199-1235.
13. R. V. Jensen, R. (1984) Urban, Economics Letters 15, 235-240.
14. Z. Artstein, (1983) Economics Letters 11, 15-17.
15. A. Matsumoto, (1997) Discrete Dynamics in Nature and Society 1, 135-146.
16. C. Chiarella, (1988) Economic Modelling 5, 377-384.
17. A. J. Lichtenberg, A. Ujihara, (1989) Journal of Economic Dynamics and Control 13, 225-246.
18. H. G. Schuster, (1994) *Deterministisches Chaos* (VCH, Weinheim).
19. Kay Hamacher, (2005) Impact of accelerated technological change on price dynamics (submitted).

Mathematical and Spiritual Models: Scope and Challenges

Jose-Korakutty Kanichukattu

Department of Statistics St Thomas College PALA Arunapuram P.O. Kerala - 686574, India, kkjstc2002@yahoo.co.in

Mathematics and spirituality are two main disciplines which help man to answer and respond to the mysteries of the universe. The successful completion of the genome projects, the brilliant breakthroughs in neurobiology research, information theory, biotechnology and the revolutions made by chaos theory strike at the roots of many fundamental questions that have occupied human thought for centuries. Mathematical models developed for the study and exploration of nonlinear phenomena and studies in the theory of chaos have contributed a major paradigmatic shift in physical sciences. Eminent scientists like Galileo, Kepler, Newton, Einstein, Hawking, Wolfrang, etc had developed many mathematical models to describe the complexities of the universe. All these advances in science via mathematics led to the revelation of the mysteries of the universe to some extent. However these models are not fully successful in explaining the complex realities of the universe. Another direction of advancement was based on the various religious and spiritual philosophies and models proposed by many thinkers. The talk will focus on various mathematical models developed by scholars in the past and present. A discussion on true spirituality and the Indian way of life based on the principle 'unity in diversity' will be made. The limitations of scientific theories and religious philosophies will be pointed out. The need for evolving a true scientific and true spirituality culture will be stressed. It is suggested that a joint action by scientists and religious leaders is needed to ensure the better future of the humanity and the universe.

V. Burdyuzha (ed.), The Future of Life and the Future of Our Civilization, 217.

Future of our Civilization: Benefits and Perils of Advanced Science and Technology

Ting-Kueh Soon

Malaysian Scientific Association, Room 2, 2nd Floor, Bangunan Sultan Salahuddin Abdul Aziz Shah, 16, Jalan Utara, 46200 Petaling Jaya, Malaysia, malsci@tm.net.my

1. Introduction

The world today is undergoing rapid and major changes. There is a significant increase in world population. Great strides were made in scientific discoveries and technological advancements. Advances in telecommunications have made the world a global village and there is growing concern on the depletion of natural resources, increasing disparity in wealth distribution and deterioration in the quality of our environment.

2. Major Changes in the World

At present, the major changes in the world are as follows:

- Population growth
- Accelerating industrialization and urbanization
- Globalization and growing disparity
- Depleting resources
- Environmental degradation
- Civil disorders and natural calamities

V. Burdyuzha (ed.), The Future of Life and the Future of Our Civilization, 219–230.

2.1 Population Growth

After the second world wars, world population has been growing at a steadily rate. In 1900, world population stood at 1.6 billions and increased by about double to 3.0 billions in 1960. This had doubled again to 6.0 billions in 1998. World population is expected to reach 10 billions in 2050. The question is, “Is such a population growth sustainable?” “Can the world feed, cloth and shelter 10,000,000,000 people?”

Even at the current situation, there is increasing incidence of poverty, hunger, famine and lack of basic amenities. Population growth is highest in the developing countries; and yet famines, civil wars, natural disasters, HIV – AIDS and other debilitating diseases are killing millions of the poorest every year. At the same time, the rich are getting richer and a small upper class is getting more affluent. And all these gains are at the expense of the poorest!

This picture definitely does not look promising, in fact, downright worrying.

2.2 Accelerating Industrialization and Urbanization

The industrialized world has advanced technologically with great scientific discoveries and cutting-edge technologies. Many of the rapidly developing countries are also trying very hard to catch up. Yet majority of the world’s population are still lacking in basic amenities and more than one billion of the world’s population is living at a subsistence level of less than one US dollar a day.

Rapid industrialization also means rapid resource consumption. There is a tremendous increase in the demand for energy, food and materials. At the same time, there is increased urbanization. It is expected that by the year 2008, 50 per cent of the world population will live in the urban areas, placing great strain on transportation, water consumption, health care and sanitation. In addition, majority of the most populated cities in the world will be found in the developing world as shown in Table 1.

2.3 Globalization and Growing Disparity

In the second-half of the twentieth century, there were tremendous developments in transportation and communication. Land transport by road and rail developed rapidly all over the world and traveling has been made so easy within landmasses. However, increasing motor vehicles on the road

Table 1. Urbanization – Urban Population by 2015. *Source:* UNITED NATIONS World Urbanization Prospects.

Urbanization as percentage of world population:
1800 (3%) 1900 (13%) 2008 (>50%)

1950	Mil	2000	Mil	2015	Mil
New York, USA	12.3	Tokyo, Japan	26.4	Tokyo, Japan	26.4
London, England	8.7	Mexico City, Mexico	18.4	Mumbai, India	26.1
Tokyo, Japan	6.9	Mumbai, India	18.0	Lagos, Nigeria	23.2
Paris, France	5.4	Sao Paulo, Brazil	17.8	Dhaka, Bangladesh	21.1
Moscow, Russia	5.4	New York, USA	16.6	Sao Paulo, Brazil	20.4
Shanghai, China	5.3	Lagos, Nigeria	13.4	Karachi, Pakistan	19.2
Essen, Germany	5.3	Los Angeles, USA	13.1	Mexico City, Mexico	19.2
Buenos Aires, Argentina	5.0	Calcutta, India	12.9	New York, USA	17.4
Chicago, USA	4.9	Shanghai, China	12.9	Jakarta, Indonesia	17.3
Calcutta, India	4.4	Buenos Aires, Argentina	12.6	Calcutta, India	17.3

exerted strong demand on the traffic infrastructure and energy consumption. As of now, major cities and urban centers are faced with huge traffic congestion and other related problems. Sea transport also expanded rapidly with shipment of fuel, raw materials, food and manufactured goods to destinations all over the world.

One of the most significant developments in transportation is air transport. With commercial planes flying all over the world, ferrying passengers and goods, "the world has virtually grown smaller". Instead of taking weeks to travel from Australia to Europe, it is possible to fly from Sydney in Australia to London in the United Kingdom within the same day. In fact, air transport has been hailed as one of the major wonders created by mankind.

However, the most significant development is in telecommunications, both wire and wireless, and the Internet. First came the telephone and followed closely by wireless radio Transmission. Then came the computer and satellite communication. The world is now "wired together" so that you can virtually talk to anyone in any part of the world. The Internet allows the world to be connected and recent development in mobile communications allows you to transmit and receive voice, picture and data from any part of the world.

The impact of such global communication is tremendous. Information is available at the tips of your fingers and you know almost instantly what happens in other parts of the world. Manufacturers, business and ordinary people can source for raw materials, spares and products from all parts of the world and home keepers can purchase their daily needs over the computer. We have finally become a "global village".

The arrival of the information age has brought vast benefits to major population of the world. It has elevated the standard of living of the poor and improved the quality of life of many. However, it has also resulted in great disadvantages to those who have not has access to such facilities. In fact, this digital divide between "have" and the "have-not" has created even greater disparity between the poor and the rich. This widening disparity does not only happen among nations, but also between different levels of society within the same nation. While the affluence is enjoying the best of life, more than one billion of the world population is living in abject poverty without even basic amenities.

2.4 Depleting Resources

With increasing population and better quality of life, the world is faced with the serious problem of depleting resources. Increasing demand for energy and transportation has resulted in depleting energy resources such as fossil fuels. We are so used to cheap fuel from petroleum and coal; however, such situation will be gone soon with these resources running out after a couple of hundred of years. At the moment, we have not been able to harness other sources of energy at a price competitive to fossil fuel. There have to be a breakthrough in the acquisition of energy soon; otherwise, we will be faced with an energy crisis within the next few decades.

We are also faced with serious depletion of biological resources and extinction of species. With more land opening for agriculture, housing, manufacturing, leisure, roads and other infrastructure development, biological resources are being depleted at such a rapid rate and species are facing extinction. With increasing desertification due to climate change, the world's habitat is being threatened from two fronts, development and climate. Changes in the world habitat could have serious consequences on species survival and water resources. It is estimated that due to the clearing land for development, species of various life forms are going extinct at several thousands a year.

There is also a growing concern on depleting mineral resources. Metals such as gold, silver and copper are being mined at such a rate that these resources will run out in few hundred years. Of course, there is a increasing trend to use alternate materials to replace these metals in various applications, but one must bear in mind that the earth is finite and there is no way that new resources will come from somewhere.

Another crucial resource that is facing serious problem is that of fresh water. There are ample salt water in the oceans and the seas, but what the world needs is fresh water for human consumption. Fresh water only makes up 3 per cent of the total water available on earth, and out of this 3 per cent, only 0.003 per cent is usable by human. With rapid land clearing and climate change; these fresh water resources are further threatened and the world will be faced with a serious water crisis in the next few decades. Of course, technologies are available to convert salt water to fresh water; but can the world afford it?

2.5 Environmental Degradation

With rapid development and little care for the environment, we are now faced with a serious problem of environmental degradation. Air quality in major cities and industrial areas are rapidly deteriorating. Acid deposition in urban and industrial centers is causing serious harm and damage to human population, buildings and materials. High ambient toxic gases such as oxides of nitrogen, ozone and carbon monoxide, in the air are seriously affecting the quality of life of the urban population.

Waste generation and bad handling of these wastes also seriously affect both air and water quality. Water resources are contaminated by such wastes and subsequently this will affect water quality. Rapid urbanization leads to increasing demand for sanitation, and this places even greater stress on water quantity and quality. This will directly lead to greater pressure on the maintenance of public health system.

Environmental degradation can also be a factor in new emerging diseases that can seriously threaten world population.

2.6 Civil Disorders and Natural Calamities

The world seems to be facing greater civil disorders and conflicts. Peace between the Israelis and Palestinians seems to elude the negotiators all the time. There is still no end to the Iraqi civil conflicts. New flash points are happening in many parts of the world. There is an underlying conflict between the developed countries and the very poor; and we are all threatened by acts of terrorism.

The underlying causes of human conflicts are inequality, social injustice and lack of compassion. Unless these factors are removed, there will be no ends to social disorders and conflicts.

In recent years, we are also experiencing greater natural disasters and calamities, both in size and frequency. Earthquakes, typhoons, tsunamis and floods are creating havocs in many parts of the world. These natural disasters seem to be occurring more often than before and we are putting part of blame on climate change. However, one must remember that human activities have been identified as one major cause of climate change.

3. Scientific and Technological Advances

The world has experienced tremendous changes in terms of scientific discoveries and technological advancements, especially in the last few decades. We have come from the atomic age in 1950s, to the electronic age in 1960s, digital age in the 1980s and the present knowledge society. JJ Thomson discovered the electron in 1900 and Dalton put up his atomic theory in early 1900s. Niels Bohr's atomic model and Mendeleev's Periodic Table of elements had lead to greater understanding of atomic and molecular structures. The understanding of properties and behaviors of matters has opened up tremendous opportunities for industrial development in the '50s. The atomic era has ready come of age in the splitting of the atom, as in the atomic bomb in early 1940s.

This was rapidly followed by the electronic age in the 60s when the electrons seem to be taking over with electronic gadgets that control everything from radios, televisions, washing machines to calculators and other office equipment. Finally, it was the microchips that started the computer revolution into the digital age in the '80s.

The greatest impact of science and technology development on human is probably the digital or information revolution. In a relative short period of twenty years, the whole world has changed with the computer age. With wire and satellite communications, Internet and mobile communications, the whole world is now "wired together". You can now access to anyone in any part of the world. You can transmit voice, picture and tremendous amount of data across the world in a split of a second. This has definitely affected our lives, including the ways we do our work, conduct our business and the ways we entertain ourselves. We are now in the information or knowledge age.

3.1 The Benefits of Science and Technology

Scientific and technological advancements have definitely contributed to the well-being of the humankind. Among the areas that science and technology contributed the most in elevating human well-being are as follows:

Agriculture and Food Production

Science and technology have contributed to increased agricultural and food production. With new varieties and hybrids, better and new agricultural practices including pre- and post-harvest technology and management, and control of diseases and pests, we have been able to produce more and better quality agricultural products including foods and other materials. With modern agricultural methods and intensive farming, we are now producing enough food to feed the world population. New varieties of foods with special characteristics are also being produced through genetic engineering. This has definitely increased food production and improved food qualities.

The major problem is food security and distribution of foods. While many countries have surplus of foods, there are many more countries with inadequate food supply. In some of the poorest countries, people, including children, are faced with hunger and famine. And there is no solution to the problem insight yet.

Manufacturing, Automation and Robotics

The industrial revolution has enabled us to produce materials and goods in a large scale. Factories now churn out large quantities of materials and products for our daily needs such as foods, clothing, building materials, household products, hygiene and personal care products, pharmaceuticals and medicines, electric and electronic appliances, and many others including motor vehicles and new products for entertainment and leisure. This has definitely increased the standard of living and the quality of life of many.

Recent developments in science and technology also lead to such new technologies as computer-aided design and computer-added manufacturing (CAD/CAM), automation and robotics that further improve efficiencies and increase productivities

Development in Material Science

There were also significant developments in material science and technology. New materials are being invented and manufactured that have better and improved properties over materials that were displaced. Plastics, polymers, fibers, glasses, ceramics, new building materials and alloys, and composite materials are replacing metals and other materials from natural sources. These new materials have improved properties such as light-weight, durable and strong, fire-proof, insect-resistant or other special properties, besides being manufactured in large quantities and having economic advantages.

Recent developments in advanced materials such as ceramics and nanomaterials further illustrate the advances made in material science and technology. With these new materials with improved properties, we have been able to make advances in other areas such as buildings, roads and other infrastructures, manufacturing, aviation and aerospace, medicine and health care, information and communication technology, biotechnology and many other areas.

Nutrition, Health and Medicine

Better understanding of nutrition and the functions of foods have contributed to better health and social well-being of the global community. Malnutrition and diseases due to deficiencies in certain nutrients can now be arrested. As a result, the standard of living has improved considerably and people are living healthier and longer.

Advances in medicine and health care also improve the quality of life. We have been able to control the spread of many infectious and communicable diseases. With vaccination, many diseases can be prevented. New drugs and technologies are being constantly introduced into the market to fight and cure different types of illness and diseases. Better nutrition and health care have definitely improved the quality of life of the global community.

Life Sciences, Genetics and Proteomics

We have made tremendous advances in the understanding of many important life processes such as metabolism, respiration and aging. With better understanding of these processes, we are able to improve health care and quality of life.

Recent advances in genetics have enabled us to understand the very basis of life. We have just completed mapping the human genomes. This is a great scientific achievement and very soon, we may be able to understand the functions of different genes in our body and proceed to make “correction” or “alteration” to defective genes.

Advances in proteomics have also great implications in health care and disease prevention. Recent breakthroughs in stem cell research have opened new possibilities of treating certain diseases that was thought "incurable" previously. And if we are able to slow down the process of aging, we may be able to live to a healthy one hundred and ten of age.

Telecommunications and Mobile Communications

One of greatest technological advancements in the twentieth century is in telecommunications and mobile technology.

Alexandra Graham Bell first invented the telephone in 1876 while Guglielmo Marconi was credited as the inventor of the first radio in 1895. The telephone was the first of wired communication that transmits and receives voice messages. This was a great invention that enables people to communicate over great distance. On the other hand, the invention of the radio has enabled voice to be transmitted across space. These two great inventions have virtually changed the world at that time. This was later further developed to include transmission of picture with the invention of the television in 1925. The television was such a great hit that you now see at least one television in each household in the developed world

Wireless communication has also advanced to include satellite communications that enable us to communicate with virtually anyone in any part of the world.

With the recent introduction of mobile communications through the hand phones, the whole world is now connected.

Computer, Internet and Worldwide Web

Charles Babbage was credited with the first "mechanical computer" in 1835. But the "electronic computer" of today really took off only after the invention of the microchip by Jack Kilby in 1958. After that, the computer age really took off with the introduction of the "mini" or "personal" computer in the late '60s and it is ready a computer world after the '80s with personal computers or "PCs" in virtually every establishment and household.

The computer age brought revolutionary changes to the world. The computer is able to receive, store, manipulate and transmit large amount of data and images all over the world. It can be used in virtual all applications in the home, office, manufacturing, aviation, weather monitoring and forecast, military warfare and many others. It could hardly be imaginable today without the computers.

The computer age took another quantum leap in 1969 with the invention of the Internet, and later with the worldwide web (www). Combining telecommunications and the computer together, and with the Internet and worldwide web (www), the whole world is ready connected and we can transmit and receive voice, picture and data from any part of the world. We are now in the information age.

3.2 Roles of Science and Technology

Science and technology are playing increasingly important roles in modern society. Among the major role of science and technology are as follows:

- Science and technology for creation of new knowledge
- Science and technology for economic development and wealth creation
- Science and technology for social well-being
- Science and technology for environmental protection
- Science and technology for future generations

Let us look at these important roles of science and technology.

Science and Technology for Creation of New Knowledge

The creation of new knowledge is one of most important and fundamental functions of science and technology. Scientific research and discoveries add new knowledge to our present understanding of matters, life and the universe. We can then use this new knowledge for the benefit of mankind.

Science and Technology for Economic Development and Wealth Creation

We have come to recognize that science and technology must also contribute to social and economic development and bring tangible benefits to mankind. We have seen rapid economic development and improvement in the quality of life in many parts of the world as a result of scientific and technological advancements. Science and technology must continue to play this important role in wealth creation and improving the quality of life.

Science and Technology for Social Well-Being

In addition to wealth creation, science and technology must also play a role in eradicating poverty, providing better nutrition and health care and improving the social well-being of the poor. Science and technology must also be used to narrow the social and economic disparity between the rich and the poor. Just enriching a small minority will further aggravate the social discontent in many societies. There is a dire need to provide food, clothing, shelter and basic amenities to the poorest in the world. This can be done through scientific research and technologies targeted at helping the poorest to produce their own food and materials.

Science and Technology for Environmental Protection

Science and technology must also play an important role in environmental protection and conservation. We are now threatened by rapid development and generation of large quantity of wastes. This has polluted our air and water, and causes our land to turn to waste or become derelict. Environmental-friendly, or so called “Green”, technologies and processes must be invented to consume minimum energy and generate little or no waste.

New methods of waste treatment and management should be put in place so that the waste can treated in such a way that will not cause air or water pollution. Perhaps waste may be considered as resource materials that can be converted into some useful products.

Science and Technology for Future Generations

Scientists and technologists must also have our future generations in mind. New processes and developments must be sustainable so that we will leave our earth intact for the many generations to come. Steps must be taken to conserve and preserve natural resources, and prevent the extinction of species.

4. Science Serving Humanity

Is science serving humanity?

The answer is yes! Science has brought better quality of life. Better nutrition and health care have improved the general well-being of people. The standard of living of the majority of population has also increased with better housing, clean water and sanitation. We are also able to enjoy sports and entertainments.

But science, or rather the applications of science, can also bring about destruction. Guns and other firearms are used to kill humans. Modern weaponries are even more destructive, aiming to kill huge masses of people. The most powerful nations in the world keep enough stockpiles of weapons of mass destruction such as nuclear and atomic bombs, to annihilate the world population many times over. Chemical and biological weapons are also being developed with the purpose of killing "your enemies". The irony is that neutron bombs are being developed so that they would kill people without harming buildings and other infrastructures.

The destructive powers of science must be checked. Science should not be used for the purpose of destruction and killing people. Scientists should be ethical in not developing weaponry and chemical or biological warfare. Instead we should unite and make the governments of the world destroy their nuclear arsenals and other weapons of mass destruction. Only then can we live with ease and peaceful co-existence.

There should also be a limit to scientific advances in not trying to medal with life. Scientists should look into the ethical issues on cloning of human embryos.

5. Conclusion

Science is neutral. We have benefited from scientific discoveries and technological advances. The world is a better place when compared with situation years ago. Science and technology must also focus on sustainable development and conservation of species so that life on earth can go on for a long period of time.

On the other hand, science and technology must not be used for destructive purposes. All weapons of mass destruction must be destroyed and conflicts between or among civilizations must be resolved peacefully.

Scientists should use new knowledge and applications of science for a better quality of life for all. The future of our civilization and of earth, is very much dependent on the development of science and technology. The earth is finite and science must be used to prepare for lives for many generations to come.

The future of life is very much dependent on the power that be – to use science and technology for humanity, rather than for destruction. To do that there must be food, education and opportunity for all.

Dark Energy and Life's Ultimate Future

Ruediger Vaas

Center for Philosophy and Foundations of Science, University of Gießen, Germany, Ruediger.Vaas@t-online.de

The discovery of the present accelerated expansion of space changed everything regarding cosmology and life's ultimate prospects. Both the optimistic scenarios of an ever (but decelerated) expanding universe and of a collapsing universe seem to be no longer available. The final future looks deadly dark. However, the fate of the universe and intelligence depends crucially on the nature of the still mysterious dark energy which drives the accelerated expansion. Depending on its – perhaps time-dependent – equation of state, there is a confusing number of different models now, popularly called Big Rip, Big Whimper, Big Decay, Big Crunch, Big Brunch, Big Splat, etc. This paper briefly reviews possibilities and problems. It also argues that even if our universe is finally doomed, perhaps that doesn't matter ultimately because there might be some kind of eternal recurrence.

Keywords: Cosmology, Universe, Dark Energy, Cosmological Constant, Quintessence, Phantom Energy, Inflation, Quantum Gravity, Far Future, Life, Intelligence.

1. Cosmic Constraints

The future is not what it used to be. There is a huge challenge and change going on in our understanding of the final fate of life and intelligence in our universe [1]. Even if all the very hard biological, psychological, sociological, and technological problems could be solved, which threaten life on Earth and perhaps also on other planets, and even if intelligent civilizations could succeed and colonize the galaxies, escaping the deaths of their parent stars and manage to use astronomical quantities of matter, energy, and information, the struggle is not won. If life and intelligence are to have a long-lasting future which is really worth to be called that way – and this

V. Burdyuzha (ed.), The Future of Life and the Future of Our Civilization, 231–247.

ultimately means quasi-eternal existence and evolution –, it has to face the cosmological constraints.

We do not know very much about these constraints. But in the last few years everything that we thought we could know – and even would know soon –, got different. In 1970 Allan Sandage published a paper called "Cosmology: A Search for Two Numbers"[2]. And until the 1990's it was thought that those two numbers would, in fact, predict the ultimate future of our universe. Since 1998 this has changed completely. Now we know those two numbers with a remarkable precision – but they turned out to be almost irrelevant after all concerning the very far future. Suddenly, a third quantity came in, which, as it seems now, rules everything.

The first of the two numbers which Sandage and his colleagues had been trying to determine for several decades is the **Hubble constant** H_0. It describes the expansion rate of our present universe. Its value is around 70 kilometers per second and megaparsec (a megaparsec is 3.26 million light-years). H_o is not a true constant, but a time-dependent variable.

The second number is called the **deceleration parameter** q_0. It describes how fast H_0 changes in the future: $q_o = 1/2\ (\rho_{m,o}/\rho_{c,o}) - \Lambda c^2/3H_o^2$, where $\rho_{m,o}$ is the present mean density of matter (dark matter included), $\rho_{c,o} = 3H_o^2/8\pi G$ is the present critical density, which makes the universe flat (or Euclidean, thus there is no global curvature), and corresponds roughly to three hydrogen atoms per cubic meter, Λ is the cosmological constant, c is the velocity of light, and G is Newton's gravitational constant.

For $q_o < 0$ the expansion gets faster (so in this case q_o should more adequately called the acceleration parameter), for $q_o = 0$ the expansion rate is constant (what holds for an empty universe – mathematically simple, but not very interesting to us), for $q_o = 0.5$ the expansion will constantly decelerate and approach (but never quite reach) zero, and for $q_o > 0.5$ the expansion will stop and reverse someday.

The third mysterious quantity, which got everything mixed up, refers to what is now called **dark energy**.

The aim of this paper is to review the new situation and its confusing many scenarios, problems, and implications for the ultimate future of life. The discussion shall be restricted to the boundary conditions of fundamental physics and cosmology, and it shall survey the different possibilities only briefly without the physical details. Here it is assumed, at least for the sake of the argument, that no entirely new physical effects will change the whole picture (and even no variations of the fundamental constants of nature or that the topology of our universe is compact), nor that there are some kinds of nonphysical entities like Cartesian souls or transcendent

Gods, which are beyond the reach of (current) science, but might alter the course of the universe in a non-predictable way. If we live in a spiritual Universe or if idealism was true and matter was just a grand illusion physical cosmology probably would be off on significance.

1.1 Dark Energy

Different methodological approaches have recently led to a consistent and coherent picture of the universe we live in [3] – a picture which is sometimes called the **concordance model**, because it is the first world-view without obvious empirical or theoretical contradictions since the beginning of relativistic cosmology in 1917. According to this model, our universe emerged from a very hot and dense state, the Big Bang, 13, 7 ± 0, 2 billion years ago, expanded ever since and cooled down to 3 Kelvin yet. This is the temperature of the cosmic background radiation, which is left over from the primordial fireball and was released some 380000 years after the Big Bang, when the atoms were formed.

Many independent measurements of distant supernovae, the Hubble constant, the matter distribution and density, the large-scale structure of galaxies, quasar spectra, gravitational lensing effects, the Integrated Sachs Wolf effect, and the temperature fluctuations in the cosmic microwave background all point to a universe which has roughly the critical energy density $\Omega_{total} = \rho_{m,o}/\rho_{c,o} + \Lambda c^2/3H_o^2 = 1$ and, therefore, is almost flat or Euclidean. The big surprise was that ordinary matter (about 4, 4 ± 0, 4 percent) and the still mysterious cold dark matter (23 ± 4 percent, probably unknown elementary particles without any electromagnetic interactions) together are adding up to only about a quarter of the total energy density. 73 ± 4 percent is made of what is now called dark energy. So if you imagine the universe as a cosmic cappuccino, the coffee stands for dark energy, the milk for dark matter, both of which we know almost nothing about; only the powdered chocolate would be what we are familiar with, namely ordinary matter made of protons, neutrons, electrons et cetera. This is the ironic success story of modern cosmology: Now we know what we don't know, and this is more than 95 percent of what the universe is made of.

Dark energy has negative pressure, therefore it is repulsive, thus it acts like antigravity (so, a better designation would be dark pressure, because it is the negative pressure which is abnormal – and not the energy density which remains positive). It is also what drives the accelerated expansion of space, which was discovered in 1998 [4]. Space does not expand any slower due to the gravitational interaction of matter within, as it was long thought, but does expand faster! Measurements of distant supernovae indicate that this accelerated expansion started about five billion years ago.

But what is about the future? What are the properties of dark energy, what physical effects will they have, and how will they affect the fate of life in the universe?

Whether space expands eternally (and how fast) or not depends on what is called the equation of state of dark energy. It is $p = \rho w$, where p stands for the pressure of the spatially homogeneous dark energy (and also for other, less exotic stuff), and ρ for its energy density. Thus, the parameter w is the ratio between the pressure and the energy density.

Here are some numbers (not only for dark energy candidates):

- $w = 1/3$ represents electromagnetic radiation
- $w \geq 0$ represents non-relativistic matter
- $w \leq 1$ represents relativistic matter (for example within neutron stars)
- $w = -1$ represents the cosmological constant Λ, which was introduced by Albert Einstein in 1917 and wrongly withdrawn some years later
- $-1/3 > w > -1$ represents a scalar-field called kosmon or quintessence (here the value of w can change with time, for instance might decrease to almost -1 when the universe grows older)
- $w = -2/3$ represents topological defects (which would be primordial relics of the early universe, but probably do not exist in significant quantities within our horizon)
- $w < -1$ represents what is called phantom energy

It is not clear yet, whether the equation of state (and perhaps its time-derivative) describes the possibilities exhaustively and sufficiently, but at the moment these are the main alternatives. Thus, w can tell us something important about the nature of dark energy. And, assuming we are on the right track, dark energy determines the ultimate fate of the universe. So what are the options?

Before surveying them, a cautious comment seems to be appropriate.

1.2 Is Dark Energy Real?

Accelerated expansion is only possible if at least one of the following assumptions is violated:

1. The strong energy condition: the density and isotropic part of the pressure seen by all observers on timelike trajectories satisfy $\rho + 3p \geq 0$. – This is violated by dark energy.
2. General Relativity as a large-scale description of our universe. – This is violated by modified theories of gravity, for example modified

Friedmann equations, string theories with large extradimensions or relativistic versions of MOND (modified Newtonian dynamics) [5]. By the way: Some theories of dark energy, some quintessence models for instance, violate not only (1) but, strictly speaking, also (2), implying modified Friedmann equations or new interactions.

3. The assumed matter-dominated, homogeneous and isotropic Friedmann-Robertson-Walker cosmological model of the universe (even beyond our observational horizon). – This is violated if our whole observable universe is an underdense "bubble" within a denser environment [6]. This large-scale inhomogeneity might be the result of very long wavelength, super-horizon perturbations generated by a period of cosmic inflation in the early universe. The observed acceleration would be a "backreaction" from these perturbations without the assumption of new physics, that is the violation of (1) or (2). However, whether a violation of (3) alone is really sufficient to cause an accelerated expansion is still very controversial [7]. Another possibility for a violation of (3), or at least a wrong application of it, is this: Our Milky Way might be located at the center of an underdense "void" (with a radius of a few dozens or hundred light-years) in the large-scale distribution of galaxy clusters. This is improbable, but it would distort the measurement of H_0 nearby versus far away and could mimic an accelerated expansion [8].

At present there is no convincing reason to drop dark energy, but one has to be aware of the alternatives. Only observations will tell which of the assumptions above is violated.

2. The Future of Life and the Universe

2.1 The Past Future: Chill or Crunch

Before dark energy was discovered, there were only two possibilities in relativistic cosmology, depending on Sandage's famous two numbers (and $\Lambda = 0$):

1. An infinite (either globally Euclidean or hyperbolic) universe is spatiotemporally open and will expand forever, but with an increasingly slower rate due to the decelerating effect of matter and energy. This scenario is called the **Big Whimper** (and sometimes the **Big Chill**), because everything finally fades away.

2. A finite (but unlimited, that is spherically curved) universe is spatiotemporally closed and will start to contract after some time due to the gravitational interactions of its matter content, finally collapsing to a high density state called the **Big Crunch**. Even if the Big Crunch were to turn into a new Big Bang, nothing could survive this transition.

By the way, there is a scenario where in some sense the direction of time is switching at the maximum size of the finite universe when the expansion turns into contraction. While some have argued that even the psychological and thermodynamic arrow of time would run backwards (from the perspective of the expanding stage), and observers would still believe to live in an expanding phase, in a quantum cosmological framework everything with classical properties is destroyed in the maximum stage due to quantum interference, and the Big Bang and Big Crunch are ultimately the same, amusingly called the **Big Brunch** [9]. So in this scenario life has to cease even earlier than in the Big Crunch model that is half way at the latest, so to speak.

Both scenarios challenge the far future of life tremendously, but there are at least some chances for long-term survival – provided, of course, that life can adapt itself to the changing conditions. It must learn to use the decreasing amount of energy and, perhaps, must even rebuild its own physical architecture (for example if protons were unstable and would decay). But if life is ultimately based on information-processing devices (that is on structure and organization, not substance), it might act like a software running on different kinds of hardware. And for this, many speculative options are conceivable. As John Desmond Bernal wrote already in 1929: "Finally, consciousness itself may end or vanish in a humanity that has become completely etherized, losing the close-knit organism, becoming masses of atoms in space communicating by radiation and ultimately perhaps resolving itself entirely into light. That may be an end or a beginning, but from here it is out of sight."

Frank Tipler [10] has argued that an advanced civilization can, in principle, colonize a closed universe entirely and should be able to manipulate its collapse to gain enough energy to live forever – forever with respect to subjective time (based on General Relativity's time dilatation) which is not to be confused with the finite objective time span such a contracting universe has.

Freeman Dyson [11] argued that in an open, eternally expanding universe with a decelerating expansion rate, life could also go on forever. "The pulse of life will beat more slowly as the temperature falls but will never stop"[12]. Given longer and longer phases of hibernation, life-bearing devices could perform infinitely many calculations with a finite amount of energy. For a society of the size and complexity of our present civilization, for instance, $6 \cdot 10^{30}$ Joule would suffice – the amount of energy radiated away by our sun within just eight hours. It is controversial however, whether such a kind of existence is really eternal. First, because deleting old information to acquire new information costs energy, so perhaps only a finite number of thoughts could be instantiated. Second, it is unclear whether life is ultimately digital or analogous, that is, based on continuous processes. Only in the latter case a finite amount of energy is really sufficient and alarm clocks for hibernation wake up calls would not necessarily break. Lawrence Krauss and Glenn D. Starkman criticized Dyson's assumptions and believe that even in a decelerated eternal expansion of the universe life is ultimately doomed [13]. John Barrow and Sigbjørn Hervik argued, however, that arbitrarily weak anisotropies of the universe suffice to harness the temperature gradients created by gravitational tidal energy, and this should be enough to drive an information-processing machine arbitrarily long [14].

Now with dark energy, the situation is not only more complicated, but also more desperate for the far future of life. But this depends crucially on the nature of dark energy which is not known yet [15]. Thus, many different alternatives have to be considered.

2.2 Resurrection and the Accelerated Future

The Big Whimper scenario is the most conservative or simple one. It is implied by the existence of a positive **cosmological constant** Λ in the framework of relativistic cosmology. Here, the expansion goes on forever, and the expansion rate is approaching a final value $H_\infty = \sqrt{(\Lambda/3)} = H_o \sqrt{(1-\Omega_m)}$ with $\Omega_m = \rho_{m,o}/\rho_{c,o}$. Because of quantum effects at the horizon (analogous to Hawking radiation at the horizons of black holes) the universe cannot cool down to 0 Kelvin but reaches within a few hundred billion years a final temperature T of $T = 1/2\pi\sqrt{(3/\Lambda)} \approx 10^{-29}$ Kelvin (corresponding to 10^{-33} electron volts). This is the end for any living system because then it cannot radiate away waste heat – and there is no life without an energy gradient [13].

The chances are better for some **quintessence** models. Here, things get very complicated for there are many different models. In most of them the accelerated expansion also lasts forever, but is slower than in the case of a cosmological constant. Depending on the models, perhaps the energy gradient remains [16]. Or the quintessence field decays, which might lead eventually to a decelerated eternal expansion [17] and a matter-dominated universe [18] or even to a revitalized universe if new matter is created out of the decaying field, and the origin and evolution of life would start again

[19]. Or w could oscillate, causing alternating periods of accelerated and decelerated expansion [20]. Furthermore, it is not known whether quintessence is exactly homogeneous and whether it couples only gravitationally or in some other weak form, for example with neutrinos. Because the density of the dark energy ~ $(10^{-3}$ electron volts$)^4$ is in the order of the neutrino mass, it was argued that both are connected [21]. These ideas are mere speculations at the moment, but not ruled out yet.

So it seems to be very unlikely that a civilization can survive forever. In the end it will run out of energy. But this is a statistical argument, since it is based on the second law of thermodynamics. Entropy can decrease due to thermal **fluctuations**, and it is in principle possible that such fluctuations can sustain a civilization for an arbitrarily long time. The probability that a civilization will survive for some time is a sharply decreasing function of that time. However, for any finite time the probability is finite, and thus many civilizations in our universe will live longer than any given time, if the universe is large enough to allow those improbable events to occur. Of course, there is almost no chance that our successors are going to be among the lucky civilizations whose life will be prolonged by thermal fluctuations in such a lottery universe.

Furthermore, in an infinite future time might not be a problem. Eventually, anything could spontaneously pop into existence due to quantum fluctuations. They would mostly result in meaningless garbage, but a vanishingly small proportion shall be people, planets and parades of galaxies. This paper will reappear again, too. And such kinds of quantum resurrection might even spark a new Big Bang. According to Sean Carroll and Jennifer Chen one must be patient, however, and wait some $10^{10^{56}}$ years (if de Sitter vacuum is the "natural ground state") [22].

2.3 Big Splat and Cosmic Collapse

Despite the accelerated expansion today, the Big Crunch scenario is also not entirely refuted. However, as far as the latest measurements of w can tell, we are safe from the collapse for at least another 25 billion years – almost twice the age of our universe today [23].

Within the framework of string cosmology, there is even a quite robust class of models, which describe our universe as a four-dimensional brane within a five-dimensional spacetime in which other 4D-branes might exist – literally parallel universes. And one of them might collide with our universe, perhaps countlessly often. Imagine two hands clapping repeatedly together. From the perspective within our universe every collision, sometimes called a **Big Splat**, looks like a Big Crunch and acts like a new Big

Bang: The extradimension temporarily vanishes, and after the other brane-universe retreats, the expansion starts again and lasts for trillions of years. Note that the branes are and remain infinite, only the inherent curvature looks like a collapse everywhere. One appeal of this Cyclic Universe [24] scenario is that it can explain the whole evolution by the action of dark energy. This is derived from string theory that is it is the manifestation of a field called dilaton with has to do with the size of the extradimension.

Whether a civilization could survive the Big Splat and spread again in the new expansion phase is not known. It seems to be unlikely, but perhaps advanced technology could create some kind of sheltering castles, for example with the help of black holes, to prevent the full brane collision locally. One could also envision a computer made of light, which would store all of our memories and information, might transmit it through the Big Splat, and recover it in the new expanding epoch of the universe. Whether this cyclic universe scenario or others are really future-eternal remains an open issue.

There are other possibilities for a Big Crunch, too.

Either dark energy leads to the contraction of our universe within a few dozens of billions of years only – a surprising possibility at least in the framework of supergravity and string theory [25].

Or something like quintessence masks a negative cosmological constant today, which will dominate the far future if the accelerating cause vanishes. A negative cosmological constant leads inevitably – and independent of whether the universe is closed or not – to a collapse.

Or the current positive cosmological constant fluctuates due to quantum effects and, ultimately and irreversibly, turns negative. It was argued that this necessarily happens under quite weak conditions [26]. This would lead to a collapse of the universe within some trillions of years.

Whether a collapsing universe ruled by dark energy allows life to continue forever, as Frank Tipler has imagined [10], is unclear. If dark energy is a homogeneous property of spacetime geometry, it probably cannot be changed and accumulated somewhere like matter. But the manipulation of anisotropies is a necessary condition for a Tiplerian eschatology.

2.4 Big Rip and Phantom Energy

Dark energy offers a third option beyond the Big Whimper and Big Crunch scenarios if $w < -1$. Then the universe is ruled by what is called **phantom energy** [27]. It tears space literally apart and leads irrevocably to a complete disruption of everything, even atomic nuclei. This **Big Rip** could happen already soon in cosmological terms, that is in the order of tenths of billions of years. There is no mechanism by which life could stop such a process, so it seems to be doomed in this case. (By the way, there is even the possibility that some kind of matter ends in a Big Rip and some other kind does not [28], and there are also models without dark energy, but with some modifications of General Relativity containing the possibility for a Big Rip – or for a "Bigger Rip" due to an even more divergent scale factor within the next few billion years [29]. On the other hand, not all phantom energy scenarios end in a Big Rip singularity [30].)

One the other hand, phantom energy could be the solution for a problem of some oscillating universe scenarios with repeated Big Bang/Big Crunch cycles: Black holes might survive the crunch and grow in each cycle until they swallow the whole universe. But according to one model, phantom energy even tears black holes apart, effectively making them boil away [31]. In a string-theory-inspired braneworld model phantom energy disturbs fields in the fifth dimension outside our universe. Those fields might turn the parameter w of dark energy around, stop the Big Rip and let the universe recollapse. Although nothing would survive the Big Rip, new structures might form during the collapse. In addition, the higher dimensional fields would let the contraction bounce back to become the expansion of a new Big Bang.

2.5 A Preposterous Universe

So what is the ultimate future? At the moment we cannot say. Further measurements have to determine w and its time derivative as precise as possible. As it seems currently, w is close to –1 (that is $-0{,}8 > w > -1{,}2$, observationally speaking) and constant over time, which is consistent with a positive cosmological constant [23]. However, if $w \approx -1$, then we have an empirical limit we probably never can go beyond. This is a simple implication of the uncertainty and intrinsic error of every measurement. Thus, we can never empirically proof that $w = -1$, only that the error bars are very small around that value. Therefore, strictly speaking, the cosmological constant could only be established by other means, that is theory (and, as it was noted [26], Λ could even fluctuate nevertheless). The same is true for other limiting cases, for instance the widely hold flat or Euclidean universe with $\Omega_{total} = 1$.

Even worse, we simply do not understand why the energy density of dark energy is about 0, 7 today. From calculations in quantum physics it should be 10^{50} to 10^{120} times higher, which is completely unrealistic and the biggest discrepancy between theoretical prediction and observation ever in the history of physics. Also we do not understand why the dark energy density has roughly the same order as the matter density today – is this just pure chance or are there deeper connections?

So there is some kind of paradoxical situation: On the one hand we are in a golden age of cosmology, measuring the fundamental cosmological parameters with higher and higher precision – and on the other hand we do understand much less than expected. We seem to live in a "preposterous universe", as Michael Turner has said.

2.6 Big Decay or Big Hit

There is another depressing possibility which might even threaten us right now:

The **Big Decay**: The vacuum state in which we live in, that is our universe is in, might not be stable, but metastable [32]. Then it would be a kind of false vacuum like the one that, according to the inflationary scenario, might have driven the very early universe to an exponential (superluminal!) expansion until its field, the inflaton, had decayed (which might have released all the energy that turned into matter). Whether or not that epoch of inflation happened, and whether it is or is not related to the current accelerated (but much slower) expansion, for example as a kind of left-over, it is possible and even plausible that our universe has not reached its ground state (the "true vacuum"). But if the vacuum is metastable, it can and ultimately will decay. This could happen spontaneously, as a kind of quantum tunnel effect, or even by accident, for example due to an high-energy experiment of a very advanced civilization. From such a phase transition a wave of destruction would spread with nearly the velocity of light in every direction. It cannot be seen, and it almost instantly wipes out everything it hits without a warning. Similar effects would occur if there are very tiny extradimensions, as string theory claims, which are compactified, that is curled up, but could unfurl or decompactify. Even if only one tiny extradimension gets large, the whole universe would be burned to death. Perhaps new forms of life would arise out of the ashes, so to speak, but because the constants of nature would have changed, nothing could be like anything we know.

Not yet excluded but very improbable is the "rather unpleasant possibility" (Alexei Starobinsky) of a **Big Hit**: that our future world line will cross a space-time singularity, for example a gravitational shock wave with an infinite amplitude, or that it hits a finite-time singularity or a space-like curvature singularity which might form as a result of sudden growth of anisotropy and inhomogeneity at some moment during expansion or due to quantum gravitational effects [33]. (The Big Rip can be interpreted as a future singularity, too.)

2.7 Wormhole Escapism and Designer Universes

All the scenarios reviewed here look more or less disappointing. If our universe is ultimately doomed, or at least the sufficient conditions for any possible information processing system disappear, the only chance for life would be to leave its universe and move to another place. There are bold speculations about **traversable wormholes** leading to other universes [34]. This seems to be possible at least in the framework of General Relativity (but like dark energy it violates some fundamental energy conditions). Perhaps wormholes could be found in nature and modified, or they could be built from scratch. If so, life could switch to another universe, escaping the death of its home. And if there is no life-friendly universe with the right conditions (physical constants and laws), an advanced civilization might even create a sort of replacement or rescue universe on its own. In fact, some renowned physicists have speculated about such a kind of world-making [35]. If such a switching of universes is possible, life might continue endlessly.

But even if our universe and every living being in it would be finally doomed, perhaps that doesn't matter ultimately. Because there could be infinitely many other universes and/or our universe might **recycle** itself due to new inflationary phase transitions out of black holes [36] or out of its high energy vacuum state, that is it creates new expanding bubbles growing to new universes elsewhere and cut their cords, metaphorically speaking [37].

2.8 Eternal Recurrence

Something strange is inevitable, if two conditions are true: Firstly, if our universe is infinite, or if there are infinitely many other universes with the same laws and constants. And, secondly, if quantum theory holds and there is, therefore, a finite number of possible states (that is due to Heisenberg's uncertainty relation there is no continuum of states and perhaps not even of space and time). If those two assumptions are valid, then according to Alexander Vilenkin every combination of discrete finite physical states are realized arbitrarily or infinitely often [38]. (Imagine a lattice built randomly out of zeros and ones: Every finite combination of zeros and units that is every local pattern occurs infinitely often.) Thus, there is a kind of spatial eternal recurrence.

This also implies that we would have perfect copies: **Doppelgänger** which are identical to us as far as quantum physics allows, and also Doppelgänger biographies, Doppelgänger earths, solar systems, Milky Ways and even Hubble volumes. Their distances are vast, but not infinite, and they could even be estimated, as Max Tegmark has shown: Our personal neighboring Doppelgänger should be $10^{10^{29}}$ meter apart, and another Doppelgänger Hubble volume, that is a region of space exactly like our observable universe, $10^{10^{115}}$ meter [39]. This means spatial eternal recurrence, but it could extend in time, which seems to be true either in a future-eternal inflationary or cyclic scenario with a flat universe. Thus, even if the history of our universe (and/or every universe) might lead to a global death, everything and every life-form might reappear over and over again, infinitely often both in space and time. Then, it is true that there is no personal eternal life, because every organism is doomed, but life as such could not be driven out of existence completely everywhere and everywhen. It would be truly eternal.

On the other hand, eternal recurrence seems to be absurd. And it is not only exact duplication – it is also every possible alternative, because all variations are equally real. As Alexander Vilenkin has said, some people "will be pleased to know that there are infinitely many [...] regions where Al Gore is President and – yes – Elvis is still alive". Thus, physical potentiality and actuality would ultimately be the same. If so, the search for options and the struggle for life doesn't matter globally. Everything that might happen will happen at one place or another – in fact, it will happen infinitely often. This might be disappointing or encouraging. Perhaps this is only a matter of personal taste. However, it seems very strange and for many people even insulting, that we are not unique, and that everything we try might succeed here but not elsewhere and vice versa – infinitely often. However, as Steven Weinberg reminded us, "our mistake is not that we take our theories too seriously, but that we do not take them seriously enough"[40].

Conclusion

In conclusion, we can safely say that the future is not, what it used to be. There is a diverse range of alternative cosmological scenarios (both for the furture and for the past [41]), some of which look really eery. It is premature, however, to announce the ultimate and unavoidable end of everything in the very far future. But undoubtly huge challenges are imminent. As Niels Bohr once joked, it is difficult to make predictions, especially about the future. But it would be of no surprise, if big surprises out there are still waiting for us to be discovered. Perhaps the universe, as John Burdon Sanderson Haldane once said, is "not only queerer than we imagined, it's queerer than we can imagine".

Acknowledgments

It is a pleasure to thank Hans-Joachim Blome, Rob Caldwell, Freeman Dyson, Gia Dvali, Katherine Freese, Gerson Goldhaber, Renata Kallosh, Claus Kiefer, Robert Kirshner, Lawrence Krauss, John Leslie, Andrei Linde, Mario Livio, Saul Perlmutter, Wolfgang Priester, Adam Riess, Antonio Riotto, Subir Sarkar, Larry Schulman, Glenn Starkman, Paul Steinhardt, Michael Turner, Neil Turok, Carsten Van der Bruck, Alex Vilenkin, Christof Wetterich, and H. Dieter Zeh for comments and discussion, André Spiegel for his kind support, and Vladimir Burdyuzha as well as Claudius Gros for the invitation to "The Future of Life and the Future of Our Civilization" symposium at the Johann Wolfgang Goethe University in Frankfurt am Main, Germany, where an earlier version of this paper was presented on May, 5th 2005.

References

1. F. C. Adams, G. Laughlin, A dying universe, *Rev. Mod. Phys.* 69: 337-372; astro-ph/9701131. – F. C. Adams, G. Laughlin, *The Five Ages of the Universe* (Free Press, New York, 1999). – M. M. Cirkovic, A Resource Letter on Physical Eschatology, *Am. J. Physics* 71: 122-133 (2003); astro-ph/0211413. – P. Davies, *The Last Three Minutes* (New York, Basic Books, 1994). – G. F. R. Ellis (ed.), *The Far-Future Universe* (Templeton Press, Radnor, 2002). – J. N. Islam, *The Ultimate Fate of the Universe* (Cambridge University Press, Cambridge, 1983). – A. Loeb, The Long-Term Future of Extragalactic Astronomy. *Phys. Rev.* D65: 047301 (2002); astro-Ph/0107568. –N. Prantzons, *Our Cosmic Future* (Cambridge University Press, Cambridge, 2000 [1998]). – R. Vaas, Die fernste Zukunft, in: U. Anton: *Die Lebensboten* (Heyne, München, 2004), 255-320. – R. Vaas, Die ferne Zukunft des Lebens im All, in: S. Mamczak, W. Jeschke (eds.): *Das Sceince Fiction Jahr 2004* (Heyne, München, 2004), 512-594. – R. Vaas, Ein Universum nach Maß?, in: J. Hübner, I.-O. Stamatescu, D. Weber (eds.): *Theologie und Kosmologie* (Mohr Siebeck, Tübingen, 2004), 375-498.
2. A. Sandage, Cosmology: A Search for Two Numbers, *Phys. Today* 23: 34-41 (1970).

3. D. N. Spergel et al., First Year Wilkinson Microwave Anisotropy Probe (WMAP) Observations: Determination of Cosmological Parameters, *Astrophys. J. Suppl.* 148: 175-194 (2003).
4. S. Perlmutter et al., Measurements of Omega and Lambda from 42 High-Redshift Supernovae, *Astrophys. J.* 517: 565-586 (1999); astro-ph/9812133. – A. G. Riess et al., Observational Evidence from Supernovae for an Accelerating Universe and a Cosmological Constant, *Astron. J.* 116: 1009-1038 (1998); astro-ph/9805201. – S. Perlmutter, Supernovae, Dark Energy, and the Accelerating Universe, *Phys. Today* 4: 53-60 (2003).
5. P. D. Mannheim, Alternatives to Dark Matter and Dark Energy; astro-ph/0505266. – S. M. Carroll et al., Is Cosmic Speed-Up Due to New Gravitational Physics?, *Phys. Rev.* D70: 043528 (2004); astro-ph/0306438.
6. E. W. Kolb, S. Matarrese, A. Riotto, On cosmic acceleration without dark energy; astro-ph/0506534. – E. W. Kolb et al., Primordial inflation explains why the universe is accelerating today; hep-th/0503117. – E. Barausse, S. Matarrese, A. Riotto, The Effect of Inhomogeneities on the Luminosity Distance-Redshift Relation: Is Dark Energy Necessary in a Perturbed Universe?, *Phys. Rev.* D71: 063537 (2005); astro-ph/0501152. – J. W. Moffat, Late-time Inhomogeneity and Acceleration Without Dark Energy; astro-ph/0505326. – D. L. Wiltshire, Viable inhomogeneous model universe without dark energy from primordial inflation; gr-qc/0503099.
7. C. M. Hirata, U. Seljak, Can superhorizon cosmological perturbations explain the acceleration of the universe?; astro-ph/0503582. – E. R. Siegel, J. N. Fry, Effects of Inhomogeneities on Cosmic Expansion; astro-ph/0504421. – E. E. Flanagan, Can superhorizon perturbations drive the acceleration of the Universe?, *Phys. Rev.* D71: 103521 (2005); hep-th/0503202.
8. K. Tomita: A Local Void and the Accelerating Universe, *Mon. Not. Roy. Astron. Soc.* 326: 287-292 (2001); astro-ph/0011484. – A. Blanchard et al., An alternative to the cosmological ‚concordance model', *Astron. Astrophys.* 412: 35-44 (2003); astro-ph/0304237.
9. C. Kiefer, H. D. Zeh, Arrow of time in a recollapsing quantum universe, *Phys. Rev.* D51: 4145-4153 (1995).
10. F. J. Tipler, *The Physics of Immortality* (Anchor Books, New York, 1994).
11. F. J. Dyson, Time without end, *Rev. Mod. Phys.* 51: 447-460 (1979).
12. F. J. Dyson, *Infinite in all directions* (Penguin, London, 1990 [1985]), 111.
13. L. M. Krauss, G. D. Starkman, Life, The Universe, and Nothing, *Astrophys. J.* 531: 22-30 (2000); astro-ph/9902189
14. J. D. Barrow, S. Hervik, Indefinite Information Processing in Ever-expanding Universes, *Phys. Lett.* B566: 1-7 (2003); gr-qc/0302076.
15. P. J. E. Peebles, B. Ratra, The Cosmological Constant and Dark Energy, *Rev. Mod. Phys.* 75: 559-606 (2003); astro-ph/0207347. – R. P. Kirshner, Throwing Light on Dark Energy. *Science* 300: 1914-1918 (2003). – J. C. N. de Araujo, The dark energy-dominated Universe, *Astropart. Phys.* 23: 279-286 (2005); astro-ph/0503099.

16. K. Freese, Cardassian Expansion: Dark Energy Density from Modified Friedmann Equations. *New Astron. Rev.* 49: 103-109 (2005); astro-ph/0501675.
17. U. Alam, V. Sahni, A. A. Starobinsky, Can dark energy be decaying?, *JCAP* 0304: 002 (2003); astro-ph/0302302.
18. J. Barrow, R. Bean, J. Magueijo, Can the Universe escape eternal acceleration?, *Mon. Not. Roy. Astron. Soc.* 316: L41–L44 (2000); astro-ph/0004321.
19. J. P. Ostriker, P. Steinhardt, The Quintessential Universe, *Sci. Am.* 284 (1), 47-53 (2001).
20. S. Dodelson, M. Kaplinghat, E. Stewart, Solving the Coincidence Problem: Tracking Oscillating Energy, *Phys. Rev. Lett.* 85: 5276-5279 (2000); astro-ph/0002360. – K. Griest, Toward a Possible Solution to the Cosmic Coincidence Problem, *Phys. Rev.* D66: 123501 (2002); astro-ph/0202052.
21. R. D. Peccei, Neutrino Models of Dark Energy; hep-ph/0411137.
22. S. M. Carroll, J. Chen, Spontaneous Inflation and the Origin of the Arrow of Time; hep-th/0410270.
23. A. G. Riess et al., Type Ia Supernova Discoveries at z>1 From the Hubble Space Telescope: Evidence for Past Deceleration and Constraints on Dark Energy Evolution, *Astrophys. J.* 607: 665-687 (2004); astro-ph/0402512.
24. P. J. Steinhardt, N. Turok, Cosmic Evolution in a Cyclic Universe, *Phys. Rev.* D65: 126003 (2002); hep-th/0111098. – P. J. Steinhardt, N. Turok: The Cyclic Model Simplified, *New Astron. Rev.* 49: 43-57 (2005); astroph/0404480.
25. R. Kallosh, A. Linde, Dark Energy and the Fate of the Universe, *JCAP* 0302: 002 (2003); astro-ph/0301087. – R. Kallosh et al., Observational Bounds on Cosmic Doomsday, *JCAP* 0310: 015 (2003); astro-ph/0307185.
26. J. Garriga, A. Vilenkin, Testable anthropic predictions for dark energy, *Phys. Rev.* D67: 043503 (2003); astro-ph/0210358.
27. R. R. Caldwell, M. Kamionkowski, N. N. Weinberg, Phantom Energy and Cosmic Doomsday, *Phys. Rev. Lett.* 91: 071301 (2003); astro-ph/0302506.
28. M. Gasperini, Towards a future singularity?, *Int. J. Mod. Phys.* D13: 2267-2274 (2004); gr-qc/0405083.
29. P. H. Frampton, T. Takahashi, Bigger Rip with No Dark Energy, *Astropart. Phys.* 22: 307-312 (2004); astro-ph/0405333.
30. P. Wu, H. Yu, Avoidance of Big Rip In Phantom Cosmology by Gravitational Back Reaction; astro-ph/0407424; – R. Curbelo, T. Gonzalez, I. Quiros, Interacting Phantom Energy and Avoidance of the Big Rip Singularity; astro-ph/0502141.
31. M. G. Brown, K. Freese, W. H. Kinney, The Phantom Bounce: A New Oscillating Cosmology; astro-ph/0405353.
32. J. Leslie, *The End of the World (*Routledge, London and New York, 1998 [1996]), 108-122.
33. A. Starobinsky, Future and Origin of our Universe: Modern View, in: V. Burdyuzha, G. Khozin (eds), *The Future of the Universe and the Future of Our Civilization* (World Scientific, Singapore etc., 2000), 71-84; astro-ph/9912054. – J. D. Barrow, C. G. Tsagas, New Isotropic and Anisotropic

Sudden Singularities, Class. Quant. Grav. 22: 1563-1571 (2005); grqc/0411045.
34. M. Visser, *Lorentzian Wormholes* (American Institute of Physics Press, Woodbury, 1996). – R. Vaas, *Tunnel durch Raum und Zeit* (Franckh-Komos, Stuttgart, 2005).
35. E. Farhi, A. H. Guth, An obstacle to creating a universe in the laboratory, *Phys. Lett.* B183: 149-155 (1987). – V. P. Frolov, M. A. Markov, M. A. Mukhanov, Through a black hole into a new universe?, *Phys. Lett.* B216: 272-276 (1989). – A. Linde, Hard Art of the Universe Creation, *Nucl. Phys.* B372:421-442 (1992); hep-th/9110037. – E. R. Harrison, The natural selection of universes containing intelligent life, *Quart. J. R. astr. Soc.* 36: 193-203 (1995).
36. L. Smolin, Did the universe evolve? *Class. Quant. Grav.* 9: 173-191 (1992). – R. Vaas, Is there a Darwinian Evolution of the Cosmos?; grqc/0205119.
37. K. M. Lee, E. J. Weinberg, Decay Of The True Vacuum In Curved Space-Time, *Phys. Rev.* D36: 1088-1094 (1987). – J. Garriga, A. Vilenkin, Recycling universe, *Phys. Rev.* D57: 2230-2244 (1998); astro-ph/9707292.
38. J. Garriga, A. Vilenkin, Many worlds in one, *Phys. Rev.* D64: 043511 (2001); gr-qc/0102010. – J. Knobe, K. D. Olum, A. Vilenkin, Philosophical Implications of Inflationary Cosmology; physics/0302071.
39. M. Tegmark, Parallel Universes, in: J. Barrow, P. C. W. Davies, C. L. Harper jr. (eds.): *Science and Ultimate Reality* (Cambridge University Press, Cambridge, 2004), 459-491; astro-ph/0302131.
40. S. Weinberg, *The First Three Minutes* (Basic Books, New York, 1977), 131.
41. R. Vaas, Time before Time; physics/0408111.

Digital Aspects of Nature and Ultimate Fate of Life

Hoi-Lai Yu

Institute of Physics, Academia Sinica, Taipei, Taiwan,
hlyu@phys.sinica.edu.tw

Assuming our physical Universe processes and registers information to determine its dynamical evolution, one can put serious constraints on the form of cosmology that our Universe can bear. We predict that our Universe at present can create at most 10^{74} registers to perform at most 10^{104} computational operations every second by integrating in new degree of freedom (galaxies) through expansion. When the degrees of freedom of the Universe will grow beyond some critical value the computation capacity of the Universe will become insufficient to determine its evolution. The Universe will inflate away its degrees of freedom within its horizon to regain dynamical evolution. The communication time required by the Universe to aware the dropping in degrees of freedom below some critical value by inflation is proportional to its Hubble radius. We predict that next inflation era will stop after inflating for a period of 10^{19} sec if the past inflation period is 10^{-33} sec.

In summary, to begin with, the Universe in the preinflationary phase was simply and bears very low computational capacity and degrees of freedom. As time evolved, the universe computational capacity grew linearly with time by expansion, however, the degrees of freedom within the Universe horizon grew as $t^{3/2}$. The number of degrees of freedom grew faster then its computational capacity. The computational cost for the Universe to determine its "physical it" exceeded its computational capacity, the Universe became lack of sufficient computational power to determine its evolution. Not being able to increase its computation power without violating any physical laws, the Universe stopped expanding its horizon but inflated its physical size to dilute its contents of degrees of freedom. To resume, the Universe requires some communication time to get aware of the dropping in the degrees of freedom below some critical value across

V. Burdyuzha (ed.), The Future of Life and the Future of Our Civilization, 249–250.

its horizon. This time is determined by the Hubble size which is about 10^{-33} sec in the past inflation. After that, the Universe resumed expansion. It is conceivable, that the Universe will have more that one chance to end up in such a situation as long as it expands to integrate in more and more degrees of freedom faster then its growth in computational capacity. That means it is conceivable that the Universe will inflate more than one time. Increasing evidences do indicate that our Universe is entering a new inflationary era. It is a hot debate problem if this new inflationary era will stop or not. From our point of view we predict that this new inflationary era will stop. Using data obtained from the past inflation, we predict that this new inflation era will stop after 10^{19} seconds. Before passing, we want to stress again that our prediction using $\Delta t_i \approx 10^{-33}$ sec is for demonstration purpose. Future experiments will measure Δt_i, H_i (here H_i is the inflationary era Hubble constant), H_f (final value of H) precisely, then our point of view will give an unambiguous prediction. The solid prediction at the moment is that this new coming inflation era will not inflate forever but will stop after some period of time.

Spirals of Complexity - Dynamics of Change

Don-Edward Beck

Box 797, Denton, Texas 76202, USA, drbeck@attglobal.net

Debates over globalization are but the surface-level collisions of the deeper tectonic plate-like cultural fault lines that remain hidden from view. The failure to both understand and deal with these evolutionary core value systems result in needless clashes over worldviews, constant threats of "us" vs. "them" or class-based violence, and expensive, politicized solutions that are both inappropriate and ineffectual. The WTO debates and conflicts in Seattle exposed these fault-lines. But where are the integral, cohesive principles and processes that can bridge over the great, global divides? Who can untie the global knot? How can the positive elements within both capitalistic thinking and socialistic goals be meshed for the common good? Consider the Twelve Postulates, an integral initiative based on an understanding of the complex dynamics that forge and transform human cultures, communities, and countries.

V. Burdyuzha (ed.), The Future of Life and the Future of Our Civilization, 251.

IV

HOW CAN WE IMPROVE OUR LIFE?

Nutrition, Immunity and Health

Ranjit Chandra

312 Quatab Plaza DLF Phase 1, Gurdaon Haryana 122002, India, rkchandra2004@yahoo.com

Nutrition is fundamental to human health and survival. The immune system is a critical intermediate proxy for relationship between nutrition and health, including infections, atherosclerosis, cancer, Alzheimer's disease, and eye problems.

V. Burdyuzha (ed.), The Future of Life and the Future of Our Civilization, 255.

Human Races and Evolutionary Medicine

Bernard Swynghedauw

Hôpital Lariboisière. 41 Bd de la Chapelle, 75475, Paris Cedex 10, France, Bernard.Swynghedauw@larib.inserm.fr

Data from the Human Genome Programme have clearly established that the human race is unique. Attempts to identify separately Black, Caucasian, Asian is not supported by biological data but is only based on socieconomic conditions. Evolutionary medicine takes the view that many contemporary diseases are likely to result from the incompatibility between our contemporary lifestyle and dietary habits and the conditions under which the evolutionary pressure had shaped our genetic inheritance. Assuming that human race is unique, it is now possible to identify new pathophysiological pathways through this paradigm. Human race has always been a social construct reflecting in a given country, at a given period of time the status of the society, even the nomenclature has always been problematic with a mixture of phenotypic, geographical and even religious qualification, including black, African, Negroes, or Semite, white, Caucasian, or yellow, Asian, Asiatic, Hispanic…[Swynghedauw, 2003]. Even self-identification is problematic, and in the last US Census 7 million persons identified themselves as member of more than one race; and 800,000 respondents said that they were both black and white [Schmitt 2001]. Finally, it is clear that race qualifies THE OTHER, and has been used throughout our history to justify the predominance of a group of persons on another. Our recent, even very recent, history is full of examples which all illustrate how humankind may use such a vocabulary to cover ignorance or, still worst, tyranny.

Human race is inherently imprecise, based on appearance, surname, language and has always been a social construct reflecting in a given country,

V. Burdyuzha (ed.), The Future of Life and the Future of Our Civilization, 257–261.

at a given period of time the status of the society[12]. From a zoological point of view, animal races, as opposed to human "races", are a subtype of a species, inter-reproductive, phenotypically extremely well-characterized (with competitions) and isolated by geography or human wishes. By contrast, human race can't be accurately defined, are, in current practice, poorly-defined by vague phenotypic traits with many confounding factors, including socioeconomic status, educational level, religion, culinary customs, tribal affiliation… Nevertheless, despite of the lack of clear definitions «race» is a commonly utilised distinction utilised in most of biomedical publications.

Is the Geographical Distribution of a Disease Racial?

This is an important issue when studying evolutionary medicine, because it could not be possible to envisage medicine under these aspects if human race is not unique [Swynghedauw, 2004].

Monogenic Inherited Diseases

The well-documented example of an overlap between race and a monogenic disease is that of Black Africans (considered as a race), and the geographical distribution of several inherited blood cell disorders, such as sickle disease and the Glucose-6-phosphate Dehydrogenase deficiency (caused by allele (A-)). Sickle disease is an inherited disease due to a mutation located on hemoglobin, which protects against the severe forms of malaria when heterozygous but have reduced fitness compared to wild-type when homozygous, as a result of the clinical consequences of the red blood cell disease [reviewed in Jobling et al., 2004]. Clearly the world map of malaria superimposed to that of the genetic disease, and the best way to explain the diffusion of the genetic mutation is selective pressure, but, clearly also, while sickle disease predominates in Black Africans, it also

[12] It is very common, for example, to say that most of the winners in the American team, during Olympic games, were Black and to attribute their physical condition to genetic predisposition, while it should be very simple to make a genome wide analysis to identify the corresponding genes or loci, or to refute the genetic thesis… which have never been done or even tried. Obviously the best explanation for the success of Black minority (like that of North-African minority in France for soccer players), is determined by the strong wish of those persons to climb into the social hierarchy

exists, with several historical birth places, in India, Spain and Arabia, and is not specific for the “Black” race.

AIDS, the most devastating new infectious disease, is most prevalent in several parts of Africa, and there is evidence that a few individuals show significant inherited resistance to the viral infection. One of the first loci conferring resistance to be identified was a chemokine receptor, CCR5, a cell surface protein that acts as a cofactor for HIV entry into the T cells. One of the variant of CCR5, D32, is not functional and, in homozygous form, is associated to resistance to HIV infection. The distribution of this variant over the world is now well documented the variant predominates in northwest Asia and northern Europe, and is not found in subsaharian Africa.

Multigenic Diseases

Multigenic diseases include cancer, cardiovascular diseases which are consequences of atherosclerosis, hypertension or diabetes, but which also have strong frequently predominant environmental components. Tobacco smoking is one of these components, but there also nutritional, psychological, socioeconomic and genetic determinants, which all play a role. Epidemiologists have published tables that allow to calculate a global risk by weighing and measuring these different factors. In spite of the absence of indisputable and clear definition, it is now well-accepted that Black Americans[13] have more risk than White of suffering from cardiovascular diseases (mortality for the total population, in number of deaths per 100,000: 247.8; Black population: 316.9; White: 243.5), including myocardial infarction (177.8; 211.6; 176.5) and stroke (57.9; 78.8; 56.4) [Mensah et al., 2005].

Can Human «Race» be a Proxy for Genetic Variations?

The answer is clearly no. The genetic diversity is too unequal distributed, the skin colour is a multigenic trait that can’t be linked with any genetic locus and, finally, a given “race” did not correspond to genetic markers.

Apportionment of Genetic Diversity

The genetic diversity is not equally distributed throughout the world. The first major point is that 83-88% of autosomal variations (the same is true

[13] The inclusion in this category was based on self-assignation

for mitochondrial markers, Y chromosome or microsatellites) is found within population between individuals, not between populations or races. A programme called STRUCTURE allow to identify four different groups of people based on the apportionment of genetic polymorphisms [Wilson et al., 2001] and showed that the two classifications, that obtained with STRUCTURE, and the usual race classification do not clearly superimposed. Studies of the genetic of the populations have definitely demonstrated that the origins of humankind (homo sapiens) was in West Africa 100,000 years ago, followed by one, probably two migrations through the actual Suez to disseminate in Europe, Asia and America (reached 10-15,000 years ago). There are many confirmations of this view, including considerations based on languages classification. Finally, genetic diversity, as calculated using F_{st} is higher in Africans than in any other continents in the world [Cavalli-Sforza and Feldman, 2003].

Skin Colour, Can't be a Proxy for Human Race

The first reason is that the biological determinants of skin colour are diverse and far from being simple and monogenic, the trait is multigenic, and includes melanosome density, eumelanine and/or pheomelanin content, at least, 13 genes for the regulation of the expression of these genes, including MC1R, are highly polymorphic receptor with more than 30 variants with 5 synonymous haplotypes in Africa and 13 non synonymous out of Africa, which indicates a strong selective functional pressure in Africa [Parra et al., 2004]. To establish a genetic link between skin colour and behaviour or any given multigenic disease is an impossible and hopeless task due to the highly complex nature of the two groups of partners. A study comparing the degree of skin pigmentation in five populations of various ranges of pigmentation has shown that pigmentation was weakly associated to genetic markers [Parra et al., 2004].

The Socioeconomic Level

The major confounding factor when studying « race » inequalities is the socioeconomic status. For example, black American community when compared to no black showed major differences in terms of average household income (+ 25.3% in favour of non black) and high-school diploma (+ 50%), the two most commonly employed markers of the socioeconomic status.

Indeed, the death-rate ratio, workers + employees/management + liberal professions, in men 25-54 years, in a French study was on average above 1 for every cause of death, including cancer, cardiovascular disease, workers and employees having a risk of death 2.9 higher than the others [Lecklerk et al., 2000]. These results have been confirmed by many studies.

Suggested Guidelines for using Race/Ethnicity in Biomedical Publication

As a consequence several authors have suggested a sort of guideline for biomedical authors [Kaplan and Bennett, 2003, Swynghedauw, 2003] which are as follow:

The reason for categorization should be specified.

The way in which individuals were assigned to a given category should be specified, even when for self assignment.

Racial/ethnicity is not a proxy for genetic variation, but more a proxy for socioeconomic status.

To distinguish between race/ethnicity as a risk factor or as a risk marker.

All relevant confounding factors should be considered (insurance status, diet, education, religion, tribal affiliation, psychological factors…) and principally socioeconomic status.

To avoid terminology (as Caucasian) that implies an immutable attribute of an individual.

References

Cavalli-Forza L. L., Feldman M. W. (2003) The application of molecular genetic approaches to the study of human evolution. Nature Genetic Suppl, 33, 266-275.

Jobling M. A., Hurles M. E., Tyler-Smith C. (2004) Human evolutionary genetics. Garland Science, New York.

Kaplan J. B., Bennett T. (2003) Use of race and ethnicity in biomedical publications. JAMA, 289, 2709-2716.

Leclerc A., Fassin D., Grandjean D., Kaminski M. (2000) Lang T et Ribet C. Ch. 14. Les maladies cardiovasculaires. In Les inégalités sociales de santé. Lang T eds. INSERM. La Découverte pub. Paris. pp 223-238.

Mensah G. A., Mokdad A. H., Ford ES et al. (2005) State of disparities in cardiovascular health in the United States. Circulation, 111, 1233-1241.

Parra E. J., Kittles R. A., Shriver MD. (2004) Implications of correlations between skin colour and genetic ancestry for biomedical research. Nature Genetics, 36, S54-S60.

Swynghedauw B. (2003) Human races and evolutionary medicine. European Rev., 11, 437-447.

Swynghedauw B. (2004) Evolutionary medicine. Acta chir belg; 104: 132-139.

Wilson J. F., Weale M. E., Smith A. C. et al. (2001) Population genetic structure of variable drug response. Nature Genet, 29, 265-269.

Bacteria in Human Health and Disease: From Commensalism to Pathogenicity

Helena Tlaskalova-Hogenova

Institute of Microbiology, Academy of Sciences of the Czech Republic, Department of Immunology and Gnotobiology, Videnska 1083, Prague 4, Czech Republic, tlaskalo@biomed.cas.cz

Continuous, reciprocal interaction between the organism and the environment is a basic feature of the life. Starting from first hours after delivery from sterile uterine environment microorganisms colonizes most of the mucosal surfaces and skin. Number of autochtonous bacteria exceeds the number of cells forming the human body. Mucosal microbiota belongs to a complex of innate, natural mechanisms that safeguard resistance of organism against pathogenic microorganisms. Immune system associated with mucosal surfaces covering the largest area of the body (200-300m^2) evolved mechanisms discriminating between harmless antigens, autochtonous microorganisms and dangerous pathogens. Numerous chronic diseases may occur as a result of changes in mechanisms regulating mucosal immunity and tolerance, which leads to impaired barrier function of the mucosa and increased penetration of microbial antigens into circulation and consequently to exaggerated and generalized immune responses. Participation of pathologically increased immunological activity to autologous microbiota and/or their components in induction of inflammatory and autoimmune processes was demonstrated. Regulation of microflora composition offers the possibility to develop new approaches in prevention and therapy.

V. Burdyuzha (ed.), The Future of Life and the Future of Our Civilization, 263.

Is There a Solution to the Cancer Problem?

Jarle Breivik

Interactive Life Science Laboratory and Section for Immunotherapy, University of Oslo at the Norwegian Radium Hospital, 0310 Oslo, Norway. jbreivik@medisin.uio.no

Cancer research has led to tremendous advances in life science, and we are currently seeing important developments in therapy. Still, the overall incidence of cancer continues to increase, and a final solution to the problem seems more distant than ever. So is it possible that we are running towards the end of the rainbow, trying to solve the unsolvable? Based on a new model of cancer development, I will argue that cancer research needs a fundamental change of perspective. The model links genetic instability and cancer development to the principles of molecular evolution, and cancer appears as a logical and inevitable cost of our multicellular composition. The solution to cancer may therefore be very different from the magic cure we are striving for.

References

Review of model: Semin. Cancer Biol. 2005, 15, 51-60

V. Burdyuzha (ed.), The Future of Life and the Future of Our Civilization, 265.

Are Embryonic Stem Cells Research and Human Cloning by our Future?

Lyubov Kurilo

National Research Center for Medical Genetics Russian Academy of Medical Sciences, Moskvorechie str.1, Moscow 115478, Russia genreprlab@mtu-net.ru

The Human Embryo and Fetus Intrauterine Development

The elaboration of different biomedical technologies, the use in medicine of reproductive technologies (RT) such as in vitro fertilization (IVF), embryo transfer (ET) into the human uterus, and freezing of germ cells and human embryos inevitably entail manipulations with germ cells and human embryos, ethical and legal problems. One of the principal ethical and legal problems is a question about status of human embryo, i.e. the stage in the intrauterine development of a human embryo (or fetus) at which a "person" in the ethical and legal sense comes into human being (that is the question of person inviolability and of human dignity' respect).

The positive or negative attitudes towards RT methods and manipulations with gametes and embryos depend mostly on when human embryo development has to be taken as the beginning of a new life: with conception; zygote (fertilized ovum); the embryo on the stage of cleavage or blastocyst' implantation; the embryo being on the initial stage of the nervous system formation or the shaping heart beats, etc. Some of specialists call the period from the beginning of fertilization until the 14th day of embryo development («pre-embryo») as a permissible period of embryo experimentation.

V. Burdyuzha (ed.), The Future of Life and the Future of Our Civilization, 267–275.

The new organism (with a diploid karyotype, 46 chromosomes) is developing after fertilization, after the reunion of the oocyte and spermatozoon chromosomes.

There is a natural selection in the process of gametogenesis, during stages of the preimplantation and intrauterine development. During natural fertilization and implantation the major part of embryos (up to 70%) gets lost due to various reasons, one of them - chromosomal aberrations. Among new-borns and grown-ups, 0.5-0.7% discloses the chromosomal anomalies (Jacobs, 1972; Kuleshov et al., 1978).

Fertilization causes a new process - of embryo development with a unique genetic development program obtained from parents. Data concerning the intrauterine human embryo and fetus development have been taken from monographic guides (Zavarzin, 1939; Patten, 1958; Bodemer, 1968; England, 1985; Moore, Persauder, 1997).

During 20 hours after natural fertilization the developing egg undergoes cleavage through a series of mitotic divisions, sinks down to the uterus. On the third day after fertilizaion morula composed of a large number of cells-blastomeres appear. By the fourth day blastula (blastocysta) is formed. At the blastocysta stage the cavity - blastocele is formed. On one of its poles an accumulation of cells is noted: the inner cell mass (ICM) giving origin to the germ, amnion, yolk sac and allantois. The outer one-cell layer of the blastocysta - a trophoblast (trophectoderm, chorion) is formed. Further on, the trophoblast, together with neighboring uterine cells and blood vessels, form the placenta. On the sixth day the implantation begins, from the process when the blastocysta is attached to the endometrium.

On the 7th-8th day the amniotic cavity, ectoderma and entoderma are formed. During the 3rd week of development gastrulation establishes all three germ layers in the human embryo. On the 14th-16th day the primitive streak is formed.

The period between 21 and 32 days is marked by commenced organogenesis. The unidirectional circulation of the blood through the developed heart is established. The separation of the heart to chambers is close to completion, and the heart is beating.

At forty to 42 days - the long bones, axial skeleton and skull are ossified. Five parts of the primary brain are formed. The histogenesis of the digestive tract primordial is going on. Sex differentiation of the gonads is quite obvious. The beginning of motor behaviour of human foetus to the 6th week has been observed (Reinold, 1981). The end of the 8th week is characterized by termination of the embryogenic period. Basic structures and organ systems are differentiated.

The fetal period of the human intrauterine development is beginning from the 9th week and finished with the birth of the foetus (38-40th week).

During the 10th week the reflex of movements of lips (reflex sucking) is established. Evidence of an operating sensory mechanism involving higher levels of the central nervous system is revealed by contraction of arms and legs in response to tapping on the chorion.

The further development of the foetus during 20-38 weeks is characterized by the further differentiation of systems and organs. The senses are functioning: the foetus reacts to light, noise, gustatory and olfactory irritations.

When discussing **the bioethical aspects of the age limit of embryo experimentation**, embryologists determine as a rule the period from fertilization (the zygote stage) up to the 14th day (until the primitive streak is formed and nervous system elements appear) (Edwards, 1986 a,b,c; McLaren, 1989) or up to the 30th day (beginning of brain differentiation (Cohen, Edwards, 1986). One of the crucial points in determining the human embryo as a personality is the question, at what time does the foetus acquire the ability to sense? The first movements of the fetus were fixed on the 6th week. By this time the fetus reacts to a touch, and synapses are found in the spinal cord. The first neuromediators are determined in the neural fibres in the 10-week fetus, and the activity of the brain stem is registered. Judging on electrophysiological and irnmunohistochemical investigations of the central and peripheral nervous system, it was concluded that the human fetus begins to feel at the age of 18-25 weeks, but until the 30th week there is no proof that the fetus is capable of processing its perceptions (Tawia, 1992). Thus the author believes that the 30th week presents the lower boundary between fetus and human.

Some scientists express the opinion that the fetus becomes a personality after seven (or fifth?) months of intrauterine development, when it acquires the ability to live independently out of the mother's body.

Thus the problem of determining at what age the embryo must be regarded as a personality having its own rights and protected by law emerges when researchers discuss the possibility to perform various manipulations with the embryo. The definition "manipulations with human embryo" implies such principal operations as in vitro culture, carrying out investigations and experiments, as well as deep freezing and storage of embryos in this condition for subsequent use for practical IVF and embryo transfer purposes, carrying out investigations with human embryos for embryonic stem cells technology or scientific research.

Stem Cells' Technologies

Definitions

- **Differentiation** (Df) - appearance of the difference between originally similar cells with formation of the specialized cells and tissues, in other words, the differentiation means the development of specialized cell types from the single fertilized egg.
- **Embryonic** stem cells (ESC) - primitive cells arising during the embryo's development. ESC give rise to all cell and tissue types of adult organisms.
- **Gene Targeting** - the possibility of directly integrating specific gene mutations into ES cells by a homologous recombination between a mutant construct and an endogenous gene. Targeting experiments yield important information about the functional role of targeting gene, gene structure, and the function of specific genes for developmental biology.
- **Pluripotent Stem Cells** (PSC) derived from the early embryo. PSC have capable of giving rise to most tissues of an organism. PSC undergo further differentiation (specialization) into SC that are committed to transfer to cells differentiation (hematopoietic lineage, cardio-myocytes, vascular endothelial cells, nerve cells, skeletal muscle and other types of cells). They are not capable of producing either totipotent SC or embryo. PSC are found in adult organisms.
- **Somatic Cells** are cells of the body others than germ cells.
- **Stem Cells** are cells that have the unlimited (self-renewal) capacity to divide (by mitosis) and to produce more differentiated cells.
- **Totipotent SC** are the early cells of embryo which have the unlimited capability to differentiate into embryo, extraembryonic membrans and tissues, as well as into different postembryonic tissues and organs of the body.
- **Transgenic Organism** realizes when alien genes have included into the cell nuclei. Cloned DNA was microinjected into the pronuclei of fertilized mouse eggs. Then these eggs were transfered into the oviducts of pseudopregnant foster mothers.These transgenic animals usually carry the introduced gene in both somatic cells and germ cells.

During of human development (as well as mammal animals) the primitive cells arise within the embryo that gives rise to all cell types of an adult organism. These cells called embryonic stem (ES) cells (Cs).

The initially of two reports were published of isolation of ESCs from inner cell mass (ICM) of mouse blastocysts (Evans, Kaufman, 1981; Martin, 1981). These ESCs cultivated in vitro appeared to be capable to self-renewal as well as producing more differentiated daughter cells. ESCs in vitro give rise to the stem cells of adult tissues (squamous epithelia, fibroblasts, haematopoietic, muscle, nerve etc.). Now the manipulation of animal ESCs has become a routine procedure.

The differentiation can be prevented by growing the ESC on feeder layer cells (for example an immortalized fibroblast cell line). These feeder layer cells produce a factor or factors that prevent differentiation and they maintain ES cell's proliferation and state of pluripotency.

There are several embryonic sources of pluripotent ESCs:

1. ESCs are derived from inner cell mass of a preimplantation embryo.
2. Embryonic germ cells (GC) are derived from primordial GC.
3. The embryonic carcinoma is also considered as the source of ESC.

There are others sources of SCs: from the blood cells of the human umbilical cord at the time of birth. They are the potential for clinical use of viable pluripotent progenitor cells.

The stem cells could be derived also from some adult tissues (for example - bone marrow cells) and from organs of aborted human embryo or fetus.

The technology of ESCs nuclear transfer into an ovum from which the nucleus had been removed (enucleate ovum) has been worked out. ES cell'nucleus and enuclear ovum had been fused by different methods. This approach is applied for organism's cloning.

Molecules that maintain the stem cells state are beginning to be identified (for example, ligands of Notch family receptors, factors like PIE-1, Kit-ligand, factor bfgf or fgf2, etc (Morrison et al., 1997; Donovan, Gearhart, 2001). The markers' ESCs and SCs are beginning to be investigated.

The principal difference between pluripotent ESCs and multipotent SCs from embryos or adult animals is in the number of types of differentiated cells that can be produced. This may reflect the different origin of the SC (Donovan, Gearhart, 2001).

The success in deriving of ES mammalian cells can be used for solving of many problems of developmental biology, cell and molecular biology, gene therapy, biotechnology, etc. The differentiation and morphogenesis of embryonic development, genetic regulation of development and mechanism of malignization are related to these problems. The application of various types of human stem cells could present a breakthrough in pharmaceutical research, drug and teratogen testing. This SC technology may be adopted for transplantation of corneal SC, epidermal grafts, hair follicle, for tissue repair generally. The applications of transgenic organisms and ESCs technology are indispensable to modeling of human diseases, oncogenesis and cell lineage analysis.

The first reports concerning human ESCs were published in 1998. Thomson with colleagues has given information on generated long-term pluripotent cell lines from early human embryos (blastocyst-derived) (Thomson et al., 1998). These blastocysts produced by in vitro fertilization for clinical purposes were donated by individuals after informed consent and after institutional review board approval. Embryos were cultured to the blastocyst stage, 14 ICM were isolated, and five ES cell lines originating from five separate embryos were derived. Four of the cell lines were cryo-preserved after 5 to 6 months of continuous undifferentiated proliferation (Thomson et al., 1998).

Dr. J. Gearhart with colleagues (Shamblott et al., 1998) obtained human fetal tissue from terminated pregnancies after informed consent. They took cells from the gonadal ridges and mesenteries of the fetus 5-8 weeks of gestation in order to maintain these cells in culture in vitro. The authors have used 5 immunological markers and alkaline phosphatase reaction that are routinely applied to characterize ESCs and EG cells (Shamblott et al., 1998). These reports showed the possibility to derive SCs for cell-based therapies in different human diseases. The use of human embryo to derive SC is widely debated in the public (Kentenich, 1999; Press Dossier EC, 2000; McLaren, 2001b; 2002; Matthiessen, 2002; Mieth, 2002; Tudge, 2002). The ES cell technology gave rise to ethical and legal problems.

Some Biological Aspects of Human Cloning

Definitions:

- **Genomic Imprinting** - a biochemical phenomenon that determines (for certain specific genes) which a gene (mother's or the father's one) from pairs of identical genes will be active in that individuum.
- **Cloning** - means making genetically identical copies of a single cell or an organism (by means of cell nuclear transfer).
- **Clone** is a complex of molecules, cells or organisms. Cloned organisms are developed by asexual reproduction from one ancestor. They must be genetically identical to a donor's nucleus genome.

Organisms cloning are created by means of replacement of the nucleus of unfertilized egg by the nucleus of somatic cell with of diploid number of the chromosomes (by means of cell nuclear transfer). After cell nucleus transfer and several cell divisions in vitro an embryo is introduced into the uterus (womb). The aim of this cloning is to create an organism with certain genetic properties (with a desirable set of genetic traits). It is described as the reproductive cloning.

However 100% genetically identity with a genome's donor may not be because there are a number of biological moments:

1. Approximately 1% DNA is mitochondrial, into cytoplasm (outside the cell nucleus). Thus a clone is not absolutely genetically identical to a nucleus' donor.
2. It is known there are different types of DNAb alterations in somatic cells. This phenomenon reduces genetically identity between a clone's genome and the genome of a nucleus' donor.
3. External and internal factors have considerable influence on the physical and mental individual development of each generation. Also this phenomenon reduces genetically identity between a clone's genome genome and the genome of a nucleus' donor.
4. There is the mechanism of function of telomere DNA sequences. The telomere shortening is due to suppression of telomerase at early stages of embryogenesis. This phenomenon (the telomere shortening) has led to cellular senescence.

The cloning organism with somatic adult cell genome will be have a more weak vitality and premature biological aging. Cloning animals do not produce a pregnancy and offspring. If pregnancies occur then there are a large number of abnormalities. Therefore it would be completely irresponsible to use this procedure on human at the present time.

For humanity there is the main purpose to maintain and to conserve of genetic variety (polymorphism) of the humane genome.

Most lawmakers believe that making babies by cloning should be outlawed.

Some Ethical and Legal Aspects of Human Cloning and Human Embryonic Stem Cell Technologies

In Russian Federation a Moratorium on human cloning (for five years) was proclaimed (19 of April 2002).

The human cloning and human embryo' stem cells technologies issues involved have been considered by a number of international bodies:

UNESCO Universal Declaration on the Human Genome and Human Rights (November, 1997).

Article 11 Practices which are contrary to human dignity, such as reproductive cloning of human beings, shall not be permitted. States and competent international organizations are invited to co-operate in identifying such practices and in taking, at national or international level, the measures necessary to ensure that the principles set out in this Declaration are respected.

Council of Europe Convention For the Protection of Human Rights and Dignity of the Human Being with regard to the Application of Biology and Medicine (Convention For the Protection of Human Rights and Biomedecine). Oviedo, 4 April 1997.

Article 18 - Research on embryos in vitro

1. Where the law allows research on embryos in vitro, it shall ensure adequate protection of the embryo.
2. The creation of human embryos for research purposes is prohibited.

Additional Protocol to the Council of Europe Convention For the Protection of Human Rights and Dignity of the Human Being with regard to the Application of Biology and Medicine, on the Prohibition of Cloning Human Beings (ETS No.168) (Strasbourg , 1998).

Article 1 - Any intervention seeking to create a human being genetically identical to another human being, whether living or deathing is prohibited...

United Nations Declaration on Human Cloning (6th Committee), January, 2005.

(a) Member States are called upon to prohibit any attempt to create human life through cloning processes and any research intended to achieve that aim;

(b) Member States are called upon to ensure that, in the application of life science, human dignity is respect in all circumstances and, in particular, that women are not exploited;

(c) Member States are called upon to adopt and implement national legislation to bring into effect paragraphs (a) and (b) above;

(d) Member States are father called upon to adopt the measures necessary to prohibit applications of genetic engineering techniques that may be contrary to human dignity.

Steering Committee on Bioethics of Council of Europe: The protection of the human embryo in vitro. Report by the Working Party on the Protection of the Human Embryo and Fetus (CDBI-CO-GT3), Strasbourg, 19 June 2003; 44P).

«This report aimed at giving an overview of current positions found in Europe regarding the protection of the human embryo in vitro and the arguments supporting them.

It shows a broad consensus on the need for protection of the embryo in vitro. However, the definition of the status of the embryo remains an area where fundamental differences are encountered, based on strong arguments. These differences largely form the basis of most divergences around the other issues related to the protection of the embryo in vitro.

Nevertheless, even if agreement cannot be reached on the status of the embryo, the possibility of reexamination certain issues in the light of the latest developments in the biomedical field and related potential therapeutic advances could be considered. In this context, while acknowledging and respecting the fundamental choices made by the different countries, it seems possible and desirable with regard to the need to, protect the embryo in vitro on which all countries have agreed, that common approaches be identified to ensure proper conditions for the application of procedures involving the creation and use of embryos in vitro. The purpose of this report is to aid reflection towards that objective».

It was pointed out that: «Obtaining SC from human embryo cannot be ethical because it necessarily involves destroying those embryos. The destruction of human life in the name of medical progress must be prohibited». Most researchers cannot accept the idea of creating human embryo just to destroy them (therapeutic cloning, cloning for research).

According to the World Medical Association's (WMA) Declaration of Helsinki (1964; 2000) we have: «10. It is the duty of the physician in medical research to protect the life, health, primacy, and dignity of the human subject». According to the WMA's Declaration of Helsinki and European Convention on Human Rights and Biomedicine, Article 2 - Primacy of the human being «The interests and welfare of the human being shall prevail over the sole interest of society or science. It is proof that it is not even necessary to obtain ES cells by destroying human embryos in order to treat diseases». A growing number of researchers believe that adult stem cells may soon be used to develop treatments of diseases such as cancer, immune disorders, orthopedic injuries, heart failure, degenerative diseases, tissues of many kinds or umbilical cord blood transplantation for hematologic diseases.

Defeat of Aging - Utopia or Foreseeable Scientific Reality

Aubrey de Grey

Department of Genetics, University of Cambridge, Downing Street, Cambridge CB2 3EH, UK, ag24@gen.cam.ac.uk

If we ignore reductions of mortality in infancy and childbirth, life expectancy in the developed world has risen only by between one and two years per decade in the past century. This is an impressive acceleration relative to previous centuries. Curiously, while many demographers consider that the current rate of increase may be maintained in coming decades, none have explored in detail the possibility that the *acceleration* will continue. This is probably because further acceleration would indeed be highly unlikely in the absence of major technological advances in the postponement not only of age-related diseases but of "aging itself" – the progressively accumulating molecular and cellular changes that underlie those diseases and also underlie the aspects of age-related functional decline that we usually do not call diseases, such as loss of muscle strength and immune response. In this essay I explain why technological advances of that sort are indeed foreseeable – indeed, I estimate a 50% chance that they will arrive within 25 years given adequate funding. I also explain why, even though these therapies will initially confer only perhaps 30 additional years of healthy life, those who benefit from them will very probably also benefit from progressively improved therapies, thereby maintaining "life extension escape velocity" and avoiding death from age-related causes at any age, with the result that (barring global catastrophe) many of them will attain four-digit lifespans.

Introduction

The approach to postponing aging that I shall describe in this essay is one of maintenance and repair. Those who like to claim that aging is intrinsically immutable are often inclined to start by asserting, ex cathedra, that

V. Burdyuzha (ed.), The Future of Life and the Future of Our Civilization, 277–290.

living organisms are qualitatively unlike machines and therefore cannot be maintained beyond their "warranty period" in the way that typical machines can. Even leaving aside the absence of any justification of the "therefore" in that assertion, there is a conspicuous fragility in the idea that organisms (even humans) are in any relevant way unlike machines. The property of living organisms that is most often suggested as distinguishing them from machines is their capacity for self-repair, and indeed that is undoubtedly something at which organisms are vastly superior to any machine currently in existence. But to consider it a qualitative difference is clearly incorrect: as a simple example one need only consider household robots that plug themselves into the mains when their batteries run low, or photocopiers that suspend operation to clean their wires when they automatically detect the need.

The pessimist often retorts that, even if this is not a qualitative difference, the difference of degree is so astronomical that the practical feasibility of maintaining an organism as one does a machine is far too distant to be worth considering. But here again we see a crass logical error, because the idea is to *augment* our natural maintenance systems: thus, the fact that they are so good already means that there is that much *less* for us to do to make them good enough to work indefinitely. There is much more to this question, as will emerge below, but the crux of the argument is as just stated.

In the next section I will describe in rather abstract terms the sort of maintenance that I believe we should be working towards in the quest to postpone aging as much as possible as soon as possible. In the following section I will go into more concrete biological detail, giving an overview of the specific types of maintenance and repair that humans need to do better in order to maintain our health and youth for a lot longer and the methods already under development to implement those required improvements. The concrete and detailed nature of those prospective interventions leads me to the view that we are potentially within only a decade of developing them all in laboratory mice, and that once we have done so we have perhaps a 50% chance of developing them in humans within only 15 years thereafter. Then, in the final section I will explain why this should be enough to put us beyond "life extension escape velocity" – the point at which we are improving these technologies faster than the remaining imperfections in them are catching up with us. Once we reach that point, and presuming we can stay there (which, I will argue, is virtually certain), no one need die of old age ever again, whatever age they attain.

Metabolism Causes Damage Causes Pathology

What is aging, actually? It is often suggested that aging is very hard to define. That is true if one requires a definition that suits all purposes, but when discussing interventions an altogether uncontroversial definition is easily found. A typical one is as follows:

> *Aging is the set of side-effects of metabolism that alter the composition of the body over time to make it progressively less capable of self-maintenance and thereby, eventually, progressively less functional.*

This definition emphasizes the ubiquitous presence of "threshold effects" in the progression of aging. In a nutshell, metabolism (the immensely complex network of processes that keep us alive from one day to the next) has *ongoing* side-effects, which we can reasonably refer to by the collective term "damage", but this damage only *eventually* causes functional decline (pathology). This allows us to identify three very distinct strategies for postponing aging and thereby extending healthy and total lifespan. Curiously (at least in retrospect), only two of them have historically been pursued. They are depicted in Figure 1, in which the flat-headed arrows are used in the conventional genetics sense to mean "inhibits".

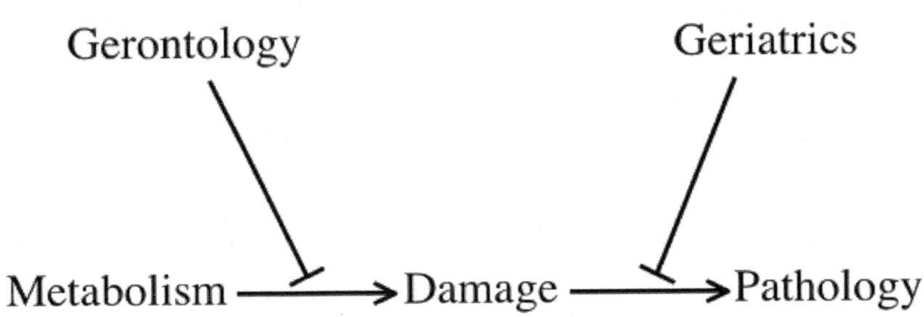

Fig. 1. The two traditional approaches to postponing aging.

It is important to stress that I will use the term "damage" in this very precise sense throughout this essay: for present purposes it is defined as the entire set of changes of bodily composition that (a) are side-effects of metabolism and (b) are eventually pathogenic. In particular, the reader should not infer any implication concerning how this damage is laid down, such as whether it could reasonably be called "wear and tear".

What is the prognosis for the gerontology and geriatrics approaches, in the foreseeable future? It is easy to see that the geriatrics approach is short-term almost by definition: as damage accumulates, its natural pathological consequences become progressively harder to avert. Besides, even if we could in principle develop geriatric medicine so sophisticated that pathology was slowed, that would be a distinctly mixed blessing, as it would constitute an extension of the frail period of life rather than of healthy life.

The gerontology approach initially seems much more promising. If one can retard the rate at which metabolism lays down damage in the first place, one will certainly extend the healthy part of life, which would seem unambiguously desirable. (Possibly the frail part would be extended too, but probably less so.) However, it has two daunting shortcomings. Firstly, damage that has already been laid down before the treatment begins will not be affected: hence, those who already have enough of it to be starting to suffer functional decline will not have that loss of function restored by such therapies. Secondly, the practicality of the gerontology approach is determined by the extent to which we understand metabolism, because altering the workings of a system that we understand only very poorly tends either to have no effect at all on its behaviour or to do more harm than good. And unfortunately, that is the case with metabolism: though we certainly understand far more about it than we did only a few decades ago, we are regularly reminded by the discoveries of fundamental new aspects of metabolism (such as RNA interference, discovered only a few years ago) [1] that in reality we have still hardly scratched the surface of its complexity. This bleak conclusion is reinforced by the failure of the rational but evidently oversimplistic approaches to extending mammalian lifespan that gerontologists have attempted over the past 50 years: it remains the case that, apart from a scattering of reports that were never reliably reproduced, the only way to extend a mammal's lifespan without making genetic changes in its germ line is to elicit a response that metabolism already has available to it, namely the intensification of repair and maintenance that results from moderate deprivation of nutrients [2]. (Genetic manipulations may presage non-genetic ones, however, so we should not ignore reports of life-extending genetic changes unrelated to manipulation of nutrient sensing or utilization [3-4]. However, these remain isolated results at the time of writing.)

If we wish to postpone aging any time soon, therefore, it seems clear that we must seek a third way – something radically different from the gerontology and geriatrics approaches. Just such an approach has been the focus of my work since 2000 and has become known as "Strategies for Engineered Negligible Senescence" or SENS [5,6]. It can best be introduced by embellishing Figure 1, as shown in Figure 2.

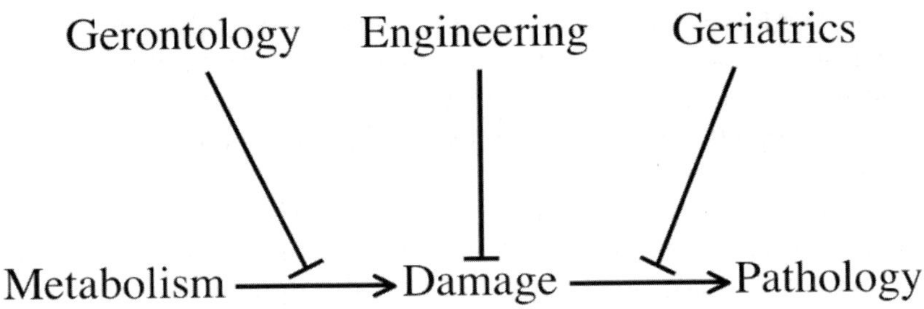

Fig. 2. How the engineering (SENS) approach to postponing aging relates to the two traditional approaches.

The key feature of the SENS approach is that it intervenes early enough to avoid being a "losing battle" like the geriatrics approach, but at the same time it does not attempt to *improve* the indescribably complex and well-honed machine that is our metabolism, but rather to *clean up after it.* The SENS approach relies on a frequently overlooked aspect of aging which is mentioned in the definition I gave earlier: that even though metabolism causes damage *ongoingly*, throughout our whole life, damage only *eventually* causes functional decline. If you live and eat essentially as your mother told you to, and if you are not particularly unlucky in terms of genetics, you will probably be able to run and think roughly as fast at the age of 40 as you could when you were 20. In short, the SENS approach does not attempt to interfere in processes – neither the process whereby metabolism causes damage, nor that by which damage causes pathology. Rather, it seeks to remove the damage that metabolism lays down, at least as fast as it is laid down, and thereby to prevent it from ever translating into pathology.

Perhaps the most obvious initial objection to the SENS approach, and certainly a common one heard from biogerontologists, is that it "must" be utterly infeasible simply because it seeks to reverse age-related decline. The idea here is that reversing a process is intuitively far harder than slowing it down, and we have made precious little progress (even in mice, let alone humans) in slowing aging down. There are two main errors in this logic. The first is that reversing a process is only necessarily harder than retarding it if one restricts oneself to using the same methods for reversal as one would for retardation and doing them so well that the retardation outstrips the progression. In reality, there are other approaches to reversing a process that do not act in this "head-on" way. Consider the predicament of a person in a small rowing-boat in the center of a large lake, which has sprung a leak. Our hero has two fundamentally different options for staying afloat until rescue arrives: he can try to plug the leak, thereby retarding the rate at which water enters, or he can bail water over the side, counterbalancing the influx. The latter process constitutes a reversal of the accumulation of the problem, but by a method that (unlike forcing the water back through the hole!) is technologically no more challenging than plugging the leak.

The other error in the idea that reversal is inherently far harder than retardation is equally important. If one has few tools available, one may only be able to plug the leak rather imperfectly, so that some water continues to enter and one will prolong one's survival but not indefinitely. In the case of bailing, by contrast, a sufficient but finite rate of removal of water will suffice to keep one afloat for as long as may be required. This has especially profound implications for life extension in the long term, as will be explained below.

From Boats to Biology: Is the Analogy Valid?

Analogies are all very well for showing that an idea makes sense in principle, but what about putting it into practice? In order to demonstrate that the SENS approach is truly foreseeable, it is necessary to describe in concrete terms what the "damage" is that SENS must repair, and also to propose specific biotechnological approaches to that repair for each type of such damage. Moreover, the proposed approaches must embody sufficient detail to give confidence that we can get there from here in a meaningfully predictable timeframe.

Without further ado, then, I offer in Table 1 what I claim is an adequately complete list of the types of side-effect of metabolism that can be considered to qualify as "damage" by the definition being employed in this essay – that is, changes that there is some reason to believe contribute to age-related pathologies of one sort or another. By "adequately complete" I mean that it includes all types of change in our molecular and cellular composition that may contribute to tissue dysfunction in a currently normal lifetime; I acknowledge that other types of such change, such as nuclear mutations that do not affect the cell cycle, may be pathogenic when we reach ages considerably exceeding our existing lifespan.

The suggestion that this list is indeed adequately complete is a bold one and is routinely challenged. However, there are two strong arguments for this contention. The first concerns the dates noted in the right-hand column, the most recent of which is 1982. The analytical sophistication available to biologists has advanced very considerably since then, so the fact

Table 1. The seven "deadly things": types of damage that SENS seeks to combat and the dates they were first suggested to contribute to mammalian aging. For details see refs. 5 - 7.

Type of age-related damage	Suggested by, in
Cell loss, cell atrophy	Brody, 1955
Senescent/toxic cells	Hayflick, 1965
Oncogenic nuclear mutations/epimutations	Szilard, 1959; Cutler, 1982
Mitochondrial mutations	Harman, 1972
Intracellular aggregates	Strehler, 1959
Extracellular aggregates	Alzheimer, 1907
Extracellular crosslinks	Monnier and Cerami, 1981

that this list has not been extended as a result constitutes a strong circumstantial argument that no "eighth sin" will be discovered in the future either (except, as noted above, in those who reach ages that the seven problems listed above currently prevent anyone from attaining).

The second argument that the above is a complete list is perhaps more attractive to the biologist: it is that the list can be derived from first principles by examining our biology systematically. The starting-point for doing this is to note (a) that the list is of types of damage, not of processes that cause that damage (which would be a much longer one – indeed, one that certainly could not be confidently completed with current knowledge) and (b) that, by definition, damage can only accumulate in long-lived structures. Intracellular proteins, for example, vary somewhat in half-life but never survive for more than a small fraction of the human lifespan: thus, any deleterious modifications that they suffer are eliminated when they are destroyed. With these two points in mind, we can then ask: what are we made of? The first-level answer is: cells and stuff between cells. Cells of a given type can become more or less numerous with age: when this is deleterious we have the first two of the seven types of aging listed in Table 1. Within cells there are only two types of long-lived molecule – DNA (which of course is long-lived in an unusual way, because it is synthesized by replication) and garbage, i.e. indigestible substances that are sequestered indefinitely, usually in the lysosome. That accounts for items 3 to 5 in Table 1. In the extracellular space, similarly, there are just two types of long-lived molecule: complex proteinaceous structures such as the lens of the eye and the artery wall that can become chemically and thus physically modified over time (item 7), and garbage, again of different composition in different tissues but collectively termed amyloid (item 6).

So far, so good: we have a satisfactorily complete description of the problem. What about solutions? Table 2 summarizes the current state of play as I see it.

Table 2. Foreseeable approaches to repair or obviation of the seven types of damage listed in Table 1. For details see refs. 5 - 7.

Type of damage	Proposed repair (or obviation)
Cell loss, cell atrophy	Stem cells, growth factors, exercise
Senescent/toxic cells	Ablation of unwanted cells
Oncogenic nuclear mutations/epimutations	"WILT" (Whole-body Interdiction of Lengthening of Telomeres)
Mitochondrial mutations	Allotopic expression of 13 proteins
Intracellular aggregates	Microbial hydrolases
Extracellular aggregates	Immune-mediated phagocytosis
Extracellular crosslinks	AGE-breaking molecules

The first point to emphasize about Table 2 is that, of the seven therapies listed, two are not strictly *repair* strategies (reversing the accumulation of the specified type of damage) but rather *obviation* strategies which make the phenomenon no longer capable of causing pathology, and thus make it cease to classify as "damage". These are items 3 and 4 in the list, addressing nuclear and mitochondrial mutations respectively. For nuclear mutations, the proposal is a treatment for cancer that does not stop cells from accumulating the mutations that allow them to divide uncontrollably, but instead gives them a time-bomb – telomere shortening – which they cannot defuse even by the hypermutation that makes cancers so versatile in eluding all contemporary therapies. For mitochondrial mutations the suggested strategy is to make such mutations harmless by allotopic expression – introducing copies of the 13 protein-coding genes of the mitochondrial DNA into the nucleus, with modifications such that they will still be targeted to mitochondria and thus maintain mitochondrial function irrespective of any mitochondrial mutation.

How far away are the therapies listed in Table 2? In order to answer that question one must ask it somewhat more precisely, by specifying two additional things: how well the therapies must work, and in what organism. It is necessary, therefore, to give several answers.

There are two main reasons why the first substantial steps in extending healthy lifespan with late-onset interventions will occur in mice sooner than in humans. Firstly there is the biological reality that organisms with longer lifespans are already avoiding aging rather well and thus are harder to improve by copying ideas (genes, in particular) from even longer-lived species. Secondly there is the sociological reason that society mostly considers the deaths of rather large numbers of mice in the quest to perfect a therapy to be much more acceptable than the death of even one human in that quest. Neither is likely to change any time soon, so we first address the extension of mouse lifespan.

As noted, we must also specify a degree of progress that can be considered an appropriate milestone. The one that I have championed in recent years, with the moniker "Robust Mouse Rejuvenation" (RMR),[7] is to treble the remaining average lifespan of a cohort of naturally long-lived mice that are already 2/3 through their natural lifespan before any intervention (whether genetic, pharmacological or dietary) is begun. Long-lived mouse strains typically live to three years of age on average, so this means initiative a protocol on such mice at the age of two years and giving them an average age at death of five years.

The last two items in Table 2 are the ones in which we are furthest advanced at present: they have progressed to clinical trials. In both cases the main work remaining to be done in mice is to apply the same principles to other major types of (respectively) crosslink and amyloid than the ones which these pioneering therapies address, but in fact it may transpire that these initial treatments, though currently restricted to one category of crosslink and one amyloid-accumulating tissue, will address a sufficient proportion of their respective categories of damage to deliver RMR (so long, of course, as the therapies for the other five classes of damage are also up to scratch).

Compensation for cell loss is also going rather well. Many tissues that lose cells during normal aging or in the context of disease are the subject of intensive research into cell replacement using growth factors or stem cell therapy, some of which has also reached the clinic. This work lags behind the two SENS strands just discussed only insofar as the differences between therapies for different tissues are probably more challenging, relying as they currently do (at least in the case of stem cell therapies) on rather precise ex vivo "pre-differentiation" of initially over-versatile stem cells that are otherwise prone to develop not only into the desired cell type but also into various unwanted ones.

Elimination of supernumerary cells in rodents varies greatly in difficulty depending on the type of cell to be eliminated. The simplest is visceral fat, which can be surgically removed from the abdominal cavity of rats and results in the abrupt alleviation of previously advanced diabetes. Potentially there is also the possibility of converting the cells in question to a benign form, but no systematic method to identify such an intervention is yet evident. There are two attractive options that do qualify as "rationally designed", however, both exploiting the identifiability of the problematic cells by their excessive expression of particular genes. In the first method, "suicide" genes [8] are introduced by somatic gene therapy: these typically enter cells of many cell types, but the gene is placed under the promoter of the excessively-expressed gene so that it is only expressed in the cells that one wishes to eliminate. In the second approach, the undesired cells are removed by the immune system as a result of stimulation by appropriate vaccines and adjuvants [9]. However, none of these strategies has yet reached the clinical trial stage.

The remaining three SENS strands may be considered the "critical path" towards RMR, as they are all some way from implementation even in mice. Allotropic expression of the mitochondrial proteins from nuclear transgenes may be closer than it seems in vitro, as recent work gives considerable confidence that the only remaining requirement is to identify amino acid changes to these proteins' transmembrane domains which makes them a little less hydrophobic and thus more readily importable by the mitochondrial protein import apparatus. The remaining issue for RMR in regard to mitochondrial mutations is delivery of these genes to affected cells, and the current state of somatic gene therapy in mice is such that this may be only moderately challenging, especially in view of the fact that introduction of these genes into mitochondrially healthy cells should be harmless.

The runner-up in the difficulty stakes for RMR is probably the removal of intracellular aggregates. Indigestible material progressively impairs cell function, not least by impairing the degradation of other substances that the cell was hitherto able to process efficiently. An approach to this problem which I introduced in 2002 is to identify microbial enzymes that can break down such compounds (or convert them to ones that mammalian metabolism already handles).

Finally we come to nuclear mutations, and specifically those which promote cancer. I recently proposed an anti-cancer strategy termed WILT (Whole-body Interdiction of Lengthening of Telomeres) that is as ambitious as it is audacious: the use of both ex vivo and somatic gene therapy to delete the genes for telomerase and (as and when they are identified) ALT (Alternative Lengthening of Telomeres) from as many of our cells as possible. This will have deleterious side-effects that are obvious and daunting: telomere shortening will irresistibly eliminate the stem cell pools that maintain all our continually-renewing tissues, such as the blood, the gut and the skin. My proposal is to avert these consequences by periodic replenishment of our stem cell pools with new cells that also lack genes for telomere elongation but have had their telomeres extended ex vivo to normal lengths with exogenous telomerase.

Life Extension Escape Velocity

Once RMR is achieved, I am convinced that society's attitude to the postponement of human aging will become unrecognisable. I have therefore predicted that there is a 50% chance of our achieving a comparable advance in human life extension within 15 years after we achieve RMR. This human milestone, which I rather unimaginatively term "Robust Human Rejuvenation" or RHR, is not in my formulation precisely proportional to RMR: rather than a trebling of the remaining lifespan of people who are already 2/3 of the way to the prevailing average age at death, I define it as only a doubling. This means roughly 25-30 years of extra healthy life for people who are 55 when treatment begins.

Why have I chosen a relatively toned-down version of RMR to define as RHR? Simply, because 25-30 years is a familiar duration in the history of technology, and specifically in that part of the history of many technologies which, in respect of life extension, I will now discuss. How long does it take, following some fundamental technological breakthrough, for that technology to progress by incremental refinements to a stage beyond that which the architects of the original breakthrough could reasonably have contemplated? The answer seems rather reliably to be in the 20-30 year range. Lindbergh flew the Atlantic 24 years after the Wright brothers' first flight. Commercial jetliners first flew 22 years after that, and supersonic airliners 20 years after that. In computing, the personal computer arrived about 28 years after the first electronic computer and the first convenient laptops arrived about 20 years later. In medicine, the discovery of antibiotics followed the publicising of the germ theory by about 30 years and was in turn followed, after another 25 years, by the development of methods to manufacture vaccines specific for a particular disease.

The implications of this pattern for the lives of people who are in middle age or younger at the time that RHR is achieved is clear, but no less dramatic for that. Put simply, there is a very high probability that the 25-30 years of good health conferred on its recipients by the first-generation panel of rejuvenation therapies (defined as those which achieve RHR) will suffice for the development of much more thorough and comprehensive therapies, capable of delivering more like a century of extra life to those who are in relatively good health at the time those therapies arrive. This is where the longevity escape velocity (LEV) concept [7] arises. The recipients of the *first*-generation therapies – the ones that gave only around 30 years of extra healthy life – will, at least if sociopolitical pressures do not intervene, mostly *also* be among the beneficiaries of the *second*-generation ones, since they will be in the same degree of health at that time as they were when the first-generation therapies arrived. The same logic of course applies indefinitely into the future, just so long as the rate of progress in improving the comprehensiveness of the therapies continues to outstrip the rate at which the remaining imperfections in those therapies allow the accumulation of eventually pathogenic damage.

What does this add up to for lifespan? Clearly the lifespans of those who live their entire lives in a period when progress is faster than LEV will be indefinite, since a given individual's risk of death at any adult age will be less than at earlier adult ages. What is less immediately clear is how to estimate the lifespans of those already alive (and at various ages) at the time RHR arrives. My estimates are depicted (for actual numbers are too speculative to estimate) in Figure 3. To summarize: I estimate that 50-year-olds who are in average health at the arrival of RHR and who are, thereafter, able to benefit from the latest and best rejuvenation therapies, will have at least a 50/50 chance of reaching their own personal escape velocity – that is, of being restored to a truly youthful state with a very low mortality risk. Most of those who are only 30 at that time will never reach a state of age-related frailty. Moreover, elite individuals – those who would naturally live to 100 or more even in the absence of these therapies – will have that 50/50 chance even if they are already in their 70's when RHR arrives.

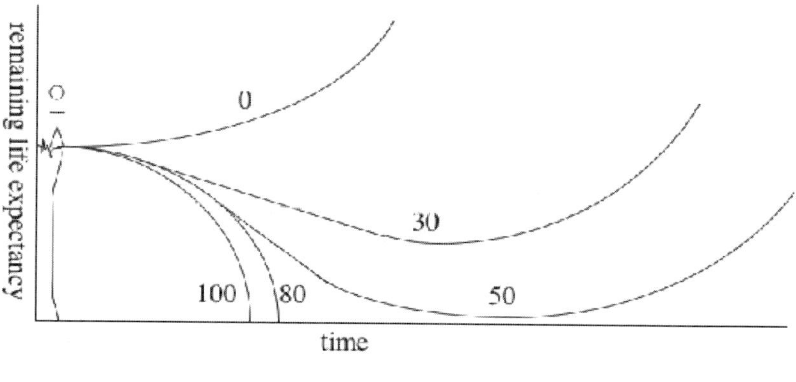

Fig. 3. Plausible trajectories of "biological age" for typical individuals of the specified ages at the time RHR arrives, presuming access to the best therapies at any time.

A key corollary of the above considerations is that there will be a stunningly sharp "cusp" in the increase in lifespans of those born in successive years. One way to quantify this is that the first 1000-year-old is probably only about ten years younger than the first 150-year-old. Another is that a whole generation will be, in the words of the Australian writer Damian Broderick [11], the "last mortal generation" – a cohort who live roughly as long as those born in 1900, but whose offspring mostly live indefinitely and die only of causes unrelated to age. The sociopolitical will surely be unprecedentedly profound.

Conclusion

I have attempted in this essay to outline the methods by which humanity will in due course defeat its greatest remaining scourge. Much of what I have written is plainly speculative in the extreme, yet I have stuck my neck out and given estimates of timeframes, with probabilities attached to them. Some feel that speculations of this sort are irresponsible, engendering unwarranted optimism about the rate of progress. I take the diametrically opposite view: I am convinced that it is irresponsible to remain silent on such matters, because doing so engenders unwarranted *pessimism*: the public are predisposed to presume that nothing can be done about aging and thus do not agitate for efforts to hasten progress, and that will only change if their sights are raised [12]. I thus have no compunction in setting out (here and elsewhere) a scenario that, after much consideration, I consider the most likely way in which, and rate at which, we will move to a post-aging world.

References

1. Fire A., Xu S., Montgomery M. K., Kostas S. A., Driver S. E., Mello C. C. (1998) Potent and specific genetic interference by double-stranded RNA in Caenorhabditis elegans. Nature; 391: 806-811.
2. de Grey A. D. N. J. (2005) The unfortunate influence of the weather on the rate of aging: why human caloric restriction or its emulation may only extend life expectancy by 2-3 years. Gerontology; 51: 73-82.
3. Migliaccio E., Giorgio M., Mele S., Pelicci G., Reboldi P., Pandolfi P. P., Lanfrancone L., Pelicci P. G. (1999) The $p66^{shc}$ adaptor protein controls oxidative stress response and life span in mammals. Nature; 402: 309-313.
4. Schriner S. E., Linford N. J., Martin G. M., Treuting P., Ogburn C. E., Emond M., Coskun P. E., Ladiges W., Wolf N., Van Remmen H., Wallace D. C., Rabinovitch P. S. (2005). Extension of murine life span by overexpression of catalase targeted to mitochondria. Science; 308: 1909-1911.
5. de Grey A. D. N. J., Ames B. N., Andersen J. K., Bartke A., Campisi J., Heward C. B., McCarter R. J. M., Stock G. (2002) Time to talk SENS: critiquing

the immutability of human aging. Annals of the New York Academy of Sciences: 959: 452-462.

6. de Grey A. D. N. J. (2003) An engineer's approach to the development of real anti-aging medicine. Science of Aging Knowledge Environment vp1.
7. de Grey A. D. N. J. (2004) Escape velocity: why the prospect of extreme human life extension matters now. PLoS Biology; 2: 7 23-726.
8. Felmer R. N., Clark J. A. (2004) The gene suicide system Ntr/CB1954 causes ablation of differentiated 3T3L1 adipocytes by apoptosis. Biology Research; 37: 449-460.
9. Berd D. (2004) M-Vax: an autologous, hapten-modified vaccine for human cancer. Expert Reviews on Vaccines; 3: 521-527.
10. de Grey A. D. N. J. (2005) Forces maintaining organellar genomes: is any as strong as genetic code disparity or hydrophobicity? BioEssays; 27: 436-446.
11. Broderick D. (2000) The Last Mortal Generation: How science will alter our lives in the 21st century. New Holland Publishers, London.
12. de Grey A. D. N. J. (2005) Resistance to debate on how to postpone ageing is delaying progress and costing lives. EMBO Reports; 6: S49-S53.

Cardiology in the XXI Century

Sergei Konorskiy

Kubanskyi Medical State University, Sedina str.4, 350063 Krasnodar, Russia, sergei.konorskiy@rambler.ru

Ischemic heart disease is a leading cause of mortality in our planet in XXI century, the greatest problem of modern cardiology. Although infarct size limitation remains a highly desirable goal, it has been extremely difficult to achieve clinically. A fundamental problem is the fact that ischemic myocardium dies quite rapidly. Infarcts achieve 80% of their potential size within 3 hours of coronary occlusion. Thus, myocardial infarction and subsequent heart failure are likely to remain major health problems.

When heart failure becomes refractory to drug treatment, more aggressive options have to be considered. Three such options are currently available: (1) replacement of the failing heart, either total (cardiac transplantation) or partial (implantation of a left ventricular assist device); (2) reshaping of the dilated left ventricle by a remodeling operation or a passive-constraint device; and (3) resynchronization by multisite packing. However, all of these techniques have limitations and contraindications, which justify the ongoing search for other therapies. The most recent technique, which is currently generating growing enthusiasm, is myocardial regeneration.

Myocardial Regeneration: What It Is

Myocardial regeneration consists of the repopulation of irreversibly damaged muscle with new contractile cells to restore functionality in the necrotic areas, thereby improving global heart function. Ideally, these cells come from the spared peri-infarct myocardium. Until recently, it was thought that adult mammalian cardiomyocytes were terminally differentiated and, therefore, could not divide. This dogma has been recently revisited in light of pathological studies performed in patients with ischemic

V. Burdyuzha (ed.), The Future of Life and the Future of Our Civilization, 291–306.

and dilated cardiomyopathies, which showed that some cells can actually reenter a mitotic cycle (A.P. Beltrami et al., 2001). The only practicable perspective is an exogenous supply of cells for effecting repopulation of injured areas.

Ideally, exogenous cells should satisfy the following criteria: (1) be easy to collect and expand; (2) form stable intramyocardial grafts; (3) be able to electromechanically couple with host cardiomyocytes so as to beat in synchrony with them; and (4) be devoid of arrhythmogenic and oncogenic effects. It is important to note, however, that the" ideal' cell fitting this description is not yet available for clinical use.

How Can the Grafted Myocytes be Kept Alive?

Cardiomyocytes

Ideally, dead myocardium should be replaced by living cardiac tissue, and, therefore, cardiac myocytes should be a first choice for cardiomyoplasty.

Fetal (and neonatal) cardiomyocytes were able to successfully engraft into infarcted myocardium, express gap junction proteins (thus allowing them to couple with host cardiomyocytes), survive for long periods of time, and improve left ventricular function (J. Muller-Ehmsen et al., 2002).

Reasoning that more grafted cardiomyocytes would give rise to larger grafts, we performed a dose-escalation study in injured rat hearts using neonatal cardiomyocytes. Disappointingly, all grafts were small (1-2% of the left ventricular mass), and there was no increase in graft size with increasing cell dose. This indicated that cell death was probably limiting the amount of new myocardium formed (C.E. Murry, H. Reinecke, 2003).

In fact, depending on the transplantation protocol, injury (cryoinjury, infarct) and animal model (mouse, rat, rabbit, dog, pig), cell survival after transplantation may range anywhere from 0% to 20%.

The next question becomes," why do the cells die?' Information on mechanisms of graft cell death is currently limited, but it appears likely that ischemia plays a major role. Cell survival is better in normal hearts than acutely injured hearts, and vascularized granulation tissue supports cell survival better than acutely necrotic myocardium, although not as well as normal myocardium. The dilemma here is that cancer are extremely sensitive to ischemic injury: this is exactly the reason they die in the ischemic heart in the first place. Similarly, an old infarct scar represents a relatively ischemic tissue and, therefore, one might expect poor survival of this highly metabolic cell type after transplantation.

Significantly better results were obtained when cardiomyocytes were heat shocked the day before grafting, i.e., a 54% reduction in cell death was observed. Heat shock is thus a simple and effective approach to enhancing graft cell survival (M.Zhang et al., 2001).

Skeletal Myoblasts: From Bench to Bedside

Strictly speaking, skeletal myoblasts are not stem cells since they are solely committed to the myogenic lineage. When muscle injury develops, these cells are rapidly mobilized from their normally quiescent state, actively proliferate, and form new myotubes, which effect regeneration of the damaged muscle fibers. In terms of clinical application, these cells exhibit several attractive characteristics, including (1) an autologous origin - which overcomes problems related to availability, rejection, and ethics; (2) a high potential for in vitro expansion; (3) commitment to a well-differentiated myogenic lineage, thereby virtually eliminating the risk of tumor development, and (4) a high resistance to ischemia, which is a major advantage given the hypovascularization of postinfarct scars in which they are implanted.

Experimentally, skeletal myoblasts injected into infarcted areas differentiate into myotubes that are embedded in scar tissue. The grafted cells retain the morphological and electrophysiological characteristics of skeletal muscle without any evidence of transdifferentiation into cardiomyocytes (H. Reinecke et al., 2002).

Interestingly, the lack of gap junctions with recipient cells does not preclude the grafted myoblasts from promoting improvement in both regional and global left ventricular function (M. Jain et al., 2001). Overall, this improvement appears to be sustained over time, is almost linearly correlated with the number of injected cells (hence the importance of large-scale production and of means of reducing the high early posttransplantation cell death rate), and is additive to the protection afforded by angiotensin-converting enzyme inhibitors.

Skeletal myoblast transplantation has also been performed in clinical trials (P. Menasche et al., 2003). Overall, the first conclusion drawn from these initial studies is that the procedure is feasible, i.e., that it is possible to grow several hundred million cells from a small muscular biopsy within 2 to 3 weeks under Good Manufacturing Practice conditions, and that the final cell yield can be surgically injected into multiple sites across and around the scar without any procedural complications.

A significant fraction of the skeletal myocytes will undergo cell death due to limited resources (nutrients, oxygen) in the graft bed. Promising strategies to improve survival of grafted myoblasts include again heat shock prior to grafting and enhancement of vascularization in the graft bed by the use of vascular endothelial growth factor. The latter is based on the hypothesis that the cellular cardiomyoplasty effect could be reinforced by improved graft survival resulting from an improved blood supply to the graft through both vasodilation and enhanced angiogenesis induced by vascular endothelial growth factor. This would be particularly beneficial in the early stage after cell transplantation, when grafted cells are subjected to various pathological processes caused by environmental stress, such as ischemic and mechanical injury known to result in both the necrosis and apoptosis of grafted myoblasts. Given that angiogenesis takes a minimum of several days to vascularize an infarct, and that cell death is most extensive in the first day after grafting, it seems that prevascularizing a graft cell bed would offer significantly more benefit than an angiogenic therapy given at the time of myoblast implantation.

A second important finding is that myoblast transplantation is potentially proarrhythmic. It is difficult to conclusively establish a causal relationship between ventricular arrhythmias and cell grafting in the absence of control groups and in a patient population prone to develop this type of events because of the underlying cardiac disease, but safety requirements make it necessary to consider this relationship as more than likely. The mechanism of these allegedly transplantation-related arrhythmias remains elusive, but could be related to differences in electrical membrane properties between donor and recipient cells (B. Leobon et al., 2002).

Skeletal muscle cells have much faster action potentials and fiber conconduction velocities than cardiac myocytes. Hence, it is possible that successful coupling of the two muscle types would result in an arrhythmogenic substrate. Nevertheless, the possible gain from coupling the two muscle types justifies additional, careful animal experimentation. As long as the potential proarrhythmic risk associated with myoblast transplantation has not been fully elucidated, it is safer to recommend the implantation of an internal defibrillator. Limited reports of the presence in human hearts of myotubes in the injected areas lend support to a potential link between engraftment of cells and improvement in function (A.A. Hagege et al., 2003; F. Pagani et al., 2003). The mechanisms of this improvement remain unclear, however.

How Can the Graft be Synchronized with the Rest of the Myocardium?

A successful myocyte graft in the heart should not only fill the void left by dead myocardium, it should also be in sync with the rest of the heart. The heart acts as a functional syncytium, meaning that all the myocytes in the heart act together as an electromechanical unit. This is in contrast to skeletal muscle, where cells are truly syncytial. i.e., have fused to form multinucleated fibers but the individual fibers are insulated from one another. Electromechanical coupling in the heart is achieved by the specialized cell-cell junctions, the intercalated disks which contain adherence junctions and gap junctions for mechanical and electrical coupling, respectively. Electrical couplings achieve by gap junctions. Connexin43 is the major gap junction protein in the mammalian left ventricle. In contrast to heart muscle cells, mature skeletal muscle fibers are electrically isolated from one another, a prerequisite for fine motor control.

Coupling of Cardiomyocyte Grafts

It is without doubt that cardiomyocytes fulfill all the requirements to couple electromechanically with the host myocardium. Indeed, several studies have shown development of intercalated disks, complete with gap junctions, between grafted fetal or neonatal cardiomyocytes and host cardiomyocytes. The principal impediment to coupling of cardiomyocyte grafts with the host myocardium is formation of scar tissue. As infarct healing proceeds, however, the grafted cardiomyocytes typically become infiltrated by scar tissue. The scar forms a physical barrier between the graft and the host cardiomyocytes. Murry and Reinecke (2003) were unable to demonstrate coupling of graft and host at later time points. Thus, any cell-based therapy that aims to induce such electromechanical coupling will need to address the issue of insulation by scar tissue.

Coupling of Skeletal Myocyte Grafts

Skeletal muscle cells in the heart appear firmly committed to their fate. This is characterized by myoblast fusion and myotube formation, maturation, expression of skeletal muscle specific myosin heavy chains, and the failure to express cardiac markers (H. Reinecke et al., 2002).

Experiments indicate that cardiomyocytes have the capacity to form electromechanical junctions with skeletal muscle cells and to use these junctions to induce synchronous beating in the skeletal muscle. Why does this coupling not occur in vivo after grafting? Skeletal muscle cells in culture are less differentiated than in vivo graft cells, and in culture the cells still have low levels of N-cadherin and connexin43. It appears that this low-level expression is sufficient to permit physiological coupling. As the graft cells mature in vivo, however, N-cadherin and connexin43 appear to be downregulated to undetectable levels, thereby precluding coupling.

Genetic modification of the graft cells may further foster survival, and may also allow skeletal myocytes to better integrate with the host myocardium.

Finally, it is also conceivable that the benefits of myoblast transplantation are not directly related to their intrinsic contractile properties, but rather to the release of growth factors capable of mobilizing a resident pool of cardiac stem cells, thereby promoting endogenous regeneration and/or rescue of reversibly damaged recipient cardiomyocytes (P. Anversa, B. Nadal-Ginard, 2002).

Clarification of the mechanisms by which myoblast transplantation favorably affects the function of the infarcted myocardium has not only a theoretical interest, but practical implications as well. Thus, if the primary effect of transplantation is to alter remodeling, it should be performed at a relatively early stage of the disease, within a given time window that needs to be specified. Conversely, if improved function is the result of additional contractile cells replacing the scar area, engraftment would theoretically be successful at any time point after infarction.

Regardless of the mechanism by which skeletal myoblasts improve postinfarct function, the lack of synchronous electromechanical coupling of these cells with host cardiomyocytes other than occasionally remains a major limitation. As previously mentioned, only fetal carbon dioxide have been shown, so far, to express junction proteins and thus really integrate into the host tissuc. The recognition that "true" allogeneic cardiac cells cannot be easily used clinically justifies the search for an alternative cell type capable of acquiring "cardiac-like" characteristics. These cells would need to have a plasticity enabling them to either jump across their lineage boundaries to become cardiomyocytes or, more realistically, differentiate into cardiomyocytes from an initially uncommitted state. And this is where the story of stem cells in the treatment of heart failure really begins.

Stem Cells: in Quest of the Holy Grail?

Broadly speaking, stem cells can be defined by two major properties: self-renewal and the capacity to generate various types of differentiated cells. From a practical standpoint, stem cells can be divided into two main categories: (1) embryonic stem cells, which are totipotent, i.e., which can generate all differentiated cell types in the body; and (2) tissue-specific stem cells, which are multipotent, i.e., which can only generate the cell types occurring in a particular tissue in embryos and, in some cases, adults.

Several studies have shown that adult stem cells of a given tissue could in fact differentiate into cells of other tissues, both in vitro and after in vivo transplantation. This process, termed transdifferentiation, enables grafted cells to overcome their developmental restriction program following exposure to signals originating from a novel environment. In the setting of heart failure, this transdifferentiation potential has mainly been recognized in bone marrow-derived cells.

Bone Marrow Stem Cells

Over the past few years, transplantation of bone marrow-derived cells has raised growing interest because, like myoblasts, they can be used as autografts, but, unlike myoblasts, their plasticity allows them to change their phenotype in response to organ-specific cues. These cells could thus specifically convert into cardiac and/or endothelial cells following engraftment into myocardial tissue, thereby resulting in true regeneration of postinfarction scars. These conclusions were derived from several experimental studies, mostly performed in rodents and frequently entailing cell injections shortly after the acute ischemic injury.

As a result, a burst of clinical studies was generated, in a totally disorganized fashion. It is therefore high time for a critical review of the literature and to highlight several key clinical aspects that need to be addressed to avoid the whole issue being jeopardized because of uncontrolled trials lacking robust and sound preclinical foundations.

Selection of Cell Type

The bone marrow is a complex medium that comprises two distinct stem cell populations: hematopoietic stem cells and mesenchymal stem cells. Our current ignorance regarding the most suitable bone marrow cells for myocardial regeneration has led some groups to advocate the use of total, unfractionated bone marrow in the assumption that this mix would necessarily contain the “good” cells (E.Tateishi-Yuyama et al., 2002). This approach immediately gained wide clinical acceptance because of its simplicity: the bone marrow is aspirated, washed of red blood cells, and promptly reinjected without an intervening period of culture (at most, the aspirate is only expanded for a few days). While this use of unpurified bone marrow may be effective, one may speculate about the advisability of administering a treatment without knowing which of its components has the expected therapeutic effect. However, a recent study by Menasche and colleagues using a sheep model of myocardial infarction has failed to document any benefit of transplantation of total bone marrow into postinfarction scars (A. Bel et al., 2002).

This concern can be addressed by sorting subpopulations on the basis of specific surface markers. The first option consists in selecting hematopoietic progenitors, with the caveat that we do not yet know which population should then be targeted.

A second option consists in selecting mesenchymal (or stromal) cells, which can usually be isolated in vitro through their adhesive properties and which, under appropriate culture conditions, generate various mesodermal-type progenies, such as adipocytes, chondroblasts, osteoblasts, and skeletal myoblasts. C.M. Verfaillie (2002) identified a fraction of these cells capable of acquiring the phenotypic characteristics of cells outside the mesoderm lineage, which they called "multipotent adult progenitor cells". However, the expectations raised by the potential therapeutic value of these cells should be tempered by the following considerations: (1) they are difficult to grow in a consistent fashion; (2) they cannot be isolated with specific markers (they are only identified by a negative staining for the most common surface antigens); and (3) there are no available data showing that they can functionally repopulate a damaged myocardium following transplantation in vivo.

Scale-Up

Regardless of the type of bone marrow-derived stem cells selected, progenitor cells are only present in minute amounts in the peripheral blood and their percentage in the bone marrow itself is also very low (1% to 2% of the total bone marrow cell population). Not unexpectedly, the degree of engraftment and the related improvement in outcome parameters are directly related to the number of injected cells. To achieve clinically meaningful benefits, it is thus mandatory to scale-up the number of stem cells, which poses a real challenge. One possibility is to try to expand them in vitro, but the risk is then that they might lose, at least in part, their pluripotentiality. Another, more conceptually attractive approach consists in mobilizing progenitors endogenously by cytokines like granulocyte colony-stimulating factor and/or granulocyte macrophage colony-stimulating factor. However, the efficacy of this strategy, first demonstrated in mice (D. Orlic et al., 2001), could not be replicated in primates. In addition, safety concerns were raised about the effects of such mobilization, and the associated leukocytosis, in patients at the acute phase of myocardial infarction. It is hoped that this as yet unsettled issue will be clarified by the results of ongoing clinical trials.

Fate of Engrafted Cells

The changes undergone by grafted cells upon exposure to their new myocardial environment are still largely unknown. In most cases, their lineage-unrelated progeny have been identified by morphology and immunohistochemistry using lineage-specific antibodies (e.g., against troponin I or α-myosin). This has usually led to the conclusion that the implanted bone marrow cells had transdifferentiated. Transdifferentiation, however, requires that at least three major criteria be met (D.J. Anderson et al., 2001): (1) the injected cells must be clonal, since otherwise it is difficult to establish conclusively that the converted cells really come from the donor; (2) the cells should be transplanted without in vitro culturing, which might alter their properties; and (3) it is mandatory that the phenotype of the presumably transdifferentiated cells be precisely characterized at the morphological, molecular, and functional levels: this is of utmost importance, as in the absence of functional criteria, it remains possible that the grafted cells have simply taken on the shape of the recipient cells or only express some of the proteins characteristic of these cells, which is unlikely to be enough to generate clinically relevant benefits. Likewise, it is important to point out that the frequency of conversion is usually very low, which raises the question of its therapeutic significance. Critical analysis of the literature shows that few studies claiming transdifferentiation really meet these criteria.

Indeed, this concept itself is challenged by an increasing number of experiments showing that the apparent plasticity of bone marrow cells may not reflect transdifferentiation, but rather their fusion with host cells (A. Medvinsky, A. Smith, 2003). This phenomenon could be therapeutically useful if the grafted cells rescue the diseased recipient cells by transferring some missing genetic information, as occurs in the liver, but the effects of genomic reprogramming in the resulting cells are still unknown and clearly raise safety issues that require clarification.

Alternatively, it is conceivable that the effects of bone marrow grafting on angiogenesis (and, possibly, myogenesis) are disconnected from phenotypic changes of the transplanted cells, but are instead solely related to the secretion of angiogenic growth factors (J. Rehman et al., 2003).

Conversion of grafted cells to a cardiomyocyte-like phenotype appears to require cocultures or coimplantations with fetal cardiac cells (C. Badorf et al., 2003), which is consistent with the previous observation that direct cell-to-cell contact is an effective means of inducing the targeted differentiation of cells with a plasticity potential (G. Condorelli et al., 2001). However, these manipulations are likely to be equally difficult to implement in clinical practice. Of note, unprocessed mesenchymal cells have been reported to improve function in a porcine model of myocardial infarction, but although the engrafted cells seemed to have converted to myogenic cells, none of them expressed markers specific for a cardiac phenotype (J.G. Shake et al., 2002). Since then, the same group has reported on the ability to track these magnetically labeled mesenchymal cells with magnetic resonance imaging following their endocardial delivery, but in the only animal that was assessed at 4 weeks after transplantation, only one fourth of the original injection sites could still be detected, and postmortem histology failed to provide any evidence for "transdifferentiation" of the engrafted cells (D.L. Karitchman et al., 2003).

Role of Host Environment

That host environment has a profound influence on the fate of engrafted cells is demonstrated by the homing phenomenon. Thus, the intravenous injection of radioactively labeled endothelial progenitors does not result in any noticeable engraftment in normal myocardium and it is only in the case of myocardial infarction that the cells are found homing in the border zone (A. Aicher et al., 2003), thus suggesting that damaged tissue emits signals that act as sensors for circulating bone marrow cells.

Repopulation of infarcted mouse hearts by hematopoietic progenitors from syngeneic animals only yields 0.02% and 1% to 2% cardiomyocytes and endothelial cells of donor origin, respectively (K.A. Jackson et al., 2001). However, recognition that local signals can drive the fate of implanted cells towards a given lineage may be a double-edged sword inasmuch as grafting cells into fibrous scars could merely convert them into fibroblasts, which has indeed been observed and is exactly the opposite of the intended goal.

Taken together, these data suggest that bone marrow cells may be electively useful, primarily for inducing *angiogenesis,* when transplanted into ischemic - and thus still living - tissue that harbors the appropriate signals for inducing lineage switching and/or reversibly injured cells that could be rescued by the increased vascularization triggered by the graft-secreted angiogenic growth factors. Conversely, skeletal myoblasts appear to be better suited for inducing *myogenesis* following engraftment into postinfarction fibrous scars, and restoring, at least partially, their functionality. Should this paradigm be validated by further studies, its clinical correlate would be that bone marrow-derived cells are best indicated in the setting of *acute coronary syndromes,* whereas skeletal myoblasts remain more effective cell substitutes in the late stage of *chronic heart failure.*

Analysis of Clinical Trials

As indicated above, in spite of the many still unresolved issues, the potential of bone marrow cells immediately generated much enthusiasm, which promptly translated into a flurry of clinical studies. Some of these studies involved intraoperative injections of total unfractionated bone marrow or $CD133^+$ progenitors concomitantly with coronary artery bypass grafting. In most trials, however, cell delivery was achieved percutaneously, using three different approaches. In patients with acute myocardial infarction, bone marrow was injected directly into the affected coronary artery shortly after it had been reopened by balloon angioplasty and stenting (B.E. Strauer et al., 2002; B. Assmus et al., 2002).

In patients seen at a later, more chronic stage, cells were delivered transendocardially (E.C. Perin et al., 2003) or transvenously through a coronary sinus catheter. This device has an extendable needle, which, under echocardiographic guidance, perforates the venous wall, and a microcatheter is then advanced into the postinfarction scar tissue where cells are dropped (C.A. Thomson et al., 2002). Finally, in patients with refractory ischemia, mononuclear cells were injected through an endoventricular catheter after electromagnetic mapping (H.F. Tse et al., 2003), with the primary objective of increasing angiogenesis and relieving ischemic symptoms.

Although these trials, as well as those still under way, should be considered as preliminary, several clinically relevant lessons can nevertheless be drawn. First, the procedure appears technically feasible regardless of the route of cell delivery. Second, the precise phenotypic characterization of the injected cells is overall rather poor and usually limited to the identification of $CD34^+$ progenitors, the percentage of which was not unexpectedly found to be very low (no more than 2%). Third, the technique, on the whole, appears to be safe. There were no apparent complications following iliac crest biopsies, nor have any posttransplantation arrhythmias been documented. However, these trials have not attempted to increase the number of progenitors by in vivo pharmacological manipulations, and ongoing studies which now involve cytokine-induced stem cell mobilization will be necessary close monitoring of their safety record. Fourth, all these studies have reported improved perfusion and function following bone marrow cell transplantation, to an apparently similar level whether cells are collected from bone marrow or peripheral blood (B. Assmus et al., 2002) and, more surprisingly, whatever the number of injected cells, which is strikingly variable from one trial to the other and even within the same trial (9 to 28 x 10^6 and 10 to 245 x 10^6) (B.E. Strauer et al., 2002; B. Assmus et al., 2002). However, in the absence of concurrent control groups and randomization, the robustness of these data is arguable and only phase 2 studies designed and powered to show efficacy, if any, will allow a definite conclusion. Importantly, this efficacy is not only dependent on cell type, but also on the method for transferring cells. Preclinical data are still scarce in this respect and additional animal experiments are required to assess which of the percutaneous approaches (intracoronary, transvenous, endoventricular) are the safest, the easiest to implement clinically, and the most effective in optimizing cell functionality, retention, and long-term engraftment.

Embryonic Stem Cells

These pluripotent cells are conceptually attractive because their capacity to generate all differentiated cell types should allow them to truly regenerate infarcted myocardium. These cells can be derived from fertilized oocytes that are no longer targeted for childbearing. This involves differen tiating them into cardiomyocytes prior to engraftment into injured areas (K.R. Boheler et al., 2002). However, several major issues remain to be addressed before considering potential clinical applications. These include: (1) identification of the factors committing embryonic stem cells to a cardiomyogenic lineage and the phenotypic characterization of the resulting cells (i.e., ventricular vs atrial vs pacemaker cells); (2) demonstration of their engraftment potential into infarcted myocardium along with establishment of electrical junctions with host cardiomyocytes and functional efficacy; (3) checking that this engraftment is not associated with an uncontrolled proliferation leading to teratomas; and (4) assessment of their immunogenic potential (J.S. Odorico et al., 2001).

Meanwhile, preliminary experimental studies have yielded encouraging results. Transplantation of mouse embryonic stem cells into infarcted areas in nonimmunosuppressed rats has been associated with differentiation into cardiomyocytes, increased angiogenesis, and improved survival and left ventricular function for up to 32 weeks after transplantation (J.Y. Min et al., 2003). Notwithstanding the ethical and political debate around embryonic stem cells, it thus seems important to pursue research in this direction to avoid missing a potentially effective means of achieving cell replacement therapy.

Clinical studies should be carefully designed to assess the safety and efficacy of cell therapy, taking great care that their methodology complies with the guidelines adopted for drug trials, as this is a prerequisite for them to yield clinically meaningful data. We will probably have to wait another 3 to 5 years before knowing whether the high expectations currently raised by myocardial cell replacement therapy are indeed borne out by actual outcomes in patients.

Stimulation of postnatal neovascularization is an important therapeutic option to rescue tissue from critical ischemia. After the discovery of growth factors promoting the migration, proliferation, and tube-forming activity of endothelial cells, recombinant growth factors or genes encoding for growth factors were used to improve tissue neovascularization.

In spite of early enthusiasm, intravascular therapy with the recombinant angiogenic growth factors basic fibroblast growth factor or endothelial cells factor has been ineffective in placebo-controlled trials (S. Yla-Herttula, K. Alitalo, 2003). Some of the studies using intravascular protein delivery may have been unsuccessful due to their short half-life. Indeed, in animal studies, repeated or sustained infusion of growth factor protein was required to accelerate collateral growth. In contrast, gene therapy transducing the vascular endothelium and/or myocardium may enable sustained production of the necessary angiogenic proteins. Recent double-blind, randomized, placebo-controlled trials showed favorable anti-ischemic effects using either catheter-based intramyocardial injection of plasmids encoding endothelial cells factor (D.W. Losordo et al., 2002) or adenoviral delivery of endothelial cells factor via intracoronary infusion (C.L. Grines et al., 2002). A potential improvement may depend on type of administration (to achieve long-term expression of the gene of interest), dosage, and selection of those patient groups likely to respond to therapy (S. Yla-Herttula, K. Alitalo, 2003). Moreover, some of the adverse events could be overcome by using better vectors for gene delivery.

The Future: Combination of Cell and Growth Factor Therapy?

Cell therapy and gene therapy have proven effective in promoting neovascularization in various animal models. Moreover, some clinical studies have yielded intriguing results. Although definitive proof from large randomized trials is not yet available, both strategies may be able to treat myocardial or peripheral ischemia by improving the formation of new blood vessels. Combinations of these two strategies may have additional advantages.

References

1. Beltrami A.P., Urbanek K., Kajstura J. et al. (2001) Evidence that human myiocytes divide after myocardial infarction. N Engl J Med; 344: 1750-1757.
2. Muller-Ehmsen J., Peterson K.L., Kedes L. et al. (2002) Long term survival of transplanted neonatal rat cardiomyocytes after myocardial infarction and effect on cardiac function. Circulation; 105: 1720-1726.
3. Murry C.M., Reinecke H. (2003) How can cellular grafts be kept alive and synchronized with the rest of the heart? Dialog Cardiovasc Med; 8: 143-147.
4. Zhang M., Methot D., Poppa V. et al. (2001) Cardiomyocyte grafting for cardiac repair: graft cell death and anti-death strategies. J Mol Cell Cardiol; 33: 907-921.
5. Reinecke H., Poppa V., Murry C.E. (2002) Sceletal muscle stem cells do not trasdifferentiate into cardiomyocytes after cardiac grafting. J Mol Cell Cardiol; 34: 241-249.
6. Jain M., DerSimonian H., Brenner D.A. et al. (2001) Cell therapy attentuates deleterious ventricular remodeling and improves cardiac perfomance after myocardial infarction. Circulation; 103: 1920-1927.
7. Menasche P., Hagege A.A., Vilquin J.T. et al. (2003) Autologus skeletal myoblast transplantation for severe postinfarction left ventricular dysfunction. J Am Coll Cardiol; 41: 1078-1083.
8. Leobon B., Garcin I., Vilquin J.T. et al. (2002) Do engrafted skeletal mioblasts contract in infarcted myocardium? Circulation; 106 (suppl II): II549. Abstract.
9. Hagege A.A., Carrion C., Menasche P. et al. (2003) Autologus skeletal myoblast grafting in ischemic cardiomyopathy. Clinical validation of long-term cell viability and dufferentiation. Lancet; 361: 491-492.
10. Pagani F., DerSimonian H., Zawadska A. et al. (2003) Autologus skeletal myoblasts transplanted to ischemia damaged myocardium in humans. J Am Coll Cardiol; 41: 879-888.
11. Reinecke H., Poppa V., Murry C.E. (2002) Skeletal muscle stem cells do not transdifferentiate into cardiomyocytes after cardiac grafting. J Mol Cell Cardiol; 34: 241-249.
12. Anversa P., Nadal-Ginard B. (2002) Myocyte renewal and ventricular remodelling. Nature; 415: 240-243.
13. Tateishi-Yuyama E., Matsubara H., Murohara T. et al. (2002) Therapeutic Angiogenesis using Cell Transplantation (TACT) Study Investigators. Therapeutic angiogenesis for patients with limb ischaemia by autologus transplantation of bone-marrow cells: a pilot study and a randomised controlled trial. Lancet; 360: 427-435.
14. Bel A., Messas E., Agbulut O. et al. (2002) Transplantation of autologus fresh bone marrow cells into infarcted myocardium: a word of caution. Circulation; 106 (suppl II): II463. Abstract.

15. Verfaillie C.M. (2002) Adult stem cells: assesing the case for pluripotency. Trends Cell Biol; 12: 502-508.
16. Orlic D., Kajstura J., Chimenti S. et al. (2001) Mobilized bone marrow cells repair the infarcted heart, improving function and survival. Proc Natl Acad Sci USA; 98: 10344-10349.
17. Anderson D.J., Gage F.H., Weissman I.L. (2001) Can stem cells cross lineage boundaries? Nat Med; 7: 393-395.
18. Medvinsky A., Smith A. (2003) Fusion brings down barriers. Nature; 422: 823-825.
19. Rehman J., Li J., Orschell C.M., March K.L. (2003) Peripheral blood "endothelial progenitor cells" are derived from monocytes/macrophages and secrete angiogenic growth factors. Circulation; 107: 1164-1169.
20. Badorff C., Brandes R.P., Popp R. et al. (2003) Trasdifferentiation of blood-derived human adult endothelial progenitor cells into functionally active cardiomyocytes. Circulation; 107: 1124-1132.
21. Condorelli G., Borello U., De Angelis L. et al. (2001) Cardiomyocytes induce endothelial cells in tras-differentiate into cardiac muscle: implications for myocardium regeneration. Proc Natl Acad Sci USA; 98: 10733-10738.
22. Shake J.G., Gruber P.J., Baumgartner W.A. et al. (2002) Mesenchymal stem cells implantation in a swine myocardial infarct model: engraftment and functional effects. Ann Thorac Surg; 73: 1919-1926.
23. Kraitchman D.L., Heldman A.W., Atalar E. et al. (2003) In vivo magnetic resonance imaging of mesenchymal stem cells in myocardial infarction. Circulation; 107: 2290-2293.
24. Aicher A., Brenner W., Zuhayra M. et al. (2003) Assessment of the tissue distribution of transplanted human endothelial progenitor cells by radioactive labeling. Circulation; 107: 2134-2139.
25. Jackson K.A., Majka S.M., Wang H. et al. (2001) Regeneration of ischemic cardiac muscle and vascular endothelium by adult stem cells. J Clin Invest; 107: 1395-1402.
26. Strauer B.E., Brehm M., Zeus T. et al. (2002) Repair of infarcted myocardium by autologous intracoronary mononuclear bone marrow cell transplantation in humans. Circulation; 106: 1913-1918.
27. Assmus B., Schachinger V., Teupe C. et al. (2002) Transplantation of progenitor cells and regeneration enhancement in acute myocardial infarction. (TOPCARE-AMI). Circulation; 106: 3009-3017.
28. Perin E.C., Dohmann H.F.R., Borojevic R. et al. (2003) Transendocardial, autologous bone marrow cell transplantation for severe, chronic ischemic heart failure. Circulation; 107: 2294-2302.
29. Thomson C.A., Nasseri B.A., Makower J. et al. (2002) Percutaneous transvenous cellular cardiomyoplasty: a novel nonsurgical approach for myocardial cell transplantation. J Am Coll Cardiol; 39 (suppl A): 75A. Abstract.
30. Tse H.F., Kwong Y.M., Chan J.K.F. et al. (2003) Angiogenesis in ischaemic myocardium by intramyocardial autologous bone marrow mononuclear cell implantation. Lancet; 361: 47-49.

31. Boheler K.R., Czyz J., Tweedie D. et al. (2002) Differentiation of pluripotent embryonic stem cells into cardiomyocytes. Circ Res; 91: 189-201.
32. Odorico J.S., Kaufman D.S., Thomson J.A. (2001) Multilineage differentiation from human embryonic stem cells lines. Stem Cells; 19: 193-204.
33. Min J.Y., Yang Y., Sullivan M.F. et al. (2003) Long-term improvement of cardiac function in rats after infarction by transplantation of embryonic stem cells. J Thorac Cardiovasc Surg; 125: 361-369.
34. Yla-Herttula S., Alitalo K. (2003) Gene transfer as a tool to induce therapeutic vascular growth. Nat Med; 9: 694-701.
35. Losordo D.W., Vale P.R., Hendel R.C. et al. (2002) Phase 1/2 placebo-controlled, double-blind, dose-escalating trial of myocardial vascular endothelial growth factor 2 gene transfer by catheter delivery in patients with chronic myocardial ischemia. Circulation; 105: 2012-2018.
36. Grines C.L., Watkins M.W., Helmer G. et al. (2002) Angiogenic Gene Therapy (AGENT) trial in patients with stable angina pectoris. Circulation; 105: 1291-1297.

Cancer Problem in the Eyes of the Skin Multiparameter Electrophysiological Imaging

Yuriy F. Babich

Center of Biomedical Engineering, Kurska 12A/38, Kiev 03049, Ukraine, babich@ua.fm

The skin multiparameter electrophysiological imaging (SMEI) is a new functional and informational imaging modality of wide application. SMEI enables non-invasive high-resolution visualization/monitoring of the skin (and sub-skin tissues) in a set of electrical parameters and, thus, provides information on metabolic and inter/intra-cellular processes. With the aid of SMEI, a novel class of spatial-temporal phenomena of the skin electrical landscape (SEL) have been *in vivo* discovered, specifically initial and EMF-induced wave-like and pulsating structures with a speed of calcium waves.

By a demonstrative example of cutaneous melanoma, we present some findings of our current project on development of a novel criteria of early cancer-diagnostics based on the multiparameter analysis of the SEL dynamic abnormalities. In order to contrast the tumour, established its malignant geometry and reveal invisible distant suspicious abnormalities, exposures of low intensive mm waves and magnetic field (MF) were used.

Keywords: Cancer, skin, melanoma, functional imaging, dynamic systems, electrical impedance, electrical potential, calcium, electromagnetic field (EMF), EMF-bioeffects.

1. Introduction

Over some decades, the incidence of cancer has escalated to epidemic proportions, now striking nearly one in two men (44%) and more than one in three women (39%). This increase translates into approximately 56% more cancer in men and 22% more cancer in women over the case of a single generation. As admitted by recent National Institute of Cancer (USA) and

V. Burdyuzha (ed.), The Future of Life and the Future of Our Civilization, 307–320.

ACS estimates, the number of cancer cases will increase still further, dramatically doubling by 2050 [1]. Yet ancient philosophers knew, that one might be cured, if only the disease got to know. But "neither classical x-ray mammography, nor any other technologies of medical images can distinguish malignant tumour from benign one, - and, sometimes - with sufficient confidence, - even from a normal tissue which has in consequence high percentage of false-positive and false-negative diagnoses. The known medical systems of images are also very inefficient in detection of micrometastasises or an early stage of relapse" [2].

The idea of cancer identification by its electrical properties is not a new idea - it was firstly proposed by Fricke and Morse in 1926 [3]. The literature reports many measurements of the impedance of cancerous tissues in comparison with normal ones [4-5]. Static electric parameters of tumours essentially differ from normal tissues. For example, the electrical conductivity and permittivity of cancerous tissue has been found to be significantly greater than the electrical conductivity and permittivity of normal tissues [6,7]. Every biological process is also an electric process and "health and sickness are related to the bio-electric currents in our body" [8]. Cells are electromagnetic in nature, they generate their own electromagnetic fields and they also harness external electromagnetic energy of the right wavelength and strength to communicate, control and drive metabolic reactions. As early as 1968, Frohlich [9] hypothesized that non-thermal EMF at GHz frequencies would be excited coherency in the case of biological metabolic processes. Excitations of this type may be significant at the basis of long-range biological interactions, but their identification has posed considerable experimental difficulties. During last decades, a large number of studies reported sharp mm-wave effects (Ca^{2+} efflux, altered enzyme activity, possible interference with apoptosis, etc.) at the weakest effective field strength (e.g., at a level of pW/cm^2 and less). It has been already assumed that EMF first interacts with cell membranes and then initiates a series of intracellular enzyme cascades [10]. Interactions of both electric and magnetic fields with cell membrane components are also well documented. Blank and Goodman [11] model proposes that gene activation by MF could be due to direct interaction with moving electrons within DNA. Clinical evidence points to the significance of DNA repair mechanisms in tumour prevention and tumour initiation. "Numerous experimental works have shown the possibility of modifying and controlling the selective permeability of the cell membrane by transmitting electromagnetic waves. This leads to the possibility of verifying the specific reactions of healthy cells compared to the reactions of pathological cells and subsequently to select target cells on which to act for clinical purposes.

Pathological cells resonate differently from healthy cells due to a different tissue composition [12].

Many experimental findings on tissue cultures, specifically, evidence that:

1. Cancer cells exhibit both lower electrical membrane potentials and lower electrical impedance than normal cells.
2. The reduction in membrane electrical field strength in turn causes alterations in the metabolic functions of the cell.
3. All cancer cells have abnormal electron transfer systems. A normal cell development involves normal energy flows [14].

The tumour determinative quality is a disordered mechanisms of intra/intercellular signalling. In general terms, most benign tumour cells keep certain level of functional intercellular communication via gap junctions, but malignant cells lose the functional ties both between themselves and with normal cells [15, 16].

4. Cells become independent of normal tissue signalling and growth control mechanisms. In a sense cancer cells have become desynchronised ones from the rest cells of a body [13].
5. Mitochondria as dynamic intracellular organelles play a central role in oxidative metabolism and apoptosis. In carcinogenesis, the mitochondrial membrane potentials undergo significant changes [17].

On the other hand, certain chemicals, viruses and bacteria create cancers by modifying the electrical charge of the cell surface. It also means that therapeutic methods which manipulate the electrical charge of cell membranes can result alterations in genetic activity. Therefore, a key component of cell repair and cancer treatment would be to reestablish a healthy membrane potential in the body's cells [13].

A natural question arises: how to explore these distinctive features for development more effective approaches to cancer earlier diagnosis and more directed treatment? During last decades a few methods of electrical bioimpedance tomography [4, 18-20] and electropotential [21] measurements have been introduced to an area of breast tumour diagnosis. They used some distinctive features of electric parameters of normal and tumorous tissues but they made certain steps in investigation *only the first of the five above mentioned distinctive features*. Differing by relative cheapness, this approach was acknowledged as a useful complementary mean for conventional X-ray mammography. At the same time, methods of functional/metabolic imaging (e.g. PET) ensure more sensitive and specific diagnostics, but much more expensive. However, as it was cited above [1, 2], neither of these and other known technologies did not meet the current and ever growing problems of early oncodiagnostics and, thus, effective therapy was absent.

In order to co-ordinate/synchronize all the functions of the integrated organism, its cells have to communicate, specifically via so-called "gap junctions" between cells (which, incidentally, were firstly discovered with the aid of electrophysiological measurements). In healthy tissues, fluctuation of local Ca^{2+} concentration (the key second messenger of intra/ intercellular signaling) and, correspondingly, variations of specific local conductivity may have a range of 3 orders and, in principle, could be in vivo registered at tissue level with the aid of conventional electrophysiological means. There is a remark - if one knew the exact time and venue of the event like in case of *in vitro* observations then conventional electroimpedance scanners cannot be used (mainly due to their limited spatial resolution) for obtaining coherent, non-fragmental images of biological tissue electrical 2-3D patterns neither in static, nor – the more so - in dynamics.

So, if we had a proper imaging technique for *in vivo* detecting and monitoring functional electrical abnormalities, i.e. yet at the level of informational disorders, it would become possible to develop both: basically earlier tumour diagnosis and effective – controlled therapy.

Our own research into the subject started in the middle of the 80-es, after an original scanner of high-enough resolution for *in vivo* non-destructive visualization of the Skin Electrical Landscape (SEL) has been developed [22] and first detailed/coherent SEL images have been obtained.

The reasons for choosing up the skin were the following. On the one hand, skin is the most appropriate model organ for ***in vivo*** studying some of the above-mentioned characteristics. On the other hand, the skin as the biggest and a multi-functional organ is of high peculiar interest, in particular:

- Traditionally, physicians regarded the skin as a mirror of the interior body, including internal malignancies;
- The skin involves a variety of sensory fields (in ontogenesis, origins from the same embrionic leaf as eyes and nervous system);
- The skin as a boundary (not a wall!) between self and the environment. (*Note*: Boundaries bring order to our lives);
- Eastern medicine alleges existence of invisible singular objects at the skin - so-called "accupuncture points (AP) and channels". (AP, as it was lately found, specifically differ from adjacent tissue by significantly more marked intercellular ties (i.e.: 4-7-fold time larger concentration of cells with gap-junctions [23]);
- And, metaphysically, even so: "~The deepest entity in us is the skin, we are all living at the skin surface" (Paul Valery).

In a sequence of earlier lab-clinical studies, which were carried out with the aid of the developed experimental devices, a set of novel spatial-temporal phenomena have been shortly revealed [24-26], specifically: initial low-conductivity autowave (soliton) activity at the AP areas; induced high-conductivity wave-like directional activity at the AP areas in response to local mechanical or remote exposure to non-thermal EMF in mm range of length waves (mm-EMF); pulsating structures, etc.

Characteristic speed of these wave-like structures was up to 7-10 mm/min- just in harmony with the diffusion conception of Ca^{2+} waves and those observed *in vitro* [27]. In order to further investigate the SEL phenomenological features and their diagnostic significance, a more sophisticated set-up has been developed.

2. Methods and Materials

The main goal of our current project was finding out some basically earlier signs of tumour malignization based on the analysis of the skin 2D electrodynamic features. To this end, the project implied also carrying out of the following preliminary steps: (i) development of upgraded imaging technology (hard and software) for visualizing the skin (and sub-skin tissues), which would enable more multi-parameter and detailed dynamic imaging/monitoring the information and metabolic inter/intra-cellular processes; (ii) study of the SEL phenomenological features in health and some relevant pathologies; (iii) selection of proper test stimuli in order to contrast the area of malignization, etc. Mostly, as the test stimuli, we used weak electromagnetic exposure of therapeutic intensity, i.e. constant and reversible magnetic field (MF+, MF-), 1-4 min of ~1-10 mT, the mm-EMF of extremely low intensity (1-4 min, 50-70 GHz, 1-100 mkW/sm^2). Some other test stimuli like mild irritants (e.g. 0.1% sol. of nicotinic acid or X-ray contrast agents) were also used in experiments with allergic subjects. A breath-holding test was of particular attention in the study of SEL dynamics at tumour areas.

An updated set-up of SMEI was made as a portable scanner, which provides non-invasive imaging of the SEL via matrix of 32 × 64 needle electrodes (bipolar method) - spatial resolution at the skin surface - 1.0 mm; measurement time interval – 4 ms/pixel; 6 electrical parameters at the frequency band 2 kHz ÷ 1 MHz (measuring current – 1 ÷ 20 μA): (i) Z_{kHz}- full electric impedance at the kHz-band; (ii) Z_{MHz}- full electric impedance at the MHz-band; (iii) φ_{2kHz} - phase angle (angle between the test current and voltage drop) at the kHz-band; (iv) φ_{MHz} - phase angle at the MHz-band; (v) E_{kHz} - electrical potential at the kHz-band (i.e. ≡ asymmetry of Z_{kHz}); (vi) E_{MHz} - electrical potential at the MHz-band (i.e. ≡ asymmetry of Z_{MHz}).

Knowledge of the data in each point of a scanning area enables to construct correspondingly both - the SEL images and calculated images, i.e.: active and capacitance components of a full impedance - $R_{кГц} = Z_{kHz} \times \sin\varphi$ (or conductivity $G = 1/R$) and $C_{kHz} = Z_{kHz} \times \cos\varphi$, which more adequately correspond to electrochemical parameters of the tissue level, which can be further subdivided into intra and intracellular levels. Specifically, the SEL data at the kHz-band mainly reflects state of cell membranes and intercellular media. At 1 MHz, cell membranes are transparent for a testing current, therefore the matrix of Z_{MHz} values provides opportunity to receive relative estimations of a spatial temporal distribution of the electrochemical characteristics of the intracellular media. Similarly, the E_{kHz} and E_{MHz} data provides additional information on cell and mitochondrial membranes, respectively. SMEI enables visualization and comparative analysis of all these parameters at every point of the scanning zone. Besides, the MHz-data gives some insight about deeper tissue layers.

The developed software for processing and presentation of primary and secondary arrays of information enables obtaining on-line rough visual and statistical analysis of the SEL dynamics. Later thorough through-frame-sequence analysis of the SEL' initial and induced features may include constructions of: difference mappings, detailed dispersion maps (σ), inter-frame autocorrelation functions $\boldsymbol{r}$ of the scan-area, histograms dynamics, plots of any SEL points and their phase planes, etc.

In order to evaluate the ultimate SMEI sensitivity, a number of experiments with such a low-contrast object as jellyfish and sea-comb have been carried out. Electrical conductivity of jellyfish is almost equal to that of the seawater, i.e. ~ 96-99% (we made measurements in a jellyfish' natural environment). Despite such low-contrast, it turned out possible to reveal the SEL of the object both in static and dynamics (e.g. in response to weak MF, which was particularly marked in the φ_{2kHz} patterns, i.e. in cell membranes. The latter is hypothesized to be the primary target of the EMF exposure [10].

Lay-out of a typical experiment. The area of visible melanoma (Fig. 1) was wetted with a hypoallergenic electroconductive gel (e.g. "Spectra-360"); then, the scan-head was placed onto the skin and lock in there to the end. The whole SEL film consists of: (i) a 6-screened sets of patterns taken mainly with a scan-time of 4 s for the left – melanoma's - half of the scan-area (in order to pick up the high-speed SEL events), and (ii) an additional set of patterns - with an 8 s scan-time for the full area ($32 \times 64\text{mm}^2$). The main part of the film is 8 series long (i.e. about 8 min), each of them consists of 15 shots (each of them consists of the 6 patterns). The 8 series correspond to the following stages:

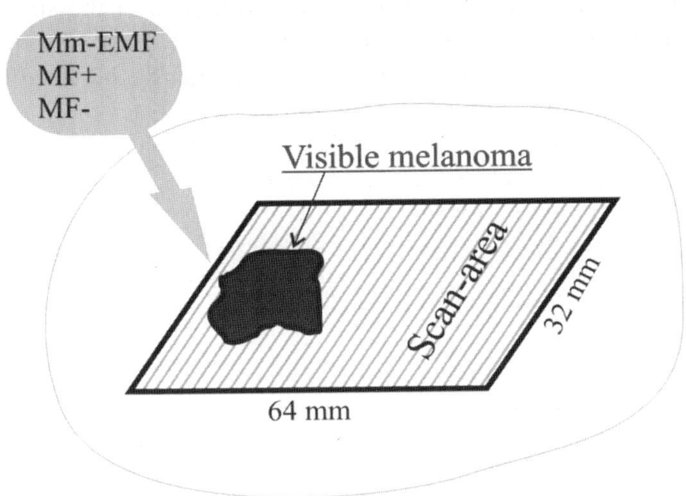

Fig. 1. Schematic sketch of the melanoma experiment.

1. No influence - in order to make certain that adaptation process is finished, as well as to assess the SEL initial dynamics; Application of mm-EMF exposure #1 (noise generator 50-80 GHz, ~5.10^{-15} Wt/cm^2, in immediate proximity to the scan-area);
2. No influence - reading out next 15 frames in order to assess the mm-EMF#1 after-effect;
3. Application of mm-EMF exposure #2 - same place, same generator, but of about 1 order higher output power than mm-EMF #1;
4. No influence;
5. Application of a weak constant magnetic field (MF$^+$) (of about 10-100 mT at the skin surface; in order to insure more homogeneous field, the polar tip was chosen 38 mm, i.e. somewhat broader, than the scan-area) to the same place at the scanner-head edge;
6. Application of a weak constant magnetic field (MF$^-$), videlicet, a fast turn over of MF$^+$ (the opposite polar tip to the same place) and reading out next 15 frames;
7. No influence.

3. Results

One of the useful methods of primary image processing is correlation analysis, which provides information in real time of the SEL dynamics.

Fig. 2a demonstrates the through-sequence dynamics of the six SEL patterns in a form of inter-frame-correlation function **r**. Here, the correlation function **r** was calculated for the whole 32×32 matrix between frame #0 and, successively, all the following 114 ones.

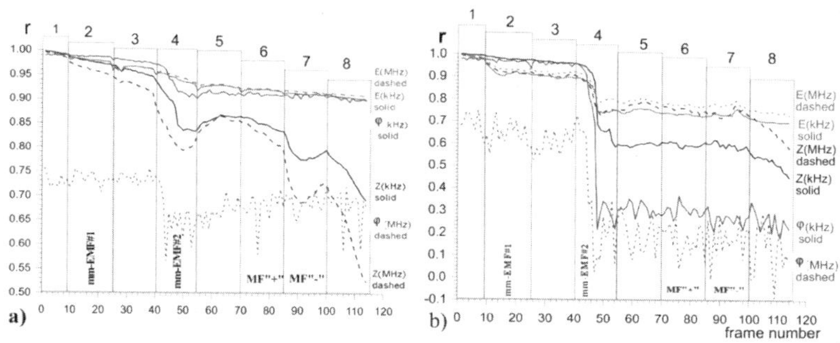

Fig. 2 a, b. Dynamics of all SEL patterns in a form of the through-sequence inter-frame correlation function **r**: a) for the whole melanoma area 32 × 32 mm (see Fig. 1); b) for only the hypersensitive area 13 × 15 mm neighbouring the visible melanoma boundary (see Fig. 3).

Ex facte, it looks like the most significant effects have happened only in response to the mm-EMF#2 and the MF' turn-over (stages 4 and 7). In order to shed more light on the curve' trend, let us ask ourselves a question – which parts of the scan-area have made most contribution to the SEL dynamics? The visible/photo image of the full scan-area is presented at Fig. 3.

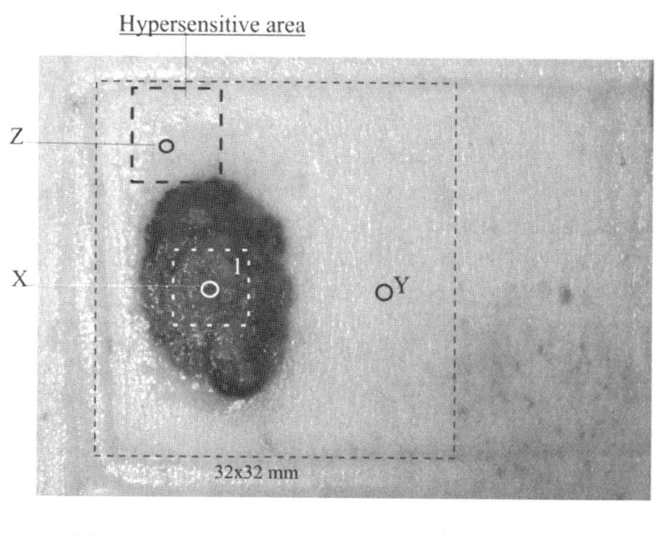

Fig. 3. A photo of the scan-area 32 × 64 mm. Pressing mark on the skin was left by the scanner-head. The hyperactive area (2) is indicated in accordance with the φ_{2kHz}- dispersion field (Fig. 4). X, Y, Z – arbitrary points chosen for phase portrait analysis (Fig. 4).

The average φ_{2kHz}-pattern (Fig. 4a), i.e. a SEL stationary representation of the 4th set of patterns, does not provide much information; it shows solely a noticeably abnormal area of higher φ_{2kHz}-values spreading well outside the boundary of visible melanoma. The corresponding functional map - the φ_{2kHz}-dispersion field of the same set (Fig. 4b) - is more informative: it displays a marked hypersensitive area of σ_{max} outside the range of the melanoma upper boundary. Another functional map, i.e. a pattern of response synchronization with regard to a melanoma boundary point (Fig. 4c), reveals not only this hypersensitive area, but also its antipodal connected environment, which noteworthy coincides with the melanoma' interior.

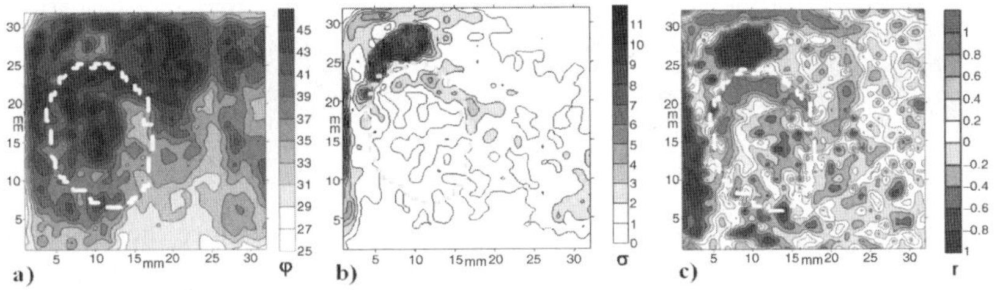

Fig. 4. φ_{2kHz}- patterns of the melanoma part (32 × 32 mm); the dashed line shows the melanoma visual boundary: a) an average φ_{2kHz} pattern of the 4th set; b) the φ_{2kHz} - dispersion field (σ in degrees) for the same stage (frames ##41-55 read out during the mm-EMF#2 test (Fig. 2)); c) a pattern of response synchronization with regard to a melanoma boundary point 10 × 22mm (marked "*"), i.e. a field of correlation function **r** between time response of the chosen point and similar graphs of all the other points. The hyperactive area neighbouring the upper boundary of visible melanoma clearly stands out against the backgrounds of b, c).

In order to assess what is going on in the hyperactive area, a φ_{2kHz}- pattern train of this area, starting from the last frame of the previous set #3 (no influence, frame #40, Fig. 2), is presented at Fig. 5. Here, the central high-magnitude structure performed a wave-like spreading with a speed up to ~10-15 mm/min starting with about 20s delay after the mm-EMF#2 was on. The speed of disappearance of the propagating structure was much faster.

Another interesting phenomenon, i.e.: the spontaneous point-like bursts of the SEL activity (supposed Ca^{2+} efflux) can be seen at the impedance dispersion fields, e.g. at those of Z_{2kHz} (Fig. 6).

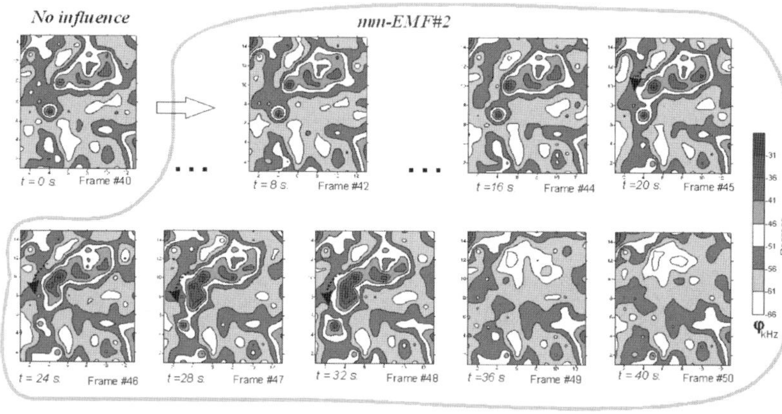

Fig. 5. A φ_{2kHz} pattern train: frame #40 was read out just before the mm-EMF#2 applied; frames ##42-50 were read out during the mm-EMF#2-test (a set #4, Fig. 2). (Frames #41 and #43 were omitted for the sake of simplicity and because of their similarity with the neighbouring ones). A response delay was about 16s – noticeable changes started at frame #45. Arrows show propagation of the tractile structure leading edge with a speed up to ~10-15 mm/min. The structure disappeared after the frame #48 faster than in 4s.

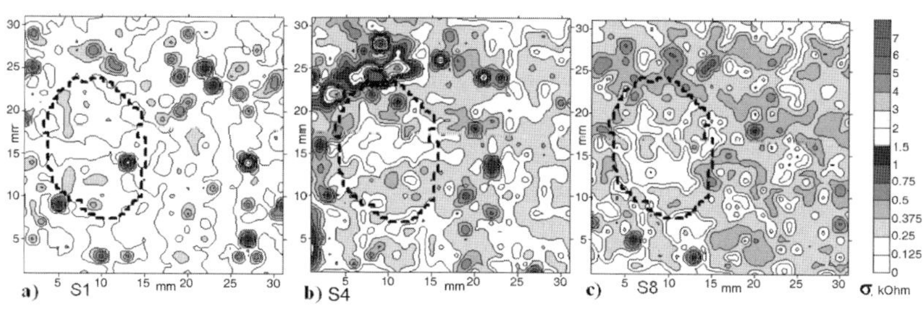

Fig. 6. Z_{2kHz}- dispersion fields of the melanoma part (32 × 32 mm) for sets #1, 4, 8.

Figure 6 shows: a) the 1st set - comparatively low-magnitude dynamics of a dozen of the bursts of chaotic topology, no signs of melanoma; b) set #4 - about an order greater dynamics at the melanoma boundary and noticeable circumference of bursts around it, marked area of $\sigma = 0$ inside melanoma; c) set #8 – significant level of the field (after the MF exposures), marked area of zero dynamics ($\sigma = 0\text{-}0.125$ kOhm) inside melanoma. Interestingly, at these sets (as well as at all those non-shown ones), the bursts don't appear inside the melanoma outline deeper than that at Fig. 6a.

Similar metabolic bursts were also observed at the E_{2kHz}-dispersion fields, but as distinct from the Z_{2kHz} patterns, maximal EMF-induced activity – gradually increasing with all the tests "on" and decreasing with "off" - was observed exactly in the middle of the melanoma area (Fig. 7a-b).

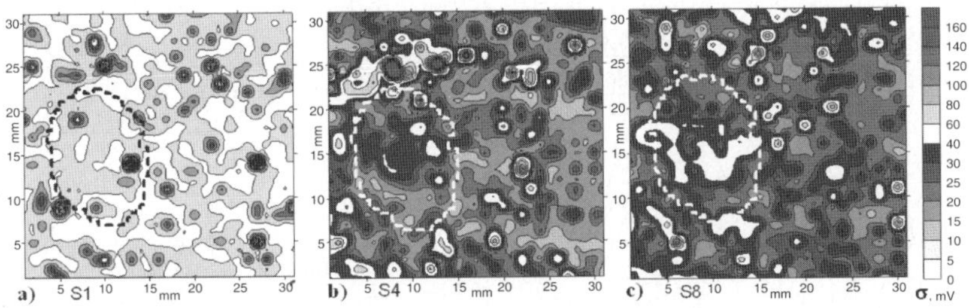

Fig. 7. E_{2kHz}-dispersion fields for sets #1, 4, 8 (compare with Fig. 6).

In both pattern trains (Fig. 6, 7):

the melanoma is indistinguishable by the SEL initial dynamics (S1);

the background' and particularly the bursts' magnitude grew up in about 3-10-fold time, correspondingly, and the upper-range bursts noticeably got enclosed the tumour in response to the mm-EMF#2 (S4).

The E_{2kHz} background dynamics (i.e. extra-cellular polarization processes) still kept growing in contrast with that of Z_{2kHz} (e.g. fluctuations of intercellular pH)(S8).

One more way to characterize the SEL patterns is phase-portrait analysis of each or at least some of more typical pixels like "X',"Y" and "Z' (Fig. 3). Figure 8 demonstrates through-sequence (1-8 stages, Fig. 2) behaviour of these points in immittance (*g*) vs susceptance (*b*) domain at 2kHz (where $b = (1/Z_{2kHz}) \times \sin\varphi$ and $g = (1/Z_{2kHz}) \times \cos\varphi$). Here, one can assess noticeable difference between the three points:

point X (inside melanoma) demonstrates 4 adjacent clusters, gradually increasing values but of (on average) stable φ character;

point Y (outside melanoma) develops itself as only 1-cluster, but with noticeably different φ;

point Z (active zone) is of most interesting behaviour. It consists of 3 – rather spaced clusters with φ similar to that of X. Cluster A shows initially reversed (to X) trend and higher sensitivity to the mm-EMF#1 exposure, i.e. a sound single Δφ response at the 2[d] stage (pre-membrane reaction), which though has not let to trigger transition into next cluster, like it happened afterwards at stage #4. During the latter and the following 5th stage, point X makes a fast transition from A to C cluster (10-fold increase of the magnitudes). Then, *in response to the MF-turnover*, it makes a magnificent jump backward to the B-cluster

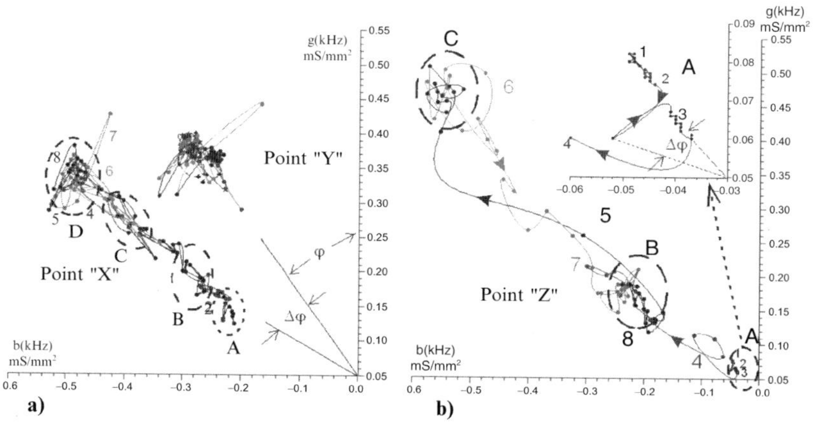

Fig. 8. Phase-portraits of points X, Y, Z (Fig. 3) in immittance (g) vs susceptance (b) domain at 2kHz. Numbers 1…8 - correspond to the 1...8 stages (Fig. 2). Letters A, B, C, D indicate clusters with some features of strange attractors (dashed ellipses).

4. Conclusion

Due to the paper limits a part of our findings in the SEL phenomenology was presented here.

We believe they may be useful as for aforesaid oncological problems, as for another broad not-far-distant field of EMF-bioeffects. Specifically, the SMEI technology enables us to *in vivo* reveal and monitor tumour/ malignancy geometry, particularly tumour microenvironment [28]. At the same time, SMEI provides a possibility to develop a very sensitive real-time biofeedback and thus to create a novel basis for tumour therapy and surgery.

The SEL heterogeneity inside melanoma outline was confirmed by histological analysis, i.e. some breaks in malignant cells were confirmed. But histology was unable to confirm other suspicious SEL abnormalities at or outside the melanoma boundary. Identification of the SEL puzzles with the aid of functional/metabolic 2-3D imaging modalities should be a most important subject of further SEL studies.

References

1. Cancer Prevention Coalition, University of Illinois at Chicago, School of Public Health, http://www.preventcancer.com/losing/nci/manipulates.htm.
2. Report of National Institute of Cancer, (1999) USA.
3. Fricke H. and Morse S. (1926) The electrical capacity of tumours of the breast. *J. Cancer Res.* 10, 340-376.
4. *Critical reviews*^{TM} *in Biomedical Engineering*. ed. J. Bourne, (1996) V.24/4-6/, 330-337.
5. Grimnes S., Martinsen O. (2000) *Bioimpedance and bioelectricity basics*, San Diego, California: Academic press.
6. Foster, K. R. and Schwan, H. P. (1996) "Dielectric properties of tissues". In: *Handbook of Biological Effects of Electromagnetic Fields*, 2nd ed., Boca Ratin: CRC Press, 25-102.
7. Jossinet, J. and Schmittm M. (1999) "A review of parameters for the bioelectrical characterization of breast tissue", Ann.New-York Acad. Sci., 873, 30-41.
8. Pekar R. (1997) Percutaneous Bio-Electrotherapy of Cancerous Tumours: A Documentation of Basic Principles and Experiences with Bio-Electrotherapy. Munich, Germany: Verlag Wilhelm Maudrich.
9. Fröhlich, H. (1968) "Long-Range Coherence and Energy Storage in Biological Systems." *International Journal of Quantum Chemistry*, 2.
10. Adey W. R. (1990) Electromagnetic fields, cell membrane amplification and cancer promotion, Extremely Low Frequency Electromagnetic Fields: The question of cancer, Wilson, B. W. Stevens, R.G., and Anderson, L.E., eds, Columbus, Ohio, Battelle Press, 211-250.
11. Blank M., Goodman R. (1999) Electromagnetic fields may act directly on DNA. J Cell Biochem, 75: 369-374.
12. Gorgun S. S. (1998) Studies on the Interaction Between Electromagnetic Fields and Living Matter Neoplastic Cellular Culture. ISSN: 1062-4767, V. 7., N. 2., Fall.
13. Steve Haltiwanger: The Electrical Properties of Cancer Cells. http://www.royalrife.com/haltiwanger1.pdf.
14. Garnett M. First Pulse: (1998) A Personal Journey in Cancer Research. New York, NY: First Pulse Projects.
15. Trosko, J., Chang, C. Madhukar, B. and Dupont, E. (1993) " Role of modulated gap junctional intercellular communication in the tumor promotion/progression phases of carcinogenesis". In: New Frontiers in Cancer Causation. O.Iversen, ed., Taylor and Francis Publishers, Wash. D.C., pp. 181-197.

16. Gap-junction-mediated intercellular signalling in health and disease, (1998) Open Meeting following Novartis Foundation Symposium No. 219.
17. Modica-Napolitano J. S., Keshav K. S. (2002) Mitochondria as targets for detection and treatment of cancer. Exp. Rev. Mol. Med. 11http://www.expert-reviews.org/02004453h.htm.
18. Brown, B. H. and Barber, D. C. (1982) "Applied Potential Tomography - a new in-vivo imaging technique". Proc. of the HPA Annual Conference, Sheffield.
19. Assenheimer M. et al. (2001) The T-SCAN™ technology: electrical impedance as a diagnostic tool for breast cancer detection. Physiol. Meas. 22, 1-8, February.
20. Cherepenin V., Karpov A., Korjenevsky A. (2002) Imaging of Breast Tissues: System Design and Clinical Testing. IEEE Trans. Med. Imaging 21(6): 662-667.
21. "Biofield" test. (1998) Cancer Press Releases, July, 31.
22. Babich, Y. (1992) "Impedanz-Bild (Introscopie) von biologischen Gewebe Verfahren"(in German). DZA. -**35**. #4, 93-97. #5, 103-109.
23. Chernilevski, V., Gudoshnikov, V. et al. (1992) "Possible intercommunication between the acupuncture points/channels and endocrine regulation mechanisms" (in Russian). Physiology of man. -18, #5, p. 171-173.
24. Babich, Y. (1998) "The skin 2D electrobioimpedance in response to a remote non-thermal mm-EMF exposure: phenomenological study". Proc.of 2d Int. Conf. On Bioelectromagnetism. p.79-80 (Melbourne, Australia).
25. Babich, Y. (2000) "Quasistationary and autowave structures of the skin electro-impedance relief". Proc. of Natl. Acad. of Sci. of Ukraine, #4, p. 199-204.
26. Babich, Y., Bakai, E. (2001) "Visualisation of the skin spatial&temporal electrical parameters to reveal their characteristic features in health and disase". Proc. of XI Int. Conf. on Electrical Bio-Impedance. Oslo, Norway. p. 131-134.
27. Berridge, M. J., Bootman, M. D., Lipp, P. (1998) Calcium - a life and death signal. *Nature* 395, 645 - 648.
28. Henning T., Kraus M., Brischwein M., Otto A.M. (2004) "Relevance of tumor microenvironment for progression, therapy and drug development'. Review. *Anti-CancerDrugs,* 15: 7-14.

Perspectives for Quantum Medicine

Volodymyr K. Magas

Departament de Fisica Teorica, Universitat de Valencia, C. Dr. Moliner 50, E-46100, Burjassot (Valencia), Spain.
Volodymyr.Magas@ific.uv.es

Quantum Medicine (QM) and more generally Physics of the Alive is based on the definition of the Alive as a fourth level of quantum organization of Nature (after nuclear, atomic and molecular levels). Each living object is considered as a quantum system with its own characteristic eigenfrequencies in the mm range wavelength, formed in accordance with genome. Such a notion is based on some theoretical considerations as well as some experiments, which allow us both to see the effect of even a few photons of specific (resonance) frequency on the macroscopic alive object and also to measure directly the radiation in the mm range from such an object. QM is using the experimental effect that the human body does not only respond to extremely low level electromagnetic radiation on the very narrow resonance frequencies - the result of such a disturbance is strongly positive! Human body can recover from many diseases (some of these cannot be effectively treated by the traditional medicine) as confirmed by clinical material. This phenomenon is called microwave resonance therapy (MRT) (discovered in 1982 by E.A. Andreev, M.Y. Belyi, S.P. Sitko [1]) and it forms a base of Quantum Medicine. The important property of QM is that MRT principally cannot cause any damage to the human body - its effect is always positive and decreasing as the body comes closer to its best shape.

At the first glance such a picture is not only in contradiction with traditional biology and medicine, but also with our common knowledge from quantum mechanics. In this paper I want to discuss the principal possibility of creation of a macroscopic quantum system such as a human body and other alive organisms from the point of view of quantum mechanics. At the moment Quantum Medicine exist as a phenomenological science and Physics of Alive as a beautiful hypotheses (more philosophical than physical), which aims to explain the observed effects of the interaction of the low energy electromagnetic radiation in the mm range with alive objects. Also I am going to discuss how, in my opinion, some theoretical physical modeling can

V. Burdyuzha (ed.), The Future of Life and the Future of Our Civilization, 321–334.

be done to describe creation of such a macroscopic quantum system through generation of a laser-like coherent electromagnetic eigenfield, and which experimental information would be necessary for such a modeling and for confirmation of the discussed hypothesis in general.

If QM will be confirmed by the future studies, it will certainly become the Medicine of Third Millennium. And also it will make us rethink our ideas about the creation of Life and the evolution of Earth biosphere in general, and about perspectives for human civilization in particular. For example, the expectations from the gene engineering might be strongly overestimated.

Introduction

In this paper I am going to talk about Quantum Medicine (QM), which is a medical application of the new notions about the nature of Life given by Physics of the Alive. What is the physical basis of the Physics of Alive? Nowadays we have a big set of experimental data showing that the interaction of the low energy electromagnetic radiation (EMR) in the mm range with alive body or single cell significantly affects its living activity. This interaction has resonance character and observed resonance in the absorption spectra of the alive objects is much narrower than could be from classical physics (much narrower than thermal spread should be [2]). And as a result of such an interaction even with very low energy EMR (with separate photons [3]) at the resonance frequencies of the human body (or other alive object) show some drastic effects [4]. In particular it can recover from many diseases – this effect gave the birth to the Quantum Medicine (Medical doctors Cherkasov, Nedzvetskii [5] first observed the medical effect of the mm range EMR).

In this work I am neither going to discuss experimental data on the interaction of EMR in the mm range with alive objects, nor its medical effects. Being a theoretical physicist I don't fill competent enough in this field. An interested reader can find more information and corresponding citations in the recent reviews [3, 4]. I only want to mention that the quality of some of these experimental results is still questionable according to some experts in the field, for example, Yu. Babich has presented his results at this Conference.

My goal is to discuss the interpretation given to this data by the Physics of Alive from the point of view of quantum physics. Thus, I accept the experimental effects listed above, as well as those, which will be discussed in the next section, as facts. Namely, I accept that the living object is sensitive to separate photons of the specific resonance frequencies in the mm range with extremely narrow width. This means that the living objects - macroscopic objects! - behave like quantum systems with characteristic energy levels. Can we understand this in some way?

We know that quantum effects sometimes can play an important role in macroscopic world - coherent laser radiation, superconductivity, zone structure of the semiconductors, Mossbauer effect etc. But this happens only under some specific conditions. Some of these we don't yet know, for example those whish allow high temperatures superconductivity. So, we cannot immediately rule out the possibility that living body or cell is a quantum system, just because it is a macroscopic object. But to justify such a statement many questions should be answered.

How does the macroscopic alive object conserve quantum properties? How is its characteristic spectrum created? The answer at the moment is - nobody knows! There are several hypotheses, but there is no really working detailed theory. The most complete and most interesting idea was proposed by S.P. Sitko with co-authors, and later it was developed by them into Quantum Medicine and Physics of Alive [2,6].

Basics of the Physics of Alive and Quantum Medicine

Physics of Alive proposes the following explanation to the quantum properties of the alive objects [2-3, 6-7]. Apart from the anatomic-morphological structure of the alive body[14] there is also its electromagnetic 'skeleton' - coherent eigenfield with characteristic frequencies in the mm range. This field is created due to electromagnetic activity of every cell, but when created, it coordinates and synchronizes of all parts and structures of the body. All the cells have one identical element - genome. It is assumed that EMR of the cell in some way reflects its genome, a since it is the same for all cells, this creates a possibility for generation of a coherent eigenfield of the body.

The existence of such a coherent electromagnetic eigenfield is a necessary condition for the quantum mechanical description of the macroscopic system. Can such a field really be generated? Frequencies of eigen-oscillations of cytoplasmic membranes of all living system should lie within $10^{10}-10^{11}$ Hz, i.e. in mm range, according to estimates in [8]. Thus, a possible resonance amplification of some modes related to genome structure in the processes of DNA replication, RNA transitions, protein

[14] At the moment we are talking only about multicell organisms. QM mostly operates with human body.

translation, etc. are also expected to be in this range. On the other hand, one should not forgot, that the water - main component of the alive bodies- intensively absorbs mm range EMR. Therefore in order to form coherent eigenfield the radiation intensity from each cell should exceed some threshold.

The basic technology of the Quantum Medicine is microwave resonance therapy (MRT), its latest modification is Sitko-MRT [9]. Doctors use generators with very low spectral density of EMR $\sim 10^{-21} - 10^{-20}\ W / Hz \cdot cm^2$ (while thermal radiation of the human body is of the order of $10^{-19} W / Hz \cdot cm^2$) to stimulate biologically active points (BAP). These points are the same as those known in acupuncture or Ancient Chinese Medicine. Interestingly, Ancient Chinese Medicine is based on the idea of that the internal organs of man are intersected by the so-called meridians, whose external tracks are situated on the body surface. There are 26 meridians and most BAPs are situated just over them.

The existence of meridians (along which BAPs, used in the QM, are situated) cannot be observed by eyes or with microscope, but it can be observed in the mm range. The following properties of the meridians have been observed as it is summarized in [3] (again I accept these as facts).

The meridians have diameter 3-5 mm (at least near BAPs).

Refraction index inside meridian is n=1 (as in atmosphere; typically for human body n = 5-6).

For the density of the external flux of $\sim 10^{-21} - 10^{-20}\ W / Hz \cdot cm^2$ BAPs completely absorb the radiation of mm range.

Then density increases to $10^{-19} W / Hz \cdot cm^2$ or more a change of behavior is observed: BAPs start to reflect the radiation.

One more important experimental point: EMR of mm range has no effect on a normal healthy body [10, 2, 6]. This is one of the key points in QM. We can introduce definition of the 'healthy' body. MRT stops producing any effect when patient reaches the 'healthy' state.

QM gives the following quantum mechanical explanation to this effect. 'Healthy' state is a stable state of the system, a ground state in our energy level system. It corresponds to global minimum of some effective potential. 'Disease' is a metastable state of the system, which corresponds to local minimum of this potential. MRT stimulates transition from metastable state first to some exited state, then cascade transition into a ground 'healthy' state, similarly as in nuclear, atomic or molecular physics. This is illustrated in Fig. 1, taking as simple example Landau-Haken potential [3].

From these positions we can find an answer to some questions, which are still unclear in the traditional biology. One is the problem of 'unnecessary' genes: it is known that up to 98% of genes are not taking part in the standard processes of the protein production. Why do we need them? Particularly took into account that the difference in genome between humans and worms is about 2%.

According to Quantum Physics of Alive the purpose of genes is not only the production of corresponding proteins, but also generation of the coherent electromagnetic 'skeleton'. And this seems to be a more complicated task, which involves all 100% of genome.

Another example - imagine a simple thing, which can happen with everyone: you have cut a finger. We all know that, if the cut is not very deep, then in a few days it will disappear. How does this happen? Why your body is restored into exactly the same form as it was before? What controls this, very complicated in details, recovery process?

Based on the Physics of Alive the phenomenon of healing of the cut finger finds the following schematic explanation. In the wound area a certain number of cells were destroyed, but electromagnetic framework - coherent eigenfield of an organism remained, since it was created by billions and billions cells of an organism. The mismatch between the structure of a coherent field of a body (the spectrum of its characteristic eigenfrequencies describes in the universal electromagnetic language all the details of a body structure and its functioning) and the deformed morphology at the injured spot initiates the standard and well-known mechanisms of cells division and generation of the particular proteins just at the injured spot. These processes proceeds under control of the electromagnetic framework until the mismatch between a framework (which gives what is necessary) and morphological structure at the injured spot becomes less than sensitivity threshold of the system realizing this mechanism of communication.

We discussed the process of healing or self-cure of the body. But what can be done if the disease becomes chronic and is not cured by itself? This corresponds to the deformation of the electromagnetic framework itself. Quantum medicine is aimed to restore the electromagnetic framework of a human body. The patented technologies of diagnostics and quantum medicine therapy ([9], see also citations in [3]) allow to determine and eliminate deformation.[15]

---

[15] Typically the course of treatment consists of 10-12 sessions, 45-70 minutes each [3]. During this time the metastable state of the framework decreases so much that not a single self-organization level can be formed here. In other words, the framework of an organism is constantly in the ground potential well.

There are to ways to put human body out of health state	
To damage the body at anatomic-morphological level: for example to cut the finger	Deformation of the electro-magnetic framework of the body - chronic disease
⇓	⇓
Some cells are destroyed, but electro-magnetic skeleton still exist, since it is created by all cell	The body is in the metastable state and cannot help itself
⇓	⇓
The mismatch between eigenfield and deformed morphology starts, controls and then stops the standard mechanism of the cell replications	MRT helps to restore the electromagnetic skeleton of a human body

After the above discussion we end up with the important conclusion, which makes QM (and its basic technology - MRT) different from the traditional medicine and all the other ways of treatment I know. Quantum Medicine is absolutely harmless![16] Indeed the MRT uses radiation of such a low intensity that it cannot make any harm to the human body as to the classical system. While it influence in quantum mechanical sense is positive - after the excitations system has a bigger chance to roll down to the global or at least dipper minimum, what corresponds to healthier state. If, by a chance, system is stopped in some local minimum, then patient does not feel better and the treatment will be continued. The treatment is going until MRT doesn't show any effect, what correspond, by definition, to a 'healthy' state.

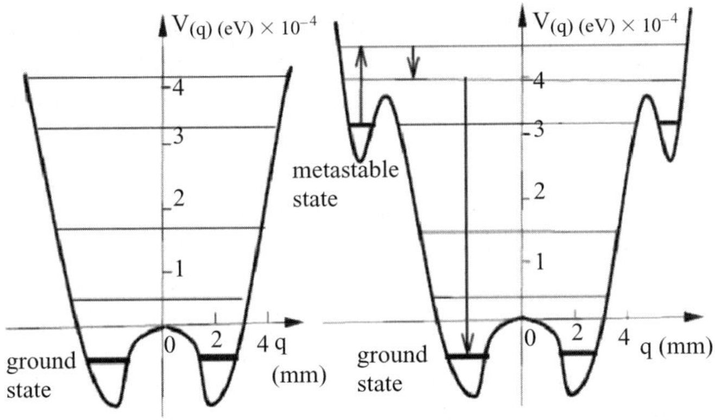

Fig. 1a Organism's ground state (health). Landau–Haken potential $V(q) = kq^1/2+K_1q^1/4$ ($k<0$, $k_1>0$)

Fig. 1b Organism's metastable state (disease). Deformed Landau–Haken potential. The way out of metastable state is shown (treatment) with use of MRT

Figure 1. Figure is from [3].

I stress once more that the QM (and more generally Physics of Alive) is still a hypothesis to explain the interaction of the low energy EMR of the mm range with alive objects. But, if it will be proven that it is true, then certainly QM will be the medicine of the 21st, 22nd and future centuries! Just due to this, in my opinion, it is worth to study and check this hypothesis in details, although most of the physicists disregard it immediately after hearing its basic assumption of coherent eigenfield. In this work I am trying to understand is the creation of such a field possible in principle from quantum-mechanical point of view.

Is where an experimental evidence of coherent eigenfield of a human body? If the picture illustrated in the Fig. 1 is true, then we should expect to see the EMR in the mm range coming from the human body - radiation during the transition of the system from excited to the ground state. A radiation was indeed measured - see Refs. [11].This appeared to be a complicated experiment because the radiation of this type has intensity similar to that used by the MRT technology, i.e. lower than the thermal radiation.[17]

This experimental data support the picture of MRT quantum influence, Fig. 1, but for me it is not clear whether they can be considered as the

[17] Please note that this experiment, similarly as the study of the meridians on the human body, was done by the same collaboration, which propagandizes the QM and Physics of Alive. Therefore in the strict sense this is not an independent confirmation. I did not find experimental results from other collaborations. Certainly more experimental studies are required here.

experimental evidence of the existence of the coherent eigenfield of a human body in the mm range, as authors claim. I don't see the reason why the intensity of such a laser-like field should be so small. If all the cells of our body are constantly radiate and absorb EMR in order to form long range correlations between all the parts of our body - then such a system should radiate strongly, not at the level of separate photons. Also it was not proven in these experiments [11] that the measured radiation is coherent.

I think that the above question can be answered in the way that this eigenfield is form in the way, that it does not exist all the time, but can be created very fast when it is needed due to some reason [12]. For example, you cut a finger as we discussed above. But, of course, such a hypothesis needs additional experimental tests. Thus, in my opinion, the present experimental situation is uncertain, even taking all the published results as true facts.

There is one more very important achievement of the Physics of Alive. We might finally have found a fundamental definition of Alive and its difference from non-Alive: the Alive object is a whole quantum-mechanical system with its own system of characteristic energy levels. Thus, this long-standing problem of biology might be finally solved, really on the fundamental level, in Physics of Alive, if this beautiful hypothesis will be confirmed.

How this Coherent Electromagnetic Eigenfield can be Created?

First of all, in this section I am going to present my ideas on how to make physical modeling in Quantum Medicine and Physics of Alive. As it was discussed in the introduction a weak point of this hypothesis is theoretical description.

Everything what we discussed in the previous section about QM and Physics of Alive was about humans and probably true for other multicell organisms, although here experimental background is even more pure [12]. If this theory of Alive is true, then already a single alive cell is a whole quantum-mechanical system with its own characteristic eigenfrequencies as it required by the fundamental definition of Alive formulated above. In Ref. [4] one can find review of the experiments on the interaction of the low energy EMR in the mm range with single cell organisms (and corresponding citations), which can be interpreted in this way, although here experimental data are very pure and such interpretation is not very convincing. Multicell objects show much wider range of different reactions to the mm range EMR, and, of course, all the coherence effects discussed above, which are the subject of QM, belong to multicell organisms. But I simply cannot imagine how the macroscopic multicell organism can be a quantum system, if its building blocks - cells - are not (contrary to our body cells themselves do not consist of almost identical building blocks).

So, in my opinion, in order to be able to create a coherent electromagnetic eigenfield of our body, we should start with the assumption that each alive cell is a quantum-mechanical system with its own eigenfrequencies. Of course, cell is also a macroscopic object - how is it created as a quantum-mechanical system? Well, I don't know. But I don't know either how it is created, even if it is not a quantum-mechanical system. Did I loose too much? Is it easier to imagine that DNA and cell with all its mechanisms is created by a chance than that some macroscopic object can have quantum properties? We cannot say we know all the possible conditions to see quantum effects in macro-world. At least, until we do not understand how Bose condensation happens in high-T superconductors.

So, I assume that each cell of the alive body is quantum-mechanical system with its own eigenfrequencies. This eigenfrequencies in some way reflect the genome of the given organism, and therefore are equal for all the cells in the same body. Then the multicell organism in the first approximation can be modeled as a homogeneous infinite matter consistent of the identical cells. Later the model can be improved to account for the particular anatomic-morphological structure. Similar type of modeling is successfully used in nuclear physics even nowadays - in the first approximation nuclei are homogeneous and infinite and consist of identical nucleons.

Since cells in our body are constantly staying together - they are bound in some way. This is primary done by chemical electromagnetic forces. But being bound is not enough for system to shown quantum properties. Earth and Sun are also bound by gravitational force. We can solve Schrodinger equation for this system, similarly as for the hydrogen atom, but due to macroscopic masses of this objects distance between energy levels will be so small, that for any practical purpose the spectrum will be continuous. Therefore in order to form a quantum system our macroscopic cells cannot be bound in some effective one-cell potential as nucleons do in the nucleus.

We should search for some other way to form an overall quantum system and what is proposed in QM is laser-like system. In such an approach our cells should form a working body of the alive-laser, similarly as molecules form working body of an ordinary laser. Let us discuss very briefly which conditions for coherent light generation are necessary. Here I will basically follow [13].

An atom or molecule (we discuss now ordinary lasers) can emit a photon via spontaneous emission or stimulated emission. Atoms and molecules in excited states randomly emit single photons in all directions according to statistical rules via spontaneous emission. In the process of stimulated emission, a photon of energy hν perturbs an excited atom or molecule and causes it to relax to a lower level, emitting a photon of the same frequency, phase, and polarization as the perturbing photon. Stimulated emission is the basis for photon amplification and it is the fundamental mechanism underlying all laser action.

Consider the simple case of a two-level system with lower level 1 of energy E_1 and upper level 2 of energy E_2. Let N_1 be the number density of atoms in level 1 and N_2 be the number density in level 2. Let $u(\nu_{12})$ be the energy density per unit frequency interval of the light at the frequency $\nu_{12} = (E_2 - E_1)/h$. The rate of spontaneous emission is independent of $u(\nu_{12})$ and proportional to N_2.

$$R_{spon.emiss.} = A \cdot N_2 . \tag{1}$$

The rate of stimulated emission depends on $u(\nu_{12})$ and can be written as

$$R_{stim.emiss.} = B_{21} u(\nu_{12}) N_2 . \tag{2}$$

The absorption rate also depends on $u(\nu_{12})$:

$$R_{absorb.} = B_{12} u(\nu_{12}) N_1 . \tag{3}$$

The proportional coefficients A, B_{12}, and B_{21} are called the Einstein coefficients. It can be shown [13] $B_{21} = B_{12} = B$, since the quantum mechanical treatment of stimulated emission is similar to that of absorption. In thermal equilibrium the probabilities that states 1 and 2 are occupied are proportional to the Boltzmann factors $\exp(-E_1/kT)$ and $\exp(-E_2/kT)$, respectively, and $u(\nu)$ is given by Plank's law,

$$u(\nu) = \frac{8\pi h \nu^3}{c^3} \frac{1}{\exp(h\nu/kT) - 1} . \tag{4}$$

Where: T is the temperature of the atoms or molecules. In equilibrium the probability of up transitions must exactly balance the probability of down transitions:

$$R_{spon.emiss.} + R_{stim.emiss.} = R_{absorb.} . \tag{5}$$

What leads to the following relation:

$$A = B \frac{8\pi h \nu^3}{c^3} . \tag{6}$$

We can study the ratio of stimulated and spontaneous emissions:

$$\frac{R_{spon.emiss.}}{R_{stim.emiss.}} = \frac{1}{\exp(h\nu / kT) - 1}. \tag{7}$$

In the ordinary lasers this ratio is much smaller than 1, since $h\nu >> kT,$ while for the quanta in the mm range it is quite opposite: at the room temperature $h\nu / kT \approx 0.01$ and correspondingly $R_{stim.emiss.}$ / $R_{spon.emiss.} \approx 100$. In the papers devoted to QM and Physics of Alive (see for example [3]) eq. (7) it is considered as giving strong support for the idea of coherent eigenfield of human body. But is this really a condition, which would allow laser action? No, the real criterion for laser action is defined by the competition not between stimulated and spontaneous emissions, but between emission and absorption.

In thermal equilibrium $N_1 > N_2$. Correspondingly, a resonant photon is more likely to be absorbed than to stimulate emission. But if $N_2 > N_1$ there is a possibility of average overall amplification for an array of photons passing through the system. This situation is called population inversion. Spontaneous emission depletes N_2 at a rate proportional to A, producing unwanted photons with random phases, propagation directions, and polarizations. Because of loss associated with spontaneous emission and other losses associated with the laser cavity, each laser is characterized by a minimum value of $N_2 - N_1 = N_{thres}$ called the inversion threshold. Only if $N_2 - N_1 > N_{thres}$ we can see a laser action. Recall that in our situation of an alive body we have to overcome the water absorption of the mm range EMR, so N_{thres} should be large. In the ordinary lasers the external pumping is used to produce a population inversion. What can be the reason for it in the multicell organisms?

I my opinion one should be very careful, when trying to apply above eqs. (1-7), which are true in the micro-world, to the processes in macro-world. They should be rather considered as guidance for what we want to achieve in order to obtain coherent EMR in our system of cells, but the exact treatment of the stimulated and spontaneous emission mechanisms, as well as of the absorption should be completely different.

The good example to feel the difference between micro and macro world is a temperature. In our case, when the coherent radiation should be emitted by cells, one cannot use the room temperature $\sim 300\,K$ as T_{cell}. In order to use temperature in principle, we assume that our radiation is in the thermal equilibrium with the heat reservoir, given in our case by the system of cells. This means that the temperature of radiation is the same as temperature of cells.[18] But what is the temperature of the cells? Temperature is our way to parameterize thermal motion, $kT/2$ is the energy per degree of freedom. But cells are macroscopic objects, they are not taking part in Brownian motion or in vibrations, as molecules in the crystal grids do. From this point of view they all are at rest and they should have an effective temperature $T_{cell} = 0$

On the other hand, cells are complicated objects with lots of internal degrees of freedom and internal energy resources. Therefore stimulated emission and absorption do not have to given by similar expressions, as in eqs. (2) and (3). For example, it is easy to imagine a cell, which from time to time absorbs the resonant photon, being already in the exited state, and then redistributes energy into motion of some its internal components, instead of always going through the spontaneous emission. Note that since cell has complicated internal structure its stimulated emission and absorption may, in general, depend on the external conditions, which have nothing to do with EMR. Similarly, since cells have internal energy resources the population inversion can, in principle, be achieved without external radiation - for example, cells might go into excited state, using their internal energy, because of some chemical signal. This may happen, for example, in the vicinity of the cut on your finger, creating thus favorable conditions for generation of the coherent EMR. Of course, this is just a principle possibility.

I don't know how the population of the cells on different energy levels can be calculated, how to estimate emission and absorption rates, etc. But these can be measured experimentally. And only after we have these data, we can discuss whether conditions for the coherent EMR generation are satisfied in the human body and other multicell organisms. Actually, if coherent field can indeed be formed in the system, then the best experimental proof, in my opinion, would be to build a bio-laser. This can be done on the completely phenomenological level, like all the QM, without knowing

---

[18] One may also ask whether cells in our body can form a heat reservoir, which in principle should be infinite, since number of cells ($\sim 10^{13}$) is many orders of magnitude smaller than $N_{Avogadro} \sim 10^{23}$. In the simulations of heavy ion collisions thermal distributions describe the spectra of the produced particles surprisingly well (see for example [14]) although the total number of produced particles is of the order of $10^2 - 10^4$. The reason for this is still unclear. In my opinion, 10^{13} is large enough number to safely apply canonical ensemble description, and the fact that cells are macroscopic objects supports this statement, since it is less likely that the radiation can influence their state.

precisely all above details. Possibly the bio-laser will be easier to build using the same techniques as in ordinary laser: external inducing EMR, resonators, if needed; just the working body should be formed by some biotissue.

Perspectives for Quantum Medicine

As we have seen in the previous sections the existence of the coherent electromagnetic eigenfield of the multicell body, what forms the basis of the QM and Physics of Alive, is not really in contradiction with our knowledge in quantum mechanics and in other fields of physics. But the present experimental data do not show any clear evidence of its existence. In my opinion this is only possible if each single cell of the body is a macroscopic quantum system. There is lots of work for the experimentalists in this field. So, the question whether Quantum Medicine is just beautiful hypotheses or really a Medicine of the Third Millennium is still open.

Just imagine for a second that it is really true. Then the perspectives for Quantum Medicine are enormous. Absolutely harmless medicine, which allows us to talk with our body on its own language of mm range electromagnetic quanta... This will make us change completely our view on medical treatment and future of the medicine.

And not only that! We will also know (at least one) fundamental difference between Alive and non-Alive. Certainly, this will affect our ideas about the creation of the Life on the Earth and, more generally, in our Universe.

According to Quantum Medicine the genes are responsible more for the electromagnetic skeleton of the Alive organism than for the production of proteins. So, the effect of mutations may be not only on the anatomic-morphological level, they can affect the eigenfield of the organism. This will change our view of the Evolution. And also our expectations from gene engineering may not come out, or at least not as soon as expected. According to QM, it is much more difficult to add, replace or modify genes than we have thought so far. Maybe because of this most of genetically modified species are not reproductive and have other defects. So, it is possible that we will need to decode human genome once more, keeping in mind coherent eigenfield of the human body.

Acknowledgments

I acknowledge fruitful discussions with Yu. Babich during this Conference.

References

1. E.A. Andreev, M.Y. Belyi, S.P. Sitko, Dokl. Acad. Nauk UkrSSR. B10 (1984) 56.
2. S.P. Sitko, Physics of the Alive, Vol. 1, Nr 1 (1993) 5.
3. S.P. Sitko, Physics of the Alive, Vol. 12, Nr 1 (2004) 5.
4. E.N. Gorban, Physics of the Alive, Vol. 9, Nr 2 (2001) 19.
5. I.S. Cherkasov, S.V. Nedzvetskii, author certificate Nr 733697, 1980 (USSR).
6. S.P. Sitko, L.N. Martchian, *"Introduction to quantum medicine"*, Kiev, "Pattern", 1994; S.P. Sitko, Physics of Alive, Vol. 9, Nr 2 (2001) 5.
7. http://www.sitko-therapy.com/
8. H. Frohlich (Ed.): *"Biological Coherence and Response to External Stimuli."*. Springer – Verlag, Berlin, Heidelberg, New York, London, Paris, Tokyo, 1988 (p. 268).
9. S.P. Sitko, *"Microwave Resonance Therapy"*, Patent Nr 2615, 15.03.1994 (Ukraine); Patent Nr 2053757, 10.02.1996 (Russian Federation); US Patent Nr 5.507.791, 16.04.1996.
10. S.P. Sitko, E.A. Andreev, I.S. Dobronravova, J. Biological Phys. 16 (1988) 71.
11. S.P. Sitko, Physics of the Alive, Vol. 6, Nr 1 (1998) 6; S.P. Sitko, Yu.O. Skripnik, O.P. Yanenko, Physics of the Alive, Vol. 7, No 2 (1999) 5.
12. Yu. Babich, private communication.
13. R. Loudon, *"The Quantum Theory of Light"*, Oxford University Press, 2000; Online course of Modern Optics, The University of Tennessee, Department of Physics and Astronomy, http://electron9.phys.utk.edu/optics421/modules/m7/lasers.htm.
14. P. Braun-Munzinger, K. Redlich, J. Stachel, *"Quark Gluon Plasma 3"*, Eds. R.C. Hwa and Xin-Nian Wang, World Scientific Publishing, 2004 Singapore, pp. 491-599. [arXiv: nucl-th/0304013].

There are 6 Million Tons of Brain Matter in the World, Why do We Use it so Unwisely

Boris N. Zakhariev

Bogolyubov's Laboratory of Theoretical Physics, Joint Institute for Nuclear Research, 6 Joliot Curie, 141980 Dubna, Russia, zakharev@thsun1.jinr.ru

There is nothing as valuable as our brains in the whole Universe. Diamonds, gold, oil, etc. which caused wars are infinitely less valuable. And what do we do with our treasure? Is it not evident that for our common wealth and happiness these brains must work most intensively and effectively? It can be achieved by **saturating the global brain matter by information**. Everybody must have high education, free access to almost any book, journal, video-, audio-products, etc., particularly, through Internet, independently of the rate of personal income. The informative interaction will evidently result in a nonlinear growth of cleverness and effectiveness of the mankind through the social many-body amplification. But now only a **negligible part of the mankind has a comparatively satisfactory level of information**. If we really understand the priority of these advantageous for everybody and noble goals, they would be easy to achieve. Instead, we lose countless genuine material and spiritual riches.

I work at the international physics institute in Dubna as a theoretician. We have taken an active part in improvement of quantum mechanics. It results in a radical simplification of this discipline "unintuitive", according to the previous opinion of the Nobel Laureate Gell-Mann. Particularly, we have unexpectedly found some kind of quantum ABC. You can read about this in our recent book "Submissive Quantum Mechanics: new status of the theory in the inverse problem approach" (in pictures). We placed its versions, Russian and English (still draft), in a free accessible form on my homepage in Internet (http://thsun1.jinr.ru/~zakharev/). It appears possible to build quantum systems with the desired spectral properties of elementary quantum 'bricks and building blocks' as in a 'children toy constructor

V. Burdyuzha (ed.), The Future of Life and the Future of Our Civilization, 335–348.

set'. It was an important step for us to make the quantum knowledge significantly more accessible to people in the world. Science must be simple! We shall return to this point somewhat later. We can even imagine that most of people will become mutual teachers and get their portion of utmost high happiness by simultaneously extending and acquiring knowledge. It is not at all an unfeasible dream. That means to have peaceful armies of information conductors instead of millions of soldiers. Also, e.g., an increased mass production of PC, powerful enough and very cheap, instead of smart bombs, etc., produced for perverted ideals of world oppressors: **liberation of their evil activity**. We all are responsible for everything that happens in the world and we need to be something like presidents (ministers of culture, etc…) of the whole mankind and our competence is much more wide than it is usually used to think.

Formally I am not an expert in the social subject, I consider here. But I am deeply concerned with it and believe that the world is objectively interested is our opinions. So it is a physicist's talk about 'lyrics' (J.W. Goethe University of Frankfurt is a very suitable place for this).

Recently I have read that Nobel Laureate Ernst Richard suggested that conferences on natural science should include in their programs reports about social problems. This idea is in resonance with my opinion.

Particularly, I have read about 1500 memoirs of different people who shared with me their most valuable life experience and improved my understanding of our life, at least in my own reference frame. I believe that anyone of us has a fundamental need to harmonize his particular professional specialty with social activity to make our common future happier. Already in ancient times Euripides said that the wealth of any person depends more on the wealth of the state than on his family. In our time of globalization the same is true for the whole world. Trying to express my dreams and feelings in physics and lyrics I would be happy, e.g., to infect you by optimism.

I believe that to solve better the world's many-body social problems, we must be more open and exchange intensively our different personal world's overviews. I myself have tried to do this and exposed my original physicist's opinions on sharp social problems, my memoirs, diaries on my homepage in Internet, although I am often in contradiction with what we hear around us. We all will gain from mutual understanding and get happier by satisfying the fundamental need of self-expression.

Many years ago I was impressed by results of investigation of the labor of 10 thousand of metal workers. It appears that 5 additional years of simple school education give some increase of their productivity. But what seemed strange at first sight was the **five times (!) decrease of defects** in their production. And this additional education gave five times acceleration in transition from processing of some detail to a new one. Is not it paradoxical that geography, mathematics, languages… etc., seemingly having nothing to do with the essence of metal work appear to be so important? No, a even simple intuition prompt us that it is quite natural. Any job is made better with the help of a more developed, more flexible, cleverer mind. And what's more, the efforts once spent on knowledge give return for the whole life. It is also easy to believe that the same effect of education must cause five times less diplomatic errors, e.g., stupid wars. You know many striking examples of such mistakes during the last years in the world. So similar generalizations can be applied to almost any kind of human activity. Particularly, education makes people of different countries mutually more attractive, creating cleverer world ecology. I like even the idea of "Universities at prisons" and its first realizations. It is, e.g., one of the wise ways to exterminate the crime. One of the previous US ministers of justice (maybe the cleverest one) said, "we invest much money in the building prisons, but it would be more effective to support the better quality of upbringing of children which are claimants to become criminals (each teacher of the first classes can easily indicate such candidates".

There are published million of different books every year in the world. And one person can hardly have time to read tens of them. I was once surprised when I understood that people of a small town like our Dubna can in principle read many times more than it is published, if everybody will read different books. And there is a tremendous ability of information exchange: to reach any interested person in the world it is enough about 6 intermediate connections only!

There is a possibility of global education without excluding a significant part of population. I want to mention two outstanding examples. The first one is the great schoolteacher, genius **V.F. Shatalov** who invented how to teach children so well that pupils of weakest class begin to learn with good marks under his guidance. He arrived at this by reducing courses to separate elementary 'bricks' each of which was accessible even for the weakest pupil. It can happen also with usual teachers that some bad pupil once prepared some lesson very well. But an ordinary teacher can give him only a low mark if he knows that nothing before was studied satisfactorily and also nothing will be so mastered in the future. But in these cases Shatalov put always an excellent mark. For the bad pupil it is a **great** event. He get enthusiastic incomparable with what the best pupils feel, to learn the next 'brick' and so on, and so on. Another principal moment of Shatalov's approach was that he checked everybody's competence in any such brick of knowledge without postponing the encouragement. This involves all pupils without exclusions into the pleasant fascinating learning process. Special care was taken of preparation of bricks to make them easy to learn, understand and to be checked quickly by the teacher.

This reminds me of the simplicity of the algorithm of the Nobel Laureate **Landau** who showed **the direct and easy way to the scientific knowledge**. My first scientific chief suggested me to pass the examinations of Landau's famous theoretical minimum. But there was an artificially created public opinion that only a narrow circle of super-students is able to do it. It appears to be a fraud spread deliberately for more comfort of the limited number of students (not a rare phenomenon in our imperfect world). Really, on the contrary, these exams were much simpler than the standard ones at the University. It was possible studying for first years to pass these examinations that were of much better quality than the standard PhD exams. There was a clearly defined, **strongly restricted** Landau's program: practically no questions outside the given textbook, from which many sections were excluded. This made the exam comparatively simple and a not too nervous procedure unlike rather indefinite exam requirements at the University. The result was much more reliable and it appeared accessible to start the process to join the famous scientific school. Further, we will also discuss the analogy of learning progress with intensification of mankind's ability to make serious discoveries. But before let us consider the importance of knowledge distribution without exceptions of some contingents of world's population. Let us introduce the notion of an 'atom of information exchange' (Fig. a). This is the circle of direct information connections of anybody with his/her information environment. Of these atoms are constructed different information complexes (Fig.b, c, d).

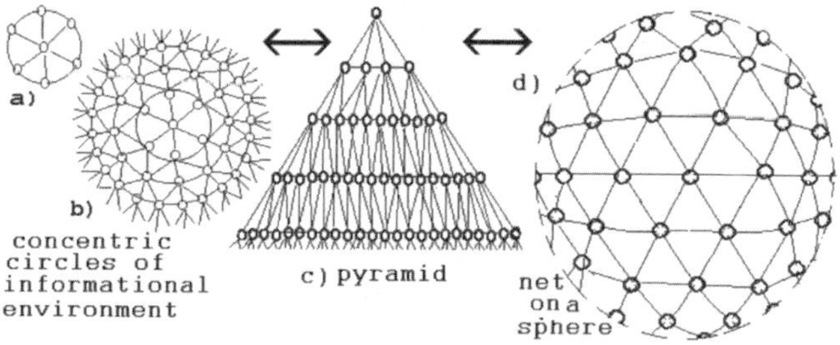

It is a widely spread opinion that for some person with concentric circles of informers around him (Fig. b) only the nearest neighbors are of crucial importance, and the further circles are for them less and less valuable (seems natural?...). So people permit themselves to be uninterested in information quality of these far circles. But the presentation of the same structure as a pyramid (Fig. c) reveals us clearly that if people at the bottom of the pyramid would be substituted by uninformed ones the whole pyramid will go one step down in information quality. The same happens with the person at the top of the pyramid. So the importance of the last circle appears to be equal to the nearest one. We all are egoists and want to be better informed. But there are different kinds of egoism from extremely narrow one to the world's wide. So to say, egoism is characterized by 'radius of its reasonability'. In the case when inside the circle of this radius there is only one person himself - it is the most primitive egoism (less than Neanderthal's one), in the limit case when all people are included in it - **most clever egoism** that **coincides with the altruism**. So the economy on education of poor nations turns out as global self-confiscation. After 11 September (the tragedy of skyscrapers destroyed in NY) the brave, clever and noble editor-in-chief of one of the most prestigious popular scientific American Journal "Science" immediately published there an editorial article. He explained that it was not a simple Ben-Ladden's crime. The main reason of this tragedy lies in that very rich USA for each dollar given, e.g., to extremely poor Bangladesh, confiscated from this country 7 $ using diversity of tricks including the trade ones. It was done according to the laws established by mighty countries ("golden billion") and accomplished by banks, trade companies located often in shining skyscrapers (with not so brilliant clean activity). To remove the basis for the terrorism, the unjust gradients of wealth distribution must be smoothed (the crime could not be excluded by military forces without radical softening the economic inequality).

One of such usual algorithms of shameful confiscation practice (really disgusting, but still not exiting sharp enough protest of the deceived part of the worlds population) is the too often strongly inadequate exchange of genuine goods for paper currency of rich countries, which continually, but not slowly are evaporated through inflation (negative! loan). And rich countries give loan to poor ones with too high interest (taking advantage of their strength and not excluding its military component). Really, it is self-damage **and spoiling of the world's social, financial, and spiritual ecology**. And the blind top of the world's pyramid of wealth must know this truth. Also, the shorter is the radius the stronger is **self-punishment** through different multiple 'channels' of social communications.

On the other hand the USA show us an excellent example of realizing during tens of years a permanent 'socialist' revolution inside their country (e.g., narrowing the gap between genders, national and racial groups, etc., particularly stimulated by the previous blacks' rebels).

So it is wrong to say that socialism was defeated by capitalism. In our country just the deviation from socialism weakened the regime. (Naive majority of people due to comparatively low level of education hoped that 'perestroika' would help them to get rid of the previous shortages, but they were unaware of the financial support (from above and abroad) of the robbery of the "new capitalist Russians" and against justice. They also believed in the attractive image of a 'free market', but did not know that it does not exist in the world and in our country the market became the exclusive property of the "capitalist sharks". Those who recently visited Moscow was surprised by the improvement of the outward appearance of the city, but it has a simple explanation that the capital is in privileged position and confiscates the wealth of the whole impoverished country.

This understanding of fundamental importance of justice inside one country (USA) must be later continued to the whole world, according to the cleverest egoism. US performed a permanent 'velvet' social revolution. I think it was to a certain degree due to an instructive example (positive influence of the opposite political camp) of the USSR's world propaganda of equality. It was wonderfully possible for our people to save some 'socialist' ideals in difficult conditions of multiple direct aggressions (against these ideas) by some countries of the golden billion and often masked partnership of other ones (their economic, political blockades, in addition to climate 'ice blockade'). So there was a partially positive mutual influence of rich and poor countries even in different political camps. Look at valuable witness of UK prime-ministers (in their memoirs) that they were obliged to speak with the leaders of the former colonial countries not 'in order form' when they could get help from the USSR with its anticolonial stand. Till now people from the countries of the third world often say that they regret the USSRs support in no longer available. So I am sure that that is not a subjective illusion.

Another praiseworthy US step in the direction of noble deeds was the initiative of their National Institute of Health (NIH) directors. Nobel Laureate Harold Varmus and later Elias Zerhouni called for free access to scientific publications. It is time for radical improvement of copyright practice, which now strongly restricts information exchange in the world. I have just found Declaration on open access to knowledge (see http://www.zim.mpg.de/openaccess-berlin/; http://www.eprints.org/berlin3/outcomes.html http://www.zim.mpg.de/openaccess-golm/index.html).

The US has achieved one of the highest education levels (there is an average 'half-university education') that must be spread further to the whole world and then to be even increased. In this case, there will be the full recoil of culture, productivity, and any wealth for everybody. So the world will gradually transit from cave values: rough strength, sly adroitness, etc., to the genuine ones worthy of the information century. Inevitably, we will combine all progressive achievements in the world's practice.

There is an opinion at the tops of the administrative pyramids that the justice on the earth is impossible because of the lack of wealth to provide everybody with the American level of income. The same I have heard from one of our previous "communist" laboratory leaders. Even before the counterrevolution he was for separation inside Russia of a 'golden minority' and for their rich live at the expense of others poverty. Such base statements of some persons from our 'elite' (better to say, 'slag') seem to me evidently unacceptable. Inside our families we usually do not allow ourselves to suppress economically our relatives defending good spiritual relations. So, the same must be true also in more wide communities and even in the whole world. This simple arithmetic of social values will be gradually mastered by humanity in future. The previous position of our country (with its positive and negative aspects) was somewhere in the middle of the world's social pyramid allows us sometimes to escape the blindness of those at its top and bottom. It is instructive in global sense due to better position to extrapolations both up and down. Even example of negative experience, may serve as useful information without which it would be impossible to improve the world. The recent unprecedented instantaneous increase of wealth gradients is absolutely evident awful injustice. The former administration used the fact that the common 'socialist' property with the help of the world wide evil web (**WWeW**) could be comparatively easily privatized. Those who were placed to defend common wealth and got accustomed to rule have gradually transformed from 'peoples servants', highly paid even at the previous regime, into the individuals capturing for nothing the peoples riches. The authorities used their administrative resources for economic counterrevolution.

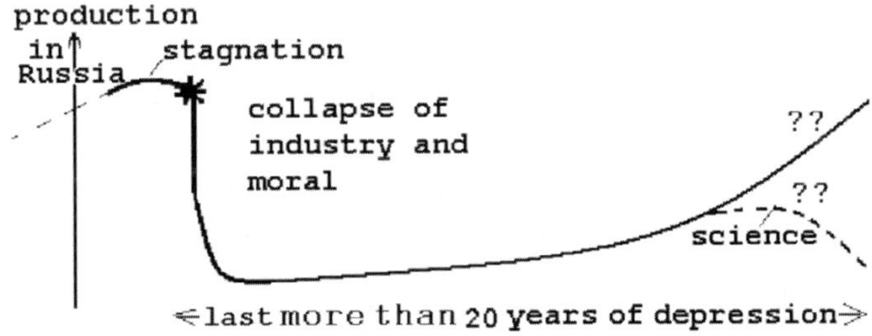

So the 'socialist' regime weakened by the capitalist propaganda of developed countries with corrupted elite appeared to be not stable in conditions of an insufficiently informed society. We were depraved from outside and from above even unaware of the influence of our unfavorable climate on economics, e.g., too short vegetation period. As a result, people hoped that they would become as rich as citizens of the USA when 'socialism' is transformed into capitalism. Really, it became the freedom mainly for rich minority to privatize what was the collective property of the majority. There began a regime of **apartheid** of 'new Russians' against the main population, supported by international web of banks, politics, PR services and different mafia structures. Our new oligarchs and the members of wwew abroad have got hundreds of billions $. And about ~10 million of 'old Russians' died in addition to a usual average death rate (more severe holocaust than that of the Hitler regime) from insufficient nutrition and medication caused by sudden robbery-privatization of almost the whole Russia's previously common property. The population of the world's 'golden billion' has easily believed that this was a success of "freedom" and "democracy" (real Orwell's newspeak). It was followed by the collapse of justice (moral) as the necessary consequence. Incomparably weaker cruelties of the previous regime in the USSR (tens/hundreds dissidents in prisons) evoked the whole day and night PR-protests of the developed countries. But there is almost no reaction to perishing of **millions now**. There was simultaneously a mighty blow to our education. Millions of pupils were thrown out of schools, for splash of narcotic, mass-prostitution. It was bad not only for those hundred millions of robbery victims, but also for oligarchs who could be expected to be now in nirvana. Really, it caused the avalanche of moral in the country and tsunami of anger (not genuine 'wealth', without the highest spiritual comfort). Oligarch Chubais said in his TV talk that he is now afraid of vengeance (the loss of main human welfare, the security).

It is important to understand different kinds of social disinformation slyness. I'll consider some Russian historical examples which seem for me very important (they have also international interest) and where I have original opinion and would be glad to share with others. I hope this information and parallels with contemporary events have some universal contents (for different countries and for future). For me it was experience of revealing historical distortions. Really they are bad for the layers themselves. Future of world civilization will sooner become better if we will learn the previous lessons.

There is a trick to hide real causes of evil by intentional distortion of notions "Empires of evil and goodness". After the Russian revolution of 1917 the previous people's oppressors got financial support from foreign banks during our Civil war. They were against the proclaimed noblest goals of highest justice by construction of a society without exploitation of man by man in the world. Without this support there would not be such a war at all. This was the opinion of the head of the government created after the Czech corps' counter-revolution putsch in Samara. This Civil War gives additional 8 million killed after the Russians defeat in the World War I with about 1,5 million lost. In spite of statements that it was against the cruelty of revolution, the world oligarchs do not take care about people killed in our country, but they were afraid of spreading the ideas which can hinder them to fleece the world. Really the rich world was not at all worried by the misfortunes of our people. The cruelties of the rulers against population were unimportant for the world's bank magnates, as is often the case now. The evidence of their unfair position was the defeat of the intervention of multiple (>10), mighty, incomparably rich, colonial countries against the exhausted people. It supported the previous slave-owners fighting for continuation of oppression of their folk. This 'brother against brother' struggle forced by foreign evil will create the so-called 'Empire of evil'. No wonder that after that criminal foreign influence with a direct intervention there appeared GULAG, mass repressions (which were a result of the distrust between people provoked by the world's oligarchs). Then followed political and trade blockades, later the German aggression of 1941, which was economically supported practically by almost the whole Europe (willingly or not). All this hindered realizing the ideal goals in the USSR what was bad even for capitalists (read Sadul's letters to his government, the most objective and informed man in the world, military adviser at the French Embassy in 1917). But the real initiators do not feel themselves responsible. "The empires of goodness" made the principal contribution to the creation of evil. Let us add also for completeness that the totalitarian regime in our country was caused long ago by the desperate situation for Russia to choose whether to be a colony of mighty rich countries or to unify people for defense with loss of some freedoms, but less than in colonies).

There was also a misunderstanding with not very clean care about our dissidents before the 'perestroika'. There were PR-protests day and night against tens (hundreds?) of dissidents in our prisons, but now when ~ten million are killed by our 'liberals', the radio and television of world's magnates and the previous dissidents and 'fighters for human rights' are not interested in their fate. Now it becomes an evident example of doubtful moral of the 'greedy' billion in the world.

It is also useful to say about one of the shortcomings of the 'socialist' USSR. Their most rich republics have got subsidy and poorer republics-donors from the Middle Asia, Russian Federation, etc. (some kind of corruption on the republics level). The rate of consumption to production per person in 1990: Tajikistan 28%, Turkmenia 53%, Russia 68%, Byelorussia 77%, Estonia 231%, Georgia 400%, Armenia 310% (publication in 1992). More developed countries now turned against Russia, supporting separation from the USSR of more favorable states (all the care about richer and against poorer people).

There is a hidden side of "liberation" of the previous Baltic Republics of the USSR. They were always in a privileged position in the Union like previous members of socialist camp (East Europe). For them our country was as a kind of colony, poorer and supplying them with rough materials for prices lower than in the world. The rich world always supported politically and financially their separation from Russia and for political and economic actions against us under the NATO protection. The same is true as concerns Georgia, Ukraine, etc. It was better especially for their elites to escape from a comparatively 'colder thermostat' (with poor Russia) to richer one with NATO. In mighty coalition they could communicate with Russia from more favorable positions (dictate prices, etc. as the USA, e.g., for Bangladesh).

Our negative example has some positive instructive side. For example, for future of the world the fortune of the huge and mighty China is important. It is natural to suppose that wwew is trying to transform their system as in our country (corruption supported from outside, etc.). Of course, the Russian experience will hinder the top of their social pyramid to rob too roughly their people who are expected to control more carefully their elite being informed about Russian's blunder.

Our ministers of education and science, are corrupted by some support that is too small for real improvement of the situation, but sufficient to stimulate their mentioned above 'care' of youth. It is like advise of Pobedonostsev, the synod's procurer and the educator of the Russian last tsar, to throw out of gymnasiums the children of simple people for stability of monarchic regime. The idea was that it is 'easier' to suppress ignorant people. Such deeds gave the opposite result: not enough educated people killed the tsar. The recent and contemporary rulers reduced our science many times being stimulated by wwew. Really, there are among the scientists not very qualitative ones, but it is an axiom that scientists on average are at the top of the qualification pyramid. So, the reduction of scientists is not a wise idea. For the better future it is cleverer to lift the qualification of the bottom, see fig.

Another 'crime' of our scientific administration is the establishment of 'age-apartheid' (artificial reduction of older collaborators in spite of damage to creative process).

Recently I have found the next figure illustrating a 'paradoxical' phenomenon that with age scientific productivity of active researchers does not decrease. This is another basic element of optimism for a cleverer future organization of the humanity (see further for new such ideas).

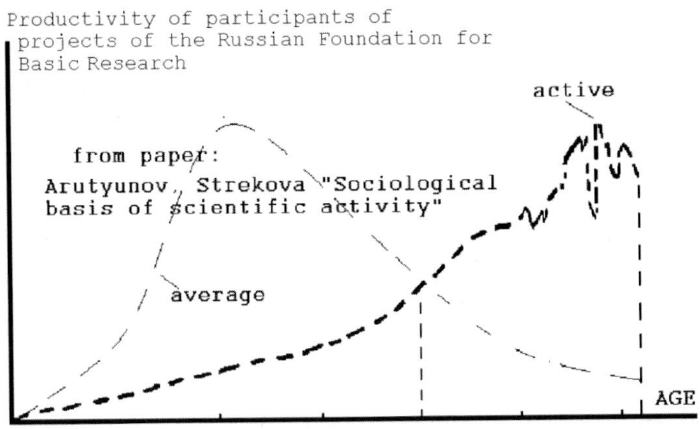

We have already discussed the dreams of better world organization. Not all will agree with me. But there is an effect of 'optical illusion' that reveals the error of many aspects of our pessimism. Here the percolation model of some important, but hidden effects in social progress can be useful. The next Fig. b shows the dependence of current I(n) on the number of cuts n of the conductive net between different electric potentials. It appeared that random cuts of different connections between knots does not significantly change I(n) from the beginning.

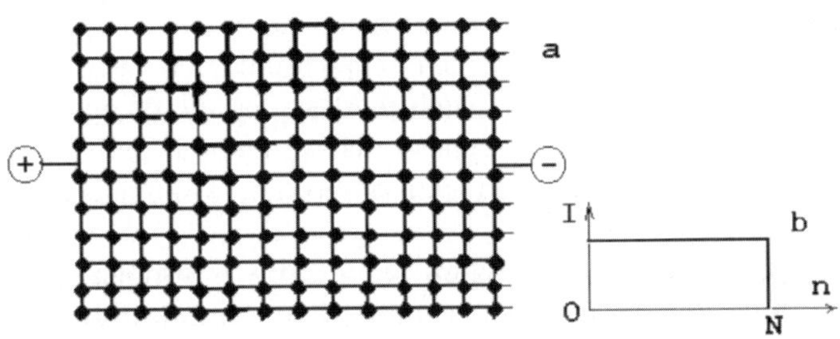

But when at some big n = N the net finally disintegrates into two parts, the current disappears abruptly. If a similar experiment is repeated, this happens almost at the same N value. An analogous effect is in opposite direction. If we make connections between different initially unbound points of the net to be constructed, then I(n) for a long time is zero and only near some big n there suddenly appears a continuous connection of two parts and the current jumps to approximately the initial value I in the previous experiment. This effect has much common with the formation of public opinion. Pair discussions do not give for a long time any significant effect on the many-body community. So the majority of people think that it is not useful to correlate mutually individual opinions because it gives no visible effect. Really, we see that these pairing contacts are of fundamental importance. It is a **necessary** preparation for future public opinion. This is one of elements of optimism: we can influence what happens with the society. There is another physical analogy: pairing correlations of electrons in metals can cause the superconductivity corresponding to the unified invincible opinion of the society.

There are two kinds of life: real one (we work, eat, etc.) and virtual one of the interpersonal conscience (ideas, models, plans). These are two parallel worlds: **"real"** and **"virtual" (mental)**. Not many people understand that the second one with its unlimited possibilities of brainpower is much more important (even more **real**) than the first one. This second one can be improved infinitely by nonlinear amplification of our possibilities due to mutual correlations (collective degrees of freedom - many-body effect). Our 'thought wavelets' can become unified waves on the global ocean of brain-matter. It will give us the combination of super-freedoms due to super-favorable environment (with micro restrictions and macro profit).

As a rule, the creative activity makes people happy. Our experience convinced us in a possibility of the genuine democratic society when everybody will take part in this most interesting and productive being. It can suit any kind of work. We can consider this problem as analogue for universal education: to 'make discoveries of elementary bricks'. Recently, I tried to discuss my personal view on this delicate subject in our institute. It was not easy to dare to give such a seminar talk. Some of my colleagues recommended me to give up this idea. They think that only opinions of great scientists can be interesting. Such modesty is one of our disappointing mistakes (complex of defectiveness that suppresses our talents, self-braking). And really, any idea, which seems important to us must be transferred to other people.

One of simple statements is: **when something is unclear, do not be ashamed of that you do not understand**. Even simple things often require significant time to be mastered. We must **not be afraid of asking simple questions**. Then we must get accustomed to manipulate these simple bricks. **Do not be ashamed if you have confused in several simple elements** (again not to be embarrassed, even quietly and slowly we can go very far). All such difficulties do not mean that you are not capable of discovering something that can wonder the world. As a rule, real science can be made clear. It usually consists of easy elementary steps as the process of climbing by alpinist on a seemingly inaccessible vertical rock.

Soon after the beginning of my scientific work I had an instructive talk with Levitan, the classic of the quantum inverse problem. I could not long dared to meet him and ask many prepared questions. Finally, I overcame the doubts and with fear made a telephone call L. He invited me to his Chair at the mathematical Department of Moscow University. I expected either immediate answers to my trivial problems or reproaches of my incompetence. It was a great surprise that L. said that he himself did not know what I wanted to understand. At first, I was somewhat frustrated? Later I understood that it was great. So L. supported the fact that I am at the boundary of what is still unknown. Any step further will be a discovery. My questions were not at all absurd and I must be braver and search for solutions that were unknown for anybody.

During the work on this text I have better understood the unified character of findings in physics and lyrics and that here the percolation can also be instructive model. All is constructed of elementary bricks (from family to the world). Every discussion of a fundamental problem is an important step to solution for many-body happiness.

For organization of the creative work the experience of the famous music composer Chaikovsky can be useful. He criticized another Russian composer Glinka who waited long for inspiration and produced little in comparison with his genius abilities. Ch. himself began every morning to work without ready ideas, but the inspiration comes during the work. The mathematician Halmosh in his memoirs said something like this: "I want to be a mathematician". He explained that although he likes most his mathematics and it gives him maximal satisfaction, but he often has difficulties (internal resistance) with the beginning of the favorite work. In order to avoid tense thinking, he began to sharpen pencils or to cut the nails. But it appeared that it is better to sit quietly down at the table, take a clean sheet of paper and to write there, e.g., a simple question, and then another one. So, gradually he drew himself into the serious work.

It is also important to communicate the competing people without hindering one another. I have even published a paper "Do we need to extinguish talents; the model of ideal creative collective". It is on my homepage in Internet. I believe that our fundamental interests are almost the same and do not contradict one another which simplifies our many-body problem cardinally. It reminds me of some clever tsar's ministers (their memoirs), who were able to convince the all-powerful monarch to proceed with his own interests. And so we can treat others as tsars and gain additional freedoms by this tactic (see the 'net on the sphere' model of community consisting of 'prime ministers').

Our future will be inevitably better and better (up to some fluctuations). It is because only everybody's understanding of his/her own genuine benefits (not perverted values) is needed. And this can be achieved without compulsion. Those who do not steel believe in better future are self punished by their social pessimism (mighty stimulus for improvement). And our efforts on the way to the ideal goal of 'interpersonal soul' will be supported with the immediate prizes of perfection of our spiritual environment. For me there is an excellent example of Schweizer whose noble deeds for Africans provided him with the so great world's sympathy, which no king could even dream to have.

Conservation of Biological Diversity

John Skinner

University of Pretoria, Faculty of Veterinary Science, Private Bag X04, ONDERSTEPOORT, 0110, South Africa Republic, john.skinner@up.ac.za

The Southern African Development Community consists of 12 countries covering 9.3 million km^2 and 195 million people live here. 12% of the region is protected while clusters of conservation areas include large (10 000 km^2) transnational Parks. Biodiversity concerns and elephants dominate conservation protocols in the region. Calls for elephant control emanate from concern for biodiversity and, beyond Parks, conflict with man. Consequently cries for elephant management are common. But conservation across southern Africa is not restricted to elephants although they, and people dominate the landscape. Several biodiversity hotspots have been recognized where high levels of endemicity occur and where man-induced fragmentation of habitats threatens their existence.

V. Burdyuzha (ed.), The Future of Life and the Future of Our Civilization, 349.

Dialogue among Civilizations as a New Approach for International Relations

Mohammad R. Hafeznia

Department of political geography, Tarbiat Modarres University (TMU), Tehran, Islamic Republic Iran, hafezn_m@modares.ac.ir

After the collapse of the bipolar system different views and theories were expressed by scholars and thinkers about the future of the world and the international system. One of them is the theory of "The clash of Civilizations" which was expressed in 1993 by the Samuel Huntington. This theory caused some anxieties in the world. In reaction to this theory, M. Khatami, the former president of the Islamic Republic of Iran, proposed the way of "Dialogue among Civilizations" as a paradigm in the international relations. 53rd General Assembly of the United Nations on 3 September 1998 accepted it and approved a resolution for the purpose of promoting dialogue among cultures and civilizations and nominated the year 2001 as "the year of Dialogue among Civilizations". This article presents the study of both mentioned theories and refers to the role of dialogue approach in creation of peaceful relations between nations and states.

Keywords: Dialogue, Civilization, Clash of Civilizations, Dialogue among Civilizations, International Relations.

Introduction

States and countries as the members of international community are enforced to make relations with each other. They have special functions in the world system. Any states and governments for being and security have aims that name national aims and interests. For achievement to these aims on the international arena the rational foreign policy is needed. Hans Morgenta, who is the prominent scholar of international relations, says the true aims of the foreign policy of any country are defense and expansion of

V. Burdyuzha (ed.), The Future of Life and the Future of Our Civilization, 351–360.

national interests (Ranney, 1993). Foreign policy of any country is defined on the basis of the constitution and is performed by the foreign affairs ministry and other related organizations (Qavam, 1991).

The correct understanding of the international situation, the position of the country in the world system and also effective diplomatic activities can enhance the national power and promote the situation of the country in the international system (Kazemi, 1994).

There is a direct relation between the national power, the political leadership and the geopolitical position of the country in the international system with the achievements of the successful foreign policy (Hafeznia, 1999). These relations can be displayed in this model:

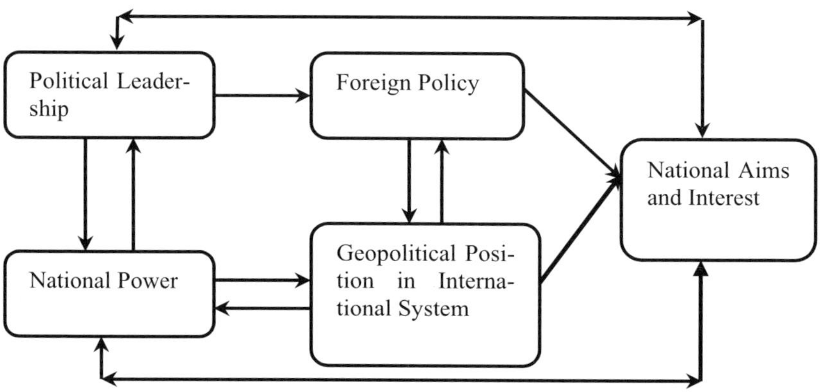

From: M.R. Hafeznia (1999)

M. Khatami, the former president of the Islamic republic of Iran, has presented his idea of the "Dialogue among Civilizations" in General Assembly of the United Nation in 1998 as a new approach to the international relations. This idea welcomed by the majority of the states and affected on the position of Iran in the world.

The Background of Dialogue among Civilizations

This idea was formed under the effect of the three factors:

as rection to the theory of "The Clash of Civilizations" which was propounded by Samuel Huntington in 1993. Professor S. Huntington is a scientific advisor of government and director of the John. M. Olin Institute for Strategic Studies at Harvard University. His theory is the product of the Olin Institute's project "The Changing Security Environment and American National Interests". The theory presented in the frame of an article was published in Foreign Affairs at summer 1993. S. Huntington expressed his theory as an analytical framework for the future of the world after the Cold War era. His statement is:

"It is my hypothesis that the fundamental source of conflict in this new world will not be primarily ideological or primarily economic. The great divisions among human kind and the dominating source of conflict will be cultural. Nation states will remain the most powerful actors in world affairs, but the principal conflicts of global politics will occur between nations and groups of different civilizations. The clash of civilizations will dominate global politics. The fault lines between civilizations will be the battle lines of the future." (Huntington, 1993)

Huntington in the framework of a spatial pattern mentions seven or eight civilizations, which include: Western, Confucian, Japanese, Islamic, Hindu, Slavic-Orthodox, Latin America and possibly African civilization.

He emphasizes that if does not prevent the clash between the components of civilizations then large conflicts between civilizations will be occurred. The most probable struggle will be happened between the Western civilization with a coalition of Islamic and Confucian civilization.

This theory caused some anxieties in the academic and political centers discussing future of the world although international changes and transformations after the Cold War enhanced a level of disputes in the world.

On the basis of this situation M. Khatami propounded his idea of "Dialogue among Civilizations" as a mechanism for prevention of civilization conflicts and as a paradigm for international relations.

1. Appearing a new approach in foreign policy of Iran took place in 1995-1997 years. The new policy had an emphasize on:
   Détente of relations between Iran and other countries;
   Making confidence, peace, cooperation and reciprocal respect;
   Mutual understanding, Dialogue and cultural policy.
2. The experience and personality of the former president M.Khatami. M. Khatami in his antecedent has been the minister of Culture and Islamic Guidance and his field is philosophy so he is a cultural person.

Moreover during his ministry he has experiences of a religion dialogue especially between Islam and Christianity. From another side the theory of dialogue and mutual understanding based on culture has a special language and differs of language of power, militarism, economic interests and diplomacy of power equilibrium. Therefore M. Khatami has not been known only a mere statesman but he is known as a reformist thinker and a humanitarian politician.

World Wide Reaction to the Dialogue among Civilizations

M. Khatami expressed the idea of dialogue among civilizations in his speech to the fifty-third United Nations General Assembly on 21 September of 1998 and proposed to declare 2001 as the year of dialogue among civilizations. He said:

"I would like to propose, in the name of the Islamic Republic of Iran, that the United Nations, as the first step, designate the year 2001 as the "Year of Dialogue Among Civilizations", with the earnest hope that through such a dialogue the realization of universal justice and liberty be initiated... Establishment and enhancement of civility, whether at national or international level, is contingent upon dialogue among societies and civilizations representing various views, inclinations and approaches. If humanity at the threshold of the new century and millennium devotes all efforts to institutionalize dialogue, replacing hostility and confrontation with discourse and understanding, it would leave an invaluable legacy for the benefit of the future generations.

Similarly, it is necessary that, as members of the United Nations, we revisit the history of the formation of this organization with a view to reform and improve the institution through a rational exchange of views" (Bekker, & Pretorius, 2001)

This proposal was welcomed and was received support by 179 states in the General Assembly (Dehghan, 1999). Besides, this led to resolution 53/22, which was formally adopted also.

This resolution welcomes to the collective endeavor of the international community to enhance understanding through constructive dialogue among civilizations on the threshold of the third millennium. The resolution had four paragraphs:

1. Expresses its firm determination to facilitate and promote dialogue among civilizations.
2. Decides to proclaim the year 2001 as the United Nations year of Dialogue among Civilizations.
3. Invites Governments, the United Nation systems including the UNESCO and other relevant international and non-governmental organizations, to plan and implement appropriate programmes to promote the concept of dialogue among civilization including through organizing conference and seminars and disseminating information and scholarly material on the subject.
4. Request the secretary-General to present a provisional report on activities in this regard to the General Assembly at its fifty-fourth session, (53rd plenary meeting: 4 Nov. 1998)

After this acceptance of these proposals General Secretaries of the UN, UNESCO and the government of Iran performed some endeavors for development of this concept by forming the conferences, roundtables, conventions and etc. Some political leaders supported the idea of the former president M. Khatami. For example: Nelson Mandela (South Africa),

Jiang Zemin (China), Mahatir Mohammad (Malaysia), Eduard Shevardnadzeh (Georgia), Bin Ali (Tunisia), Hertsuk (Germany), Banarian (India), Presidents of Italy and Austria and even Samuel Huntington.

Also some regional and international conferences have been held throughout the world. For example:

OIC (Organization of Islamic Countries) Symposium on dialogue among civilizations, Tehran, 3-5 May of 1999 (Beker & Pretoria. J, 2001;

Seminar on cultural-civilizational relations between Iran and Africa, Tehran, 1-2 May of 2001;

The conference on religion and dialogue in Harare, 12 of May 2001 (Newsletter of the conference, 2001);

International conference on dialogue of civilizations in Austria with attendance of the president of Austria and the Secretary General of UN;

130th Sessions of worldwide interassembly in Oman with presence of Mr.Pico as the representative of the secretary General of UN in dialogue among civilizations (Gozaresh goftegoo, 2001);

The conference of dialogue among civilizations in Beijing (China), 12-13 September of 2001 (Ettelaat, 9 May 2001).

Moreover, the attack of 11 September of 2001 on the America sensitively affected on relations of the Islamic and Western civilization and the attention again turned to the theory of dialogue among civilizations. This event also prepared a new bed to activate Iranian diplomacy and consultations with the governments to prevent the world of civilization and cultural conflict.

For example, travel of European delegation to Tehran in 26 Sep. 2001 and their talking with Iranian Officials especially president Khatami.

- In this visit Mr. Luee Michael the head of the European delegation, Chavier Sulana secretary General of foreign policy of European Union, Josef Pick the foreign minister of Spain were presented (Ettelaat, 27 Sep. 2001);

Jack Straw the foreign minister of U.K visited Iran and talked with Iranian officials. (Hayat-e-no, 26 Sep. 2001);

The president of South Korea (Mr. Kim day Joung) in his meeting with the editors of Asian Mass Media emphasized on the role of civilization dialogue for détente in the world (Ettelaat, 19 Sep. 2001);

Tony Beler the Prime Minister of U.K in a message requested M. Khatami to play an active role for prevention of encounter between religions and civilizations (Ettelaat, 17 Sep. 2001)

In the period of inflammation of American terrorism events, connection and consultations between president M.Khatami and UN Secretary General took place.

Especially after the tragic events of 2001 in USA tensions between the Jewish and Western states with the Islamic societies and states have increased. The importance of Civilization Dialogue Paradigm in the world enhances the geopolitical position of Iran in the international system.

The Philosophy of Civilization Dialogue

International relations in the world are under the influence of two systems or realities:

1 - a formal and legal system, 2 - an informal and geopolitical system.

The legal system comes into existence on the basis of reciprocal rights and respect between states. This system is a set of conventions, agreements, pacts, treaties, and international organizations in the regional and global scale. International organizations are a gathering of countries, which come into existence on the basis of multilateral treaty or agreement for achievement of common aims and they have a legal personality (Moghtader, 1995).

In this system states have equal rights and powers on the basis of proclamation of the principles of international laws (Mousazadeh, 1997).

The antecedent formulation of the legal system and international organizations in the international relations backs us to the Wien congress (1815), the Hague conferences (1907, 1899) and to the economical, technical and social transition in Europe (Clave, 1990; Moghtader, 1995). The evolution of trends of this process in 20th century culminated in formation of two international organizations, namely the League of Nations and UN (Colliard, 1985).

The geopolitical order is a system of relations between states which is formed on the basis of their geopolitical weight that is the source of national power. This determines the position of any country in hierarchy of the world power. Therefore the quality of international relations is the reflection of the pattern of a world wide geopolitical system.

In this system the process of a transnational political organization takes place on the pivot of a state that in the global or the regional scale is the most powerful one.

In this system a powerful state with utilization of visible and invisible tools and instruments tries to lead the regional and global relations in to the direction of its aims and interests.

The two mentioned systems have relation with each other. But the main point is this that the legal system is commonly under the influence of a geopolitical system that reflects the power relations. So in formal structures and organizations both regional and global the powerful members and states usually affect and influence on the process and decisions and partly take their leadership position on the hand (Hafeznia, 1999).

The existence of power relations paradigm between states has caused division of the world such as first world and third world, developed and under developed, core and periphery, north and south, rich and poor, etc. In other words the creation of injustice space in the international relations is a product of this paradigm. In this situation the language of the states for talking with each other is not equal.

Any dialogue requires belief to the quality of personality and respect to each other. Therefore dialogue can create possibility for mutual understanding and achievement of the peace and détente tenseness in the relations between nations and states.

Other ground which is necessary the civilization dialogue is competitions and conflicts between nations and societies on the basis of identity. This phenomenon, especially after the collapse of bipolar system, has grown up in the world. These competitions and conflicts were affected on the forming of clash of civilizations theory.

This is a reality that the struggle and conflicts between racial, religious and ethnic groups expand and humankind has the better experiences especially in Africa, south east Europe, south of Asia, Caucasus, Central Asia, south east Asia, and etc…

Recently the racist thoughts have revived in America and Europe. This evidence is a clash between some European racists with Asian families in some cities of England, Germany, France and Netherlands. Aggression and attack to the Muslims and Arab peoples in America and Europe after the 11th September of the year 2001are more important.

After the terrorist attack to America we can see some expressions and discourses in the speeches of some Western political leaders such as J. Bush – the president of USA that referred to the crusade war and clash between civilizations or the prime minister of Italy who had told about the superiority of Western civilization over Islamic civilization (Ettelaat, 3 Oct. 2001). Some scholars and political leaders of the Islamic and Arab world denied these allegations. Ayatollah Khamenai, one from leaders of Islamic Republic of Iran, was sharply against of the Bush' expression: "any body which is not with us is with terrorists". He said: "we are neither with you (USA), nor with terrorists" (Ettelaat, 27 Sep. 2001).

Till now some West and developed countries on the basis of their colonial past have racist views derogating other nations and states.

Edward Said writes: orientalists utilized the contempt of others as the instrument for legitimating of geopolitical strategy by the imperialistic countries and this manner continues from the past up to the now (Said, 1998).

Farmanfarmaaian writes: during the past two decades the options of political fighters in the middle East such as the Palestinians, Arabs and Muslim peoples interpreted as the savagery which is absent in the West civilization (Farmanfarmaaian, 1998).

O'Tuathail explains the roots of imperialism and seeks superiority of the white race over the other races. Also he notes that Roosevelt, the former president of USA, like other imperialists believed racism and preferred the white people to others (Tuathail, 1998).

The racist views and looking self-superiority is not being only in the West. They existed between other racial and ethnical groups. The sentiment of identity on the basis of one or more factors is growing. So the control of this trend and regulation of the relations between different groups require development of the culture of dialogue.

The third factor of which follows necessity of a dialogue is development of security in daily life of societies. We are not in security. Terrorism with any motive holy or not holy automatically is dreadful. Usually innocent people are suffered. There is a point that classification of terrorism needs to study.

In any case terrorism using deferent ways and tools brings to fear, destruction of homes, buildings and loses property of people.

Besides, development of terrorism especially in its political kind has a direct relation to the development of tension between nations and states. Prevention of threats of security of the societies requires the acceptance of a way of logical dialogue among states.

Therefore settlement of an international system on the basis of justice and reciprocal respect in the relations of nations and achievement to the public security and settlement of peace and peaceful coexistence and cultural and mental interactions between humankind is related to the acceptance of culture and strategy of Dialogue by the main players in the world such as governments, religions, parties, leaders, social elites, scholars and etc...

Conclusion

Samuel Huntington presented the theory of Clash between Civilizations as a paradigm for explanation of the world situation after the cold war that caused some anxieties.

In opposite to its President Khatami as a statesman with cultural personality, presented his idea of civilization dialogue as a paradigm for international relations on 53rd session of UN General Assembly.

This paradigm welcomed by the General Assembly and nominated the year 2001 as the year of Dialogue among Civilizations.

Also the heads and officials of some countries in the world welcomed to the Idea. From the other hand after the terrorist events in the USA the expectations of the world community about the role of civilization dialogue increased.

Continuation of power relations has produced social and geographical inequality between human communities. Development of identity's sentiments and competitions on the basis of religion, race, ethnicity, language, place and ect helps to the grow of misunderstanding and tensions between cultures and civilization groups. Increasing insecurity arising from multi dimensional terrorism brings to necessity of the development of a Dialogue between nations and cultures. This approach in the new world can lead peace and security for us and now we need peace and security more than in any other time.

References

Abi Saab, G. (1994) Concept of International Organization. Tehran: Elmi & Farhangi; Pub.

Amiri-Vahid, M. (1996) *The Clash of Civilization*. Tehran: IPIS.

Bekher, T. & Pretorius, J. (2001) Dialogue Among Civilizations- A Paradigm for Peace. Pretoria: UPS.

Colliard, C.A. (1985) *Institutions des Relations Internationales*. Paris: Editions Dalloz.

Dehghan, M. (1999) *Evaluation of the Détente Policy*. Tehran: Ettelaat Newspaper.

Demko, G.J. & Wood, W.B. (1994) Reordering the World. U.S.A: West view Press.

Farmanfarmaaian, A. (1998) The Geopolitics Reader (Did you Measure UP…). London: Routledge.

Ettelaat Newspaper, No.22290, 9 Sep. 2001. P. 16

Ettelaat Newspaper, No.22291, 10 Sep. 2001, P. 2

Ettelaat Newspaper, No.22297, 17 Sep. 2001, P. 2

Ettelaat Newspaper, No.22299, 19 Sep. 2001, P. 16

Ettelaat Newspaper, No.22306, 27 Sep. 2001, P. 2

Ettelaat Newspaper, No.22310, 3 Oct. 2001, P. 19

Hafeznia, M.R. (1999) *Optimal Pattern for International System*. Daneshvar, 25, 23-30.

Hafeznia, M.R. (1999) *The New Approach to foreign Policy and changing the Geopolitical Position of Iran*. Tehran: 10th Congress of Iranian Geographers-Imam Hussein University.

Hayat no Newspaper, No. 385, 26 Sep. 2001, P. 3

Huntington, S.P. (1993) The Clash of Civilizations. Foreign Affairs, 72(3), 22-50.

Huntington, S.P. (1996) *The Clash of Civilizations and The Remaking of World Order*. U.S.A: Simon & Schuster Inc.

Inis, L. & Claude, J. (1986) The Record of International Organizations in the Twentieth Century. Taipi: Tamkan University.

International Centre for Dialogue among Civilizations. (2001) Gozaresh goftego, year. 1, No.10, 23 Aug. 2001, Tehran: ICDAC.

International Centre for Dialogue among Civilizations. (2001) Gozaresh goftego, year. 1, No.11, 6 Sep. 2001, Tehran: ICDAC.

International Centre for Dialogue among Civilizations. (1998) Ketab Mah. Tehran: ICDAC.

International Centre for Dialogue among Civilizations. (2001) (Report 1). Tehran: ICDAC.

Kazemi, A.A. (1994) *International Relations in Theory and Practice*. Tehran: Ghoomes Publishing Co.Ltd.

Moghtader, H. (1995) *Public International Law*. Tehran: The Institute for Political and International Studies (IPIS).

Moussazadeh, R. (1997) *Public International Law (Vol.1)*. Tehran: (IPIS)

Qavam, A. (1991) *Principles of International and Foreign Policy*. Tehran: SAMT.

Ranney, A. (1993). *Governing: An Introduction to Political Science*: Prentice Hall.

Report Trip. (1999) – Millennium of Dialogue and Understanding. Tehran: Rasaneh Publication.

Said, E. (1998) The Geopolitics Reader (Orientalism Reconsidered). London: Routledge.

The Committee on Dialogue. (2001) Newsletter United Nations. Year of Dialogue among Civilizations. Harare: U.N

Tuathail, G.O. (1998) The Geopolitics Reader (Introduction)- London: Routledge.

New Proposals to Conserve Life and Civilization

Vladimir Burdyuzha[1], Oleg Dobrovol'skiy, Dmitriy Igumnov[2]

[1]Astro-Space Center of Lebedev Physical Institute, Russian Academy of Sciences, Profsoyuznaya 84/32, Moscow, Russia, burdyuzh@asc.rssi.ru
[2]Institute of Radiotechnics, Electronics and Automatics, Vernadskogo 78, Moscow, Russia

The preservation of life and Civilization on our planet is investigated in detail. Two proposals have been done in essence. It is shown that for the unity of stability and mutability a many poles world system is necessary. This system will be more effective than present one. The importance of the creation of center for study of the Future is noted. This center must support an ineffective activity of United Nation Organization (UNO) in some questions. The conservation of our planet for the future generations and the development of present Civilization on its must be the primary objective as a separate person, a separate state as and all mankind. But who thinks about it now.

The Earth is our Common Home

Nobody can not exactly define consequences of breaking of a fragile ecological equilibrium. These consequences can arise because of natural, technogen and social catastrophes especially in our nuclear – information century. A great number of dangers have lain now in the modern world order that force to look in the Future with trouble. Our present Civilization does not limit by Tigris or Ethrat, Dardanel or Ind. It is all World.

For last century more than 20 millions persons only in Europe were perished from Spanish influenza. The same number was perished of car accidents in all World. In Russia more than 40 millions persons were perished during civil and patriotic wars, starvation and the repression of Stalin regime. In Cambodya for a short period of Pol-Pot regime were executed

V. Burdyuzha (ed.), The Future of Life and the Future of Our Civilization, 361–364.

some millions of peaceful habitants. In Africa more than 50 millions persons died also on starvation, epidemic and military clash. Thus more than 150 millions of persons for the last century died in different situations which were not natural ones. Besides, hundreds of thousands of people were died in peace time of hurricanes, tsunamis, earthquakes and so on. The shame war was in Vietnam when napalm was used, mass murder of Kurds was in Iraq when chemical weapon was used, catastrophe in Chernobyl has also applied to mass died of people.

The full number of lost people is more than 400 millions for last century. The number of invalids is else more. If a number of unborn children takes into account then summary loses are irreplaceable. In the result a huge moral and material damage to Civilization was given. Other words 10 % of population of our planet were perished in last century (a lion part of which may be theoretically prevented). Please, think under these terrible figures.

Are we Intelligence Civilization?

It is evidently that practically every day people are killed of terrorist acts. Remember other time 11 September in USA. Imagine that among killing could be your children (Godforbid). Why no one state, no one international organization did not stop cruel murders of people in Cambodia. Iraq, Ruanda, Soviet Union. Probably the existing system of International safety does not function effectively. The main reasons of these catastrophes are good known. But we think that it is necessary to add some ones. First of all: closeness of societies, nationalism, unreadyness of conscious of people to estimate consequences of coming catastrophes, unreadyness of people to join for creation of effective system of safety. Besides, one of reasons is egocentrism of some leaders that bring to animosity and military conflicts.

All Countries are Neighbours in our Information Century

Modern catastrophes may stop life on the Earth. It is a medical fact. Which are the methods to prevent this? Which are possibilities to conserve life?

One way to provide safety is joining up of all countries under patronage of UNO in questions of forecast of our future. Other words it is necessary to create Center for research of the future. We discuss this in our declaration. (This Center will be a scientific support of UNO).

Second way to provide safety is strengthening of the modern world system. Here a mechanical example may be useful. 3 legs (better 4) is necessity for stability of a table. What will happen if a part of legs is broken? The loss of stability will take place. Excuse for this simplest example. Therefore at the beginning it is necessary to have a large number of these legs. We propose many poles organization of world. Supremacy (dictate) of one pole (state) is always fraught with serious problems (that we observe now). Even two power super states (as it was before) have not given stable world. An alternative is many poles world. But presence of a large number of poles makes such system very slow. Here mutability and development are not possible. It brings to stagnation or even to death. Probably many great empires of the past have perished on this reason. There unity of stability and mutability was broken. The presence of power states in all relations is not only desirable but it is necessary. The synchronism of components must be obligatory. But with other side all components must not be under dictate since that system is not able to development, to self-perfection. Probably here some optimum could be take place.

We propose 7÷8 poles on our planet for constant progress and development. They are: USA, West Europe, Russia, China, India, Latin America and Muslim World (here two poles may be: Sunnites, Shiites). The standard variant of poles assumes a core and some autonomics states. Of course, structures of poles may differ very much each other (a nucleus can be 2 states: China and Japan, USA and Canada et so on). A world pole is a very power state formation that has extreme man resources, military, scientific and industrial ones. In this case nobody can attack that colossus. Probably poles must have a single government. A president or a king is the head of a pole and he (she) presents it interests in the UNO.

The highest organ of authority in the world, as before, is reforming United Nation Organization that should be softly transformed in the Council Safety. This organ must possess by absolute authority in our world. The Council will consist of constant members representing poles. Each pole must have one decisive voice. Of course, exclusions may be. States like to Israel or Switzerland have given a high contribution in the modern Civilization and they may be members of this Council.

Early and now serious conflicts were and are unfortunately. The national and confessional intolerance lie in their ground. Then terrorism arises but a good decision to save world from this is absent. The presence of only one pole (like now) is not positive moment for struggle against terror. We think that terror may be decreased in many poles world. Also extreme important is struggle with drug addiction. It is very strange but opponents are here. In some small countries consumption of narcotics is permitted. This problem is not regional it is common to all mankind. It threats life on the Earth. Any portion of alcohol, tobacco, narcotics will cripple our Future. It is prolonging suicide. As and terror, narcotics are a plague of our time, unfortunately.

Which may be role of science and religions in this many poles world? Science is a fundament of any society. But and religions are very important. Practically any religion is directed on conservation of life, on healing of our souls. Orthodox Church is a soft form of religion like to homeopathy. Catholic Church is a more hard form of religion like to chemotherapy. Muslim Church is extreme hard form of religion like to surgery. All know that a knife is a tool of healing but and a tool of murder. It is necessary to remember about it.

We do not think that is necessary to proof that science and scientific researches are important since positive mutability gives only science (unfortunately and negative one also). Of course, it is necessary a power financial support for epochal scientific breaks especially in space research. Financial, energetic and people resources can be concentrated only with help of many poles world system. In modern history attempts were to put an end of some scientific searches (cloning as an example). Here it is necessary a hard expertise. Of course, religious leaders must not be outside of these processes.

Conclusions

1. We suggest to create International Center for research of our future which can support ineffective activity of United Nation Organization in some questions (Scientific Center of UNO).
2. To provide unity of stability and mutability a many poles world system is necessary. Probably this system can better preserve our Civilization and life.

V

WHAT IS OUR FUTURE

Eco-Ethics Must be the Main Science of the Future

Brian Marcotte

Eco-Ethics International Union, Strategic Analysis, Inc., 401 Cumberland Avenue, Suite 1102, Portland, ME 04101-2875, USA, brianmarcotte@thecollegium.org

Eco-Ethics is an essential part of the scientific method. Climate change will present human beings with new ecological, economic and ethical challenges in which the survival of the species and its place in "the economy of nature" is at stake. The time scales for both ecological change and economic response are now well matched a fleeting moment in the history of human beings and this argues for immediate ethically appropriate action.

V. Burdyuzha (ed.), The Future of Life and the Future of Our Civilization, 367.

The Vital Tripod: Science, Religion and Humanism for Sustainable Civilization

Bidare V. Subbarayappa

Science and Spiritual Research in India, 30, M. N. Krishna Rao Road, Basavangudi, Bangalore-560004, India, phispc@bgl.vsnl.net.in

There is no denying that the world has not only witnessed but also is a repository of amazing scientific and technological developments in the physical and biological realms alike. There is now a vast stockpile of scientific discoveries and technological innovations – a stockpile of such a magnitude that it can augment the material life of all nations in the world and provide every human being an opportunity for leading quality of life with human dignity.

In recent times there have been some path-breaking scientific endeavours in the areas of (i) Genes; (ii) Agro-technology; and (iii) Information and Communication Technology. DNA era was ushered in 1953-1954 when the epoch-making paper on the Double Helix structure of DNA was published by James Watson and Francis Crick. Thereafter came up major developments like the identification of genetic code in 1960s, recombinant DNA and DNA sequencing in 1970s, Human Genome Project as well as others since the 1980s. These have undoubtedly far reaching consequences for mankind as a whole. Their beneficial dimensions like breeding new GM varieties that have tolerance to drought, salinity, attack of pests and disease, improvement of nutritive qualities etc. cannot be ignored. That such new varieties need to be developed to face problems of the foreseeable future needs no emphasis because of the global limited resources in water. They would also increase productivity in semi-arid and dry-farming areas to augment food security in these otherwise vulnerable human settlements. The pathway of genetic engineering should have the compulsions of a futuristic vision for the augmentation of nutrition and health of the whole humanity. It is here that humanism should be the sole determinant,

V. Burdyuzha (ed.), The Future of Life and the Future of Our Civilization, 369–377.

but not the intellectual property rights; neither profit motive, nor economic nomic superiority.

In the twentieth century, human intellect scaled new heights in science and technology – multi-level harnessing of natural resources, materials science, electronics, communication, medicine, space exploration, nuclear energy and several others. However, alongside the last century has also been a harbinger of social tensions, disorders and anti-human conflicts. Behind this scenario lies what may be called human entropy – an entropy not physical, but largely mental. In essence it is an ever expanding disorder – generating selfishness – individual, communal and national. This has erected dangerous barriers between man and fellow-man. A person trained in the method of science is supposed to be an embodiment of the paradigmatic objectivity and rationality. Ironical indeed that in human affairs, between him and his fellow humans, an individual by and large is an impulsively charged ensemble of selfishness and sectarian tendencies. The blinding curtain of self-satisfaction needs to be unmasked, discarded. This is possible only through a conscious fostering of humanism in a way that the individual sees and feels the fellow – man in the same manner as he sees and feels for himself.

There is another side, rather on ugly one, of the scientific and technological developments – the production of armaments and weapons of mass destruction. Several nations are vying with one another for attaining supremacy in this destructive potential of science and technology, of killing man by fellow-man. It needs to be recognized that the killing instinct is very strong in man, stronger than that in other animals. From early times to the present, the history of mankind is more a record of onslaught of man on fellow-man through battles and wars for one reason or the other. The present is no exception. Recent studies by scientists at the International Peace Research Institute, Oslo, on the cause of armed conflicts over the past three decades, have indicated that the root cause of many of the conflicts could be traced to economic rather than ideological differences. Economic disparities in different parts of the world, the 'haves' and the 'have nots', have a long saga – social, colonial and political. These disparities have tended to generate internal tensions and external conflicts of unprecedented magnitude among nations; even within a nation, among its different societal sectors. It is unfortunate that a substantially big proportion of national GDP continues to be spent on armaments and devastating sophisticated weapons. On the other hand, the amount being spent on programs leading to poverty eradication and the fulfillment of basic needs of the vast majority of people is far less. Is it not a distressing fact that the number of children, men and women living in abject poverty ***today*** exceeds the entire human population that existed in 1900, despite the enormous and extraordinarily beneficial S & T innovations that could elevate the quality of every individual with a happy future?

Dehumanisation

In this connection, there is another dimension that merits our attention, namely, in the midst of spectacular scientific and technological achievements there has been an increasing concern over the steady deterioration of human values leading to dehumanization. This has surfaced in all cultures in several ways. A few months after the dropping of the first atom bomb in 1945, Albert Einstein observed: ***'I believe that the horrifying deterioration in the ethical conduct of people today stems primarily from the mechanization and dehumanization of our lives – a disastrous by-product of the development of the scientific and technical mentality. I do not see any way to tackle this disastrous short-coming'*** [(*Albert Einstein: The Human Side*, (eds)] Helen Dukas and B. Hofmann, Princeton University Press, N. J. 1979, p. 82). Einstein was thus emphasizing the fact that the scientific and technical mind-set or attitude alone, though it is enormously productive, does not and cannot lead to a value-based ethical, harmonious life.

About a decade later Bertrand Russell and Einstein in their *Manifesto* (July, 1955) exhorted: "We appeal as human beings to human beings. Remember our humanity, and forget the rest. If you can do so, the way is open to a new Paradise; if you cannot, there lies before you the risk of universal death" – a clarion call to scientists technologists, policy-makers and others. It needs to be recognized that human progress or what is labeled as 'development' is a two-in-one concept – first, the harnessing of natural resources without environmental degradation for elevating the material life; and the second, the refinement of man himself into an enlightened human being. Science as well as its application is one side of the coin; human refinement is another side. Each in itself is incomplete; while the blend of the two would be complete, a totality for human progress.

Scientific attitude is perhaps the best for an acceptable understating of Nature in all of its manifestations. But a reflection on the past, in the ancient period, reveals that, over the long ages, what was assiduously fostered was an integrated relation, Man **in** Nature or Man **in harmony** with Nature, in contradistinction to Man ***and*** Nature or Man ***against*** Nature that characterize modern scientific approach. Till a century and half-ago, the symbiotic relationship between man **in** nature received its sustenance, among others, through a religious mind-set – an inner experience that was part of an individual's life and, through it, of the society at large.

Religion

The word, *religion*, etymologically has the connotation of binding together or cementing. But its canvas has become so wide and complex that it eludes precise or capsuled definition because of the way in which it has evolved and diversified itself over the ages. However, as Julian Huxley has pointed out religion is a natural product of human nature and one could even say that religion is innate to the human existence itself. To a large mass of people religion generally means the acceptance and close adherence to Divine Being or supernatural force. Even so, one cannot overlook the religious impulse that is inherent ***in*** or natural ***to*** man ever since his appearance on the planet. Over the centuries, its manifestations have been molded by different traditions, attitudes and varied ritual practices that have led to certain denominations under the broad nomenclature of religion.

It needs no emphasis that the human mind, constituted as it is, has viewed with sanctity and respect fortified by religious fervour, several phenomena, especially the cosmic influences and celestial occurrences. It cannot also be denied that there is an inseparable relationship, between the life on the Earth, the Solar system in particular, and the Universe in general.

Modern cosmological ideas too have their own nuances in this respect. For example, the most current concept, namely, the Big Bang cosmology concerning the origin of the Universe has little or no explanation to offer as to **why** the 'great explosion?' occurred about 13 billion years ago. Nor does it throw light on **how** life origi-nated later on the Earth. It raises the question: Is there any unmani-fested **Power** behind the Big Bang? An emergent idea is that the seminal features of the Universe are governed by the very same constants and forces that are also the pre-requisites for the emergence of life on the Earth. If they were even slightly different from what they are now, scientists point out that it would have been impossible for not only the universe as it is now but also for the emergence of life. A cognate question then is: Is there any overarching design or a Great Designer amid all of these? Moreover, the con-stituents of all animate matter from primal amoeba to humans are made up of the same types of elements as witnessed in inanimate matter. Thus the consonance of microcosm and macrocosm con-templated by all religions looms large even from the modern scien-tific stand – point.

Today it is not in the context of the anthropomorphic idea of God or a personal God who is presumed to respond to prayers and to grant booms, that religion needs to be viewed or interpreted. But it is in the light of the celestial-terrestrial concordance, or the conso-nance of microcosm and microcosm, the purpose and values of life that one has to look at religion.

In recent times, several scientists have begun to examine the validity of the idea of a designer as well as a purpose and the associated religious attitude. There is now an increasing awareness among them that science and religion are not opposed to each other. Such an opposition, if any, is in the nature of the thumb 'opposing' the other fingers by means of which anything can be grasped. In any case, science and religion have basically one common feature – an incessant search for truth, for the origin of and the overarching design of the universe, and the emergence of life itself.

The noted astrophysicist and cosmologist, Fred Hoyle who was an atheist to start with, saw an intricate design in the Universe and wrote later in his auto-biography: 'The atheistic view that the Universe just happens to be there without purpose and yet with logical structure appears to some to be obtuse'. Freeman Dyson exclaimed: 'I do not feel like an alien in the Universe. The more I examine the Universe and study the details of its architecture, the more evidence I find that the universe in some sense must have known that we are coming'. (*God for the 21st century,* Ed. Russell Stannard, Templeton Foundation Press, Philadelphia, 2000, pp. 183-185).

Religious fervour has two dimensions: one of experiencing an undifferentiated concordance of the microcosm and the macrocosm; and the other, of fostering values of life like love, compassion, truth-fulness, non-violence and non-possession. All religions, irrespective of their labels, have been preaching and promoting these values of fundamental importance for meaningful life. Bereft of them, human life is arid; with such values the dividing labels of religions look ap-parent and not real. Man overcomes such boundaries, becomes human, engendered by human values. Religious experience stimu-lates humanism.

It is pertinent to ask such questions as: can science, the greatest enterprise of mankind today, be expected to be a source of humanism? Or can it enrich the content of humanism? Or can it reinforce other sources of humanism?

Answers to the there questions can be perceived by the following observation of the Nobel Prize winner for Physiology, Albert SzentGyorgi:

> “If my student comes to me and says that he wants to be useful to mankind and go into research to alleviate human suffering, I would advise him to go into charity instead. Research wants real egotists who seek their own pleasure and satisfaction, but find it in solving the puzzles of Nature’ (*Science Today,* May 1980, p. 35)

Einstein had his own explanation to offer in respect of these questions:

> ‘My scientific work is motivated by an irresistible longing to understand the secrets of nature and by no other feeling. My love for justice and striving to contribute towards the improvement of human conditions are quite independent from my scientific interests’ (Helen Dukas and Banesh Hoffmann, p. 18).

Though science and technology have been the most useful elements for material development and progress, they have not proved themselves to be the fountain source of humanism. On the other hand, they tend to generate dehumanization. Each nation endeavours to attain superiority in S & T to subjugate, as it were, other nations. It is important to realize that the mind-set of individuals, policy-makers, political leaders and religious exponents needs to have the entire humanity as its centrality, transcending the geographical barriers, national pride, race superiority and similar aberrations.

Avaricious ‘Outer’ Man

Life or living systems are not exactly in the nature of closed systems. They have an openness and freedom of their own, particularly human beings and communities. Though a living organism endeavours to transform as much as possible of its environmental energy into itself, man seems to be different from all other living beings in this respect. A greedy energy scavenger that he has been, he garners more energy and carelessly dissipates it in such a way that a considerable part of it becomes practically unavailable for future use. Thus he contributes to a greater disorder to the environment; also he chooses to maintain his own selfish physical satisfaction at the expense of the attendant degradation of his environment. In his avarice to be the undisputed master of biosphere, he has been instrumental for the extinction of some species, global warming and imbalance of Nature.

The energy-flow at the individual and societal levels has been of such magnitude that today it has assumed alarming proportions. This has been creating greater and greater disorder to the detriment of the much needed symbiotic relationship between man and his environment. Man is perceptibly moving now towards the peak of what may be called a high-entropy culture. Constantly engaged in augmenting his material wealth and greedily involved in his sensorial satisfactions, he has been squandering the natural energy-resources. The material man or the sensorial avaricious 'outer' man is today more in evidence than the 'inner' man embodying human values and goals. It is distressingly true that a biologically systemic living entity like man has become in several ways more harmful to the environment than the other types of living species. Though endowed with intellect and its immense capabilities, man has chosen woefully to be a maker of his own degradation, eventually leading to a none-too-happy scenario of the future of humanity, future of our civilization.

Human mind is capable of being circumscribed as well as transcendent; circumscribed by the vicarious play of senses; transcendent, beyond the senses into a realm of undifferentiated oneness with humanity. The circumscribed mind with its total disregard for humanity in general, is lured away by the material desires which are insatiable since they are governed by senses. Excessive material consumption and possessions beyond one's own needs result in the dissipation of natural wealth which otherwise could be utilized in a manner beneficial to the humanity as a whole through an enrichment of the environment. If an individual contributes more to the environment than what he takes away from it, he will leave behind an ambience that would be more fertile for the future than what it was for him.

Such an approach becomes seminal to what may be called the low-entropy culture – a culture where the individual – society relationship would be one of ethically inspired, corporate living among all societies and, more importantly, sharing the benefits of nature. In other words, man and nature become integrated entities; For this purpose, there needs to be constant endeavour to view with disdain and to discard the present day practices such as '*exploiting nature for material living* or '*taming the nature*' with a view to establishing the supremacy of man over nature. It also needs a religious vision or spirituality that transcends the insatiable material cravings which afflict man in his day–to–day existence. The inner spirit needs to be religiously aroused leading to an exalted and ennobling humanism – an experience that cements man with fellowmen without any type of distinction or differentiation. This would be a bulwark against inner enemies of man, namely, greed and hatred. This generates a new-awakening that perceives humanity as a whole; and this traverses a new pathway in which greed is discarded, need is minimised; ego is set aside, enlightened humanism is engendered; hatred is shed; universal love is nourished.

Science, with its vast potential of enriching our material life; *Religious attitude* with its emphasis on the values of life; and Humanism that knows no barriers of regions, race, color or creed, would together constitute an irresistible tripod that could resolutely sustain and elevate our civilization. Each by itself is inadequate; each needs the other for human progress; and the trinity would be a fortress against any type of destructive potential that might undermine human civilization.
human civilization.

Our civilization which has experienced a sea-change in its multi-level material enrichment, is looking for a new order – an order that sustains it by fulfilling the yearnings of man to lead a life of values in harmonious relation with nature, with fellowmen. The viable foundation for a sustainable new order is the vital tripod of Science, Religion and Humanism.

References

Barbour, Ian G.: *Religion in an Age of Science*, Harper and Row, San Francisco, 1990.

Barnett, Lincoln: *Universe and Dr Einstein*, Mentor Book, New York, 1952.

Boslough, John: (1) *Stephen Hawking's Universe*, Fontana/Calvins, 2nd impression, Glasgow, 1989; (2) *Masters of Time,* Phoenix-Orion Books, London, 1992.

Chalmers, David J.: *The Conscious Mind in Search of Fundamental Theory,* Oxford University Press, New York, 1996.

Clark, Wilson: *Energey for Survival,* Doubleday, New York, 1975.

Davies, Paul: *Accidental Universe,* Cambridge University Press, 1982; *God and the New Physics*, Simon and Schuster, New York, 1983.

Delsemme, Armand: *Our Cosmic Origins,* Cambridge University Press, Cambridge, Reprint, 2000.

Douglas, Jack D. (Ed): *The Technological Threat,* Prentice-Hall, N. J. 1971.

Dukas, Helen and Hofmann, B.: *Albert Einstein: The Human Side,* Princeton University Press, New Jersey, 1979.

Dyson, F.: 'Energy in the Universe', *Scientific American*, 225, 1971, pp. 50-59.

Georgescu-Roegen, Nicholas: *The Entropy Law and the Economic Process,* Harward University Press, Mss, Cambridge, 1971.

Heilbroner, L. R: *An Inquiry into the Human Prospect,* Norton, New York, 1974.

Huxley, Julian: *Science, Religion and Human Nature*, Watts & Co, London, 1930.

Penrose, Roger: *Shadows of the Mind,* Oxford University Press, New York, 1994.

Prigogene, Ilya and Stengers, Isabelle: *Order out of Chaos*, Fontana, Glasgow, 1984.

Russell, R. J: (1) 'Finite Creation without a Beginning: The Doctrine of Creation in relation to Big Bang and Quantum cosmologies' in *Quantum Cosmology and Laws of Science: Scientific Perspectives on Divine Action*, (Eds) Russell, R. J. et al., Vatican City State, 1993; (2) 'God and Contemporary Cosmology' in *God, Science and Humanity* (Ed) R. Hermann, Templeton Foundation, Philadelphia. 2000.

Sagan, Carl: *Cosmos*, Random House, New York, 1980.

Stannard, Russell (Ed): *God for the 21st Century,* Templeton Foundation Press, Philadelphia, 2001.

Thomas, William, L: *Man's Role in Changing the Face of the Earth,* University of Chicago Press, 1956.

HIV/AIDS and the Future of the Poor, Illiterate and Marginalized Populations

Rajan Gupta

Theoretical Division, Los Alamos National Laboratory, Los Alamos, NM 87545, USA, rajan@lanl.gov

Introduction

Today, of the global population of 6.5 billion people about 2 billion have access to modern facilities (food, water, shelter, sanitation, health care, education, jobs), who I characterize as the "haves". Of these 2 billion, roughly 1 billion live in the developed (industrialized) world and the second billion consist of the top 15-20% of the remaining world population. About 3 billion are poor (living under \$2 Purchasing Power Parity (PPP) per day) and do not have access to modern facilities [1]. Of these, roughly 1 billion people live on less than \$1 PPP per day and constitute the extremely poor. The remaining 1.5 of the 6.5 billion are in transition between the poor and the haves, i.e. they have access to some but not enough of the modern facilities. These ratios are unprecedented in human history and there are many recent success stories of development at the national level – Taiwan, South Korea, Singapore and Ireland being the most obvious – nevertheless, there remains a much larger global need.

The central message of this article is that we must act with a sense of urgency to accelerate the transition from poverty to developed modern societies. The reasons are both humanitarian and strategic – to preserve global peace, security and prosperity.

V. Burdyuzha (ed.), The Future of Life and the Future of Our Civilization, 379–400.

In this talk [2] I would like to first summarize the status of the global burden of HIV/AIDS and then use the fast spread of HIV as an example to draw attention to a much deeper and more fundamental problem – the very future of the poor – the 3 billion people living on less than $2 PPP per day. I then discuss the interconnected cycle of threats to the development of the poor and end with a prioritized list of urgently needed interventions and argue that we, for the first time in history, have the resources and the understanding (that it can be done, that it will benefit all, and that it will not decrease the wealth of the "haves") to implement the necessary and needed programs that can remove extreme poverty, if not poverty, by 2025. The purpose of this article is not to apportion blame or point fingers but to highlight the possibility that we can create a global society in which no child is denied the opportunity to develop and be part of the 21st century. If the discussion appears one-sided it is because the developed world has the most to offer and is developing most of the technology while the poor need help.

HIV/AIDS

In 1981 the world first came face to face with a new virus that destroys the human immune system when five young males checked into hospitals in Los Angeles and other major metropolis in the US with highly compromised immune systems – the AIDS (Acquired Immune Deficiency Syndrome) patients. Over the next three years researchers worked very hard to decipher the etiology of AIDS, means of transmission, and develop tests for detecting the virus. By April 1984, the teams of Dr. Robert Gallo of the National Cancer Institute, Dr. Luc Montagnier at the Pasteur Institute in Paris, and Dr. Jay Levy at the University of California, San Francisco, had identified the cause as a new retrovirus, subsequently named Human Immunodeficiency Virus, or HIV. The first HIV/AIDS antibody test, an ELISA-type test, became available in March 1985. Unfortunately, by that time the CDC had already reported 10,000 cases of HIV/AIDS in the United States, 4,942 deaths and many more were estimated to have been infected. For a timeline of the spread from 1981-88 see http://aidhistory.nih.gov/timeline/index.html

The impact of HIV/AIDS over the next ten year period (1985-95) was devastating. Even though the means of transmission, mainly through unprotected anal and vaginal sex and shared needles by IV drug users, were known very early on, the number of HIV infections grew exponentially. In

US alone the numbers grew from about 1000 cases in 1983 to 10,000 cases in 1985 to 100,000 cases in 1989 to 500,000 cases in 1995. Relatives and friends watched loved ones die and the medical community struggled without hope. Doctors could diagnose but could not treat and make whole those infected. Once the ELISA test was discovered they were able to make the blood supply safe, thus stopping the accidental spread in the developed world, but their initial hope of the quick discovery of a vaccine proved unjustified. The number of people dying annually in the developed world due to AIDS continued to rise until 1995 (reaching about 42,000 per year in the US) when triple cocktails of protease inhibitors combined with [non-] nucleoside reverse transcriptase inhibitors proved effective in bringing viral loads to negligible numbers in most patients. These drug cocktails changed the face of HIV/AIDS – a diagnosis of HIV/AIDS no longer meant imminent death. Unfortunately, these drugs were extremely expensive and toxic and the anti-retroviral therapy (ART), the aggressive form of ART is called HAART for Highly Active Anti-Retroviral Therapy, was available only in the developed world. Tragically, the burden of HIV/AIDS in the developing world has, by and large, continued to mount as if no therapy is known and in many Asian and Sub-Saharan African countries as if even the means of spread and prevention are still unknown. Even though generic manufacturers have brought down the cost of drugs to roughly $150 per person per year, nevertheless, access to ART by the poor remains very limited. The uncontrolled spread of HIV/AIDS amongst the poor continues to highlight the divide in access to health care, education, and empowerment to make safe choices between the haves and the have-nots.

By the end of 2004, roughly 40 million people were estimated to be living with HIV and 30 million had died (http://www.UNAIDS.org). In 2004 alone approximately 5 million new infections and 3 million deaths took place. Large parts of Sub-Saharan Africa are being devastated. In many regions an entire generation has been wiped out leaving grandparents to look after AIDS orphans, many of whom are HIV positive. School, hospital, security and government services are collapsing as trained staff are dying faster than new people can be trained. This need not be the case. HAART has been shown to extend the life of people in both the industrialized and developing world. Even the poorest and totally illiterate people have shown amazing fortitude in dealing with the stigma of coming forward to receive treatment, in tolerating the toxic side effects of the drugs and adhering to the demanding drug regimen. Generic manufacturers in Brazil, India and Thailand have brought down the prices to where we can provide therapy to all those diagnosed and in need (about 10 million) for about $2 billion per year. The major impediments to large-scale administration of

HAART continue to be the lack of infrastructure to deliver the medicines and the lack of financial resources (by nations and individuals) to buy the drugs in the first place. Programs in many developing countries (Botswana, Brazil, Cambodia, South Africa and Thailand) have shown that the infrastructure and delivery of drugs can be scaled up rapidly if and when drugs are made available. Unfortunately, large pharmaceutical companies holding drug patents, with the support of their governments, continue to delay access to generic versions by the poor by invoking WTO laws on patents and intellectual property rights.

Behavioral changes that would end risky sex and sharing of needles by IV drug users have been very slow. In the US the number of infections crossed the one million mark in June 2005. There continue to be 40-45 thousand new infections annually and most of the spread is in poor minorities (especially black and Hispanic women) and marginalized communities – gay men and IV drug users. This is an unacceptable burden for the richest and most developed country but dwarfs in comparison to what is happening in the developing world where roughly 90% of the infections (roughly 5 million each year) and 95% of the deaths (roughly 3 million each year) globally due to HIV/AIDS are taking place. It has increasingly become clear that poverty is a major factor in the spread [3]. The poor and the marginalized lack information, means and empowerment to protect themselves, and day to day survival (or anonymity in the case of men having sex with men and IV drug users) very often trumps safe behaviors.

Allowing people to suffer and die when medicines are available is undermining attempts to help their communities develop. Knowledge that should be handed down from parent to child is being lost, and children are growing up hungry and without supervision. The demographic impact of HIV/AIDS, killing people in the most productive stages of their lives, makes it unique in its impact on development. The long-term problems posed by AIDS orphans, already numbering 15-20 million, will need more resources to address than those needed to keep their parents alive and productive. The continued fast spread of HIV/AIDS and its devastating impact on families very clearly illustrates how hard it is to break the cycle of poverty for those living without basic provisions of health care, education, and nutrition.

The question I have repeatedly asked myself is – if we know the means of transmission of HIV (unprotected anal or vaginal sex, infected blood entering one's body mainly through reuse of needles, and from mother to child) and none of these are casual contacts, then why is HIV still spreading? Within the haves the answer is that many people continue to indulge in risky sexual behavior or are slaves to their drug addiction in spite of adequate knowledge of HIV/AIDS. Within the have-nots, poverty, despondency, day to day survival and lack of knowledge, means of protection and empowerment, make behavior change very difficult to achieve. The situation, in a much broader context, was captured amazing succinctly by Gro Harlem Brundtland, who in 2002 as the Director-General of the WHO said [4]

> *The world is living dangerously, either because it has little choice or because it is making wrong choices. On the one side are the millions who are dangerously short of the food, water and security they need to live. On the other side are the millions who suffer because they use too much [or are too indulgent]. All of them face high risks of ill-health.*

To this I will only add one thing which I elaborate on later. The haves, by not helping half the global population that lives on less than $2 (PPP) per day to break the cycle of poverty, are creating dangerous breeding grounds in which exploitation, violence, criminal behavior, terrorism, civil wars, and diseases will continue to fester. Even though the scale of the many problems and their solutions are large and daunting, my contention is that we have the knowledge and resources to address them [5, 6] and, furthermore, I add that we have no choice if we desire global peace, stability and prosperity. What we need is political and social will based on the principle of shared fate.

The first wave of HIV/AIDS has devastated Sub-Saharan Africa. This devastation did not elicit (until after 2000) much response from most people in power or policymakers sensitive to geopolitics. The cynical view was that none of these countries are important geopolitically except for their natural resources, and tapping these does not need large numbers of people. The slightly less cynical view was – yes we need to help but there is no local infrastructure through which to intervene. Past attempts at infrastructure creation have been failures and since the governments of these countries are corrupt or ineffective or both any help that is provided is abused. So how does one act to help? Fortunately, it is becoming increasingly clear that if the haves intervene at the requisite levels and holistically (as discussed at the end of this article) then the have-nots are able to rise to the challenge, work hard as partners, and are capable of breaking the cycle of poverty within a generation.

The second wave of HIV/AIDS threatens many populous, strategic and militarily powerful nations – China, India, Russia, Ethiopia and Nigeria – as cautioned by the US National Intelligence Council and the CIA in its 2001 report [7]. Recognizing that each of these countries has large populations of have-nots and marginalized people, it becomes clear that under business as usual approach HIV/AIDS will grow rapidly in these countries and if it grows to anything like the scale in Sub-Saharan Africa, that could lead to global instability. So challenges like the Millennium Development Goals, Global Fund for HIV/AIDS, Tuberculosis and Malaria, and I would like to add Kyoto protocol on climate change because natural disasters and climate change affect the poor disproportionately (the New Orleans hurricane Katrina tragedy is a glaring example), have been developed and are increasingly being supported. Assuming that the haves are willing to help the poor stand up so that they can, in time, learn to walk on their own (develop into modern societies) what I would like to do in the rest of the article is provide, based on my work in India, a list of key common threats to this development, and in light of these threats discuss what is the minimum holistic package needed to seed the transition.

It is important to stress, while making the connection between HIV/AIDS and poverty, that poverty is not the only factor. Removing poverty may not be enough to solve the HIV/AIDS problem since risky sexual behavior and IV drug use are prevalent in all socio-economic sectors. Nevertheless, experience from the developed world shows that addressing issues of poverty makes pandemics, at the very least, manageable, i.e. stabilization followed by a decline in the number of yearly infections versus uncontrolled growth. Thus, our primary concern should be to help all people achieve basic human rights and freedoms and simultaneously work to help those prone to making unhealthy choices reduce risk to themselves.

Challenges The Poor Face

The challenge we face is to provide resources and skills to the poor to jump start their transition to the 21st century. Unfortunately, there are many "sharks and barracudas" that profit from keeping them poor, exploit them, and undermine even the best attempts of very committed and innovative people to provide skills and resources to the poor. Principal amongst these sharks and barracudas are

- Despotic and/or corrupt governments
- National and transnational criminals
- Fanatics and terrorists
- Exploitative multinationals

The important question is how and why do these sharks and barracudas persist in the modern world? The answers are age old – through control and ruthless use of military power, through exploiting religious and ethnic differences, and through co-opting unscrupulous governments and influential individuals.

In this article I will not have the time to even begin to discuss bad policies by governments and the control and ruthless use of military power by them for their own self-centered purposes instead of focusing on alleviating poverty. Most people would contend that poor governance, bad policies and lack of security are the key primary factors that have to be addressed and without which nothing else will have long-term impact. This may very well be true, however, I believe that individuals can do a lot on the ground, especially facilitate the creation of a civil society. The issue I wish to highlight is that, today, non-state actors can be equally effective and important agents of destabilization or progress because they have access to significant resources, and some individuals control resources that are larger than the economies of many nations. To discuss the positive possibilities I start by highlighting the need to eliminate a major global problem that severely impacts the poor. This is the control exerted by criminal and exploitative individuals and organizations that amass vast fortunes through what I call rogue economies. These include

- Narcotics
- Alcohol
- Tobacco
- Trafficking in arms, people and human organs
- Trafficking in ill gotten goods (blood diamonds, illegal timber, counterfeit and pirated goods and medicines, banned animals and their products, etc.)

I refer to these activities using a politically loaded term "rogue economies" because they are large enough to qualify as economies and, while not all are illegal, they have illegal components. They are all insidious, are a health hazard, undermine development, undermine law and order, create mafias, warlords and drug-lords, promote corruption and prevent the formation of a civil society. I contend that as long as these rogue economies persist there will be sharks and barracudas that thrive. Today these sharks and barracudas have global reach and in some cases are richer and more powerful than governments and, in some cases, are the government.

Production and trafficking of narcotics is the prime example of a rogue economy. Even conservative estimates put the money involved at over \$500 billion per year [8]. The economies, security and development of many nations (the most significant being Columbia, Peru, Bolivia, Mexico, Nigeria, Ukraine, Russia, Tajikistan, Afghanistan, Pakistan, Myanmar,

Laos and Thailand) are overwhelmed by the drug trade. Industrialized nations spend hundreds of billions of dollars yearly to try and confront the menace but have had very limited success. Each of the countries mentioned above also illustrates the connection between drug trade, private militias, money laundering, lack of development, civil wars and terrorism. In both the industrialized and developing worlds IV drug use is a major factor in the rapid spread of HIV/AIDS. I do not believe that anyone, having given even a moment's thought to this problem, would argue that the illegal drug trade is not a global problem that needs to be addressed with a sense of urgency.

A significant fraction of the adult population (10% is a typical estimate) of many developed nations has a serious alcohol abuse problem. Nevertheless, alcoholism and its consequences for health care, lost productivity, domestic violence, crimes and road accidents are treated as an acceptable burden because of the overall prosperity. Social drinking is considered a safety valve against stress and thus a lesser of social evils. Promotion is accepted because it involves an individual's lifestyle regarding a legal activity. In the developing world alcohol abuse has become a nightmarish problem. Anyone with first hand experience of India understands that alcohol is a major impediment to development, especially in rural India. I have not visited any rural community in India where women have not listed it as the number one problem. It is the major cause of domestic violence and financial hardship. Children growing up in alcoholic homes do not have safety nets that can compensate for the stressed and unhealthy home environment and very often grow up without adequate health care, nutrition and education. A detailed analysis of the nexus between narcotics, alcohol abuse and sexually transmitted infections including HIV/AIDS in India and their implication for development and security has been presented in reference [9].

The laws on international trade in tobacco baffle me. How is it possible that the US and most of Europe have declared smoking as hazardous to health, banned smoking in public buildings, and are enacting ever more stringent laws against it and yet feel no obligation to ban export of cigarettes? What I would like to advocate is that if Americans (or citizens of any other country) wish to smoke then so be it, but we should not allow export of a health hazard (in the form of finished products like cigarettes, chewing tobacco, and cigars) to continue just because it is profitable. The case against trafficking in arms, people, human organs, exotic species and ill gotten goods is so obvious that I hope it needs no discussion.

The problem we face is that these rogue economies are so entrenched and pervasive and the profit margin is so high that they have defied control. Unfortunately, their impact on the poor continues to be devastating.

What do the poor need to develop? Generalizing from what parents offer to children, I consider the three most important things to be:

- Opportunities and Skills: These require access to heath care, nutrition, education, an environment that fosters a love for learning, job training, and activities like sports that provide healthy use of leisure time.
- Direction: This requires there be enough role models that lead by example and exemplify the virtues and payback of hard work and goal oriented focus.
- Stable, loving and nurturing environment: This provides physical, mental and emotional health and leads to the development of safe behaviors.

Rogue economies undermine each of these three. Easily seduced by the possibility of instant and easy gratification and/or money people fail to learn skills valued in the 21st century. The interdependencies between these three developmental goals highlights the need for synergy between physical and social sciences in addressing these issues, i.e. between technology and an understanding of human behavior. In this context a question I have often asked myself is why are the poor targeted by the sharks and barracudas and not by multinationals?

A simple answer is that the three billion poor are, to first approximation, not viewed by multinationals as a significant market for day-to-day commodities in the short term and for luxury goods for a long time because they do not have adequate buying power. They are, however, prime targets of national and multinational mafias and criminal organizations. These organizations are ruthless in their pursuit of profit and bring alcohol, tobacco, drugs, sex and trafficking in people, to the doorsteps of even the poorest people. Since the margin of profit in most of these activities is very large and the cost of the "quantum of indulgence" (cost of a glass of alcohol, pouch of tobacco or marijuana, price charged per sex-act, etc.) can be made arbitrarily small, criminal organizations mercilessly and ruthlessly target the poor, making them captive end users and often willing promoters and producers. To understand which economies penetrate the rural sector faster one only needs to compare the growth in these to other items that I have heard mentioned most frequently like soap, shampoo, and cosmetics that can be packaged and sold in small quantities. Therefore, if we wish human development and nation building to succeed, we need to confront the specter of rogue economies. Such focus is in the joint interest of both developed and developing nations as both pay a very high direct and indirect long-term price.

Even if we accept the depressing verdict that most of today's poor will not make the transition, what about the next generation? Unfortunately, of the 3 billion poor, roughly one billion are children under 15 who also will not get the health care, nutrition, education, skills or family support needed to develop and access 21st century jobs. My estimate of the 2005 distribution of these vulnerable children is roughly as follows: about 350 million in South and Central Asia (60% of children under 15); 30 million in West Asia (40%); 90 million in Southeast Asia (50%); 110 million in East Asia (33%); 25 million in central America (50%); 50 million in South America (45%); 290 million in Africa (80%); and 11 million in Eastern Europe (20%). These children are growing up aware of what they are missing (through TV and tourism) but will not have the skills to reach a decent standard of living themselves. Growing up with ever increasing but unfulfilled expectations, being repeatedly exploited, and without hope for betterment they could become a significant negative force. The sharks and barracudas can easily exploit and manipulate them. They can easily be recruited to serve as foot soldiers in civil wars (there are numerous recent examples in Africa, particularly in Liberia, Sierra Leone, Angola, Congo, Sudan, Rwanda, Zimbabwe, etc.) or for carrying out acts of terrorism (Middle East, Southeast Asia). It is therefore very important that the haves provide the have-nots with hope, means for steady improvement and enable them to climb the steps that will allow them to escape poverty.

The poor in different parts of the world face a number of challenges – some are specific to a region and many are common to all. I would like to focus on seven that are common.

- Population growth: Almost all the population growth is amongst the poor. With continued population growth and mechanization, each generation of the poor starts with fewer assets (particularly land) and increasing needs. Children born in poor households face a steep up hill battle for survival, are at tremendous economic and social disadvantage from birth and the norm is a life marked by neglect, deprivation, despondency, and abuse. The poor need free and easy access to modern methods of disease and pregnancy prevention and counseling to help them realize planned families as a choice.
- Lack of health care: In spite of astonishing developments in medicine and medical processes over the last 50 years and continued efforts to provide universal coverage, the poor face many impediments in accessing even basic and essential modern medicine. Starting with birth outside of registered medical facilities, partial or no immunizations, poor and inadequate nutrition throughout childhood, and constant exposure to food-, water- and mosquito-borne parasites leaves them underdeveloped.

In those that survive, the largest impact is on their mental abilities – precisely the abilities needed to be successful in the 21st century. The resulting mental deficiency is the hardest (if not impossible) to undo in later years even with the best resources. In addition, today's job markets place high value on emotional intelligence whose development requires a stable, loving and nurturing childhood. It is, therefore, unlikely that a significant fraction of the poor can make the transition to white collar jobs. Our best hope is to work with them, raise their ability to access basic health care and provide the basics so that their children can make the transition.

- Education: Many of the poor have not had any formal education so they, on their own, cannot help their children learn or provide an adequate environment for learning at home. Many of the rural schools are dysfunctional or provide very poor instruction. Thus, even when the poor make incredible sacrifices for years to send their children to school they find that the returns are very small. The children have learned very little and cannot access jobs outside menial labor. The investment of time and money leads to high expectations which are often dashed. Such children end up in no-man's land – they are aware of and know what 21st century promises and what they are missing but cannot be part of it.
- Energy: At the very minimum energy is needed in the form of light to study by at night, as kerosene to cook with so that a significant part of the day is not spent collecting firewood, for making water potable and for transportation. With increasing mechanization and reliance on technology the need for, and dependence on, energy increases in every aspect of life. Thus, easy access to inexpensive energy is essential for development. There is steady progress in connecting the poor to the energy grid, however, as discussed later there is a looming global energy crisis that needs to be addressed with a sense of urgency.
- Water, Land and Food: Access to water (irrigation and ground water to supplement rainwater) was key to the green revolution along with better seeds, fertilizers and pesticides. The majority of the poor live in rural areas and subsist on what they can grow on small patches of land. Most of this land has poor depleted soil and the output is low. Without electric power the poor cannot access groundwater. The adoption of treadle pumps, that work where the water table is close to the surface (2-5 meters), has been slow. They cannot afford good seeds or chemical fertilizers or pesticides and have few reserves to survive even one bad harvest without taking out loans. Lack of easy access to markets and mounting debts increase their vulnerability to takeover by large farmers or moneylenders. The general trend is that more and more subsistence farmers are

losing their land and are being forced to migrate to towns and cities in search of menial jobs. This migration creates a set of very difficult challenges of health care, urban sprawl, pollution and exploitation as well as many new opportunities for development [10]. Migration, in general, is a tide that cannot be stopped and, as I will argue later, essential for development. However, there is urgent need for better tools and planning to manage it both at the level of rural to urban and international migration. Similarly, there is urgent need for better tools and planning to manage rural and urban water resources.

- Environment: The poorer a person is the more she/he feels the full force of the extremes of nature. In the agrarian sector the poor are almost totally dependent on nature for survival. Lacking adequate shelter, clothing and nutrition they are ill-equipped to survive harsh winters (air-conditioning during summer and heating during winter need connection to the energy grid and are expensive and, therefore, still distant dreams). It is therefore not surprising that poverty, and especially extreme poverty is mostly restricted to within the tropics, i.e. areas with mild winters that are conducive to all life – of humans and pathogens. The poor do not have financial reserves to absorb the impact of untimely rains, extreme droughts or storms. Furthermore, air and water pollution continues to pose very severe health problems. On top of all these impediments, reduction in land holdings with every passing generation and degradation of the soil accelerates the process of forcing the poor to sell their lands and migrate to urban areas in search of menial jobs as they lack other skills.
- Poor governance, lack of modern institutions and missing civil society: Regions of poverty often have poor governance. Corruption and abuses of power become a way of life. Usually, a small minority controls most of the resources and shows little interest in developing the infrastructure (health care, education, jobs, communications, transport, etc) needed for broad-based development. All these problems are exasperated by civil unrest. Lacking modern institutions, the few people capable of bringing about change are not able to. Many of them migrate to industrialized countries or, over time, get co-opted into the corrupt system or loose their drive or are eliminated.

These seven drivers are highly interconnected and mutually reinforcing. In complex systems with highly non-linear behaviors (large couplings and feedback mechanisms) it is very hard to know in time when the threshold for runaway solutions is reached. One is often fooled by the slow initial linear rise in spread and burden and therefore reacts too late and too slowly. Such worrisome conditions exist in most, if not all, struggling countries. It is, therefore, very important to address them in an integrated holistic manner. Unfortunately, the problems are astronomical and tangled up with religious, cultural, tribal or ethnic sentiments. Thus, large-scale intervention with the commitment to build, over decades, the infrastructure and a large enough pool of dedicated and trained people to implement programs has not happened.

Why is funding so important to seed development? Historically, the poor have relied, to a very large extent, on both a formal and informal barter system for subsistence existence. To access modern amenities like education, health care, energy, sanitation and clean water requires hard currency. Accumulating adequate savings or a having a regular source of income to pay for these facilities is a significant new threshold that the poor have to cross before they can start to develop. Having steady income and enough left over after paying for basics like food are major problems for the poor and not just the extremely poor. Very often despondency, especially under the pervasive pressure of rogue economies, undermines attempts at developing sound fiscal behaviors. As a result they fail to access these amenities and cannot cross the threshold. The majority of the poor live in rural areas where the infrastructure for providing the basics is harder to develop and maintain as rural areas lack the economy of scale provided by urban areas. Since most migration to cities begins with the adult male, with the rest of the family staying in the village due to lack of housing and higher cost of urban living, villages must provide better facilities for the children. So in one way or another (through indigenous or external funding) the greatest need of the poor is access to modern basics and this requires money, people, and institutions. The obvious question is where will these funds come from if not from the haves?

Why Business As Usual Is Not Enough

Given all the recent attention on reducing poverty (see, for example, the United Nations Millennium Development Goals [11]) why should one be concerned that under business-as-usual scenarios these seven enablers will not be in place for the poor by 2025? There are many reasons:

Consider, for example, the issue of energy which is essential for modern technological societies. Assuming that by 2025 the global population is 8 billion people and the long-term desired standard of living is that of Europe (currently the per capita usage of power by Germany, France, the U.K and Japan is roughly 5.5 kilowatts, which is half that of the US), we would need three times the current primary energy used. Unfortunately, there are no proposals or plans to provide such an increase. The energy future is actually precarious as we remain dependent on fossil fuels whose reserves are dwindling and the share of solar and wind energy is small and not yet cost effective. The recent increase in oil prices and the possibility of "peak production" of conventional crude oil within this decade, followed by a global decline in production, will have major repercussions for even the industrialized world. The situation with respect to natural gas is similar but with a time delay of about 20 years [12]. Anticipating increasing competition for dwindling resources between the developed countries and growing economies like China and India, the prices of oil and gas have almost doubled in the last two years. Such fast rise in prices could easily slow the process of connecting poor households to the power grid or, worse, derail development in many parts of the world. For example, India's 2005 bill for imported oil (1.6 million barrels of crude oil per day) at $60 per barrel was $35 billion. Thus, crude oil imports alone soak up 46% of the total export revenue of $76 billion. This increasing expense, when there is a significant federal budget deficit (about 9-10% of GDP), does not bode well. It is, therefore, not clear whether even India can continue to grow and address the needs of its poor if energy prices do not stabilize. (China, for the time being, has much deeper pockets to withstand such price increases due to its large positive annual balance of trade.) The global energy infrastructure and needs are so large that we desperately need new R&D and bold thinking to ensure a stable global energy future.

History is full of examples of societies overexploiting land and water resources [13]. The results have usually been devastating – change in rain patterns, desertification, loss of biodiversity and pollution [14]. Governments find it almost impossible to advocate and enforce conservation and managed use of resources once people start viewing the resources as an entitlement to survive or make a profit. Take India as an example. To help the farmers sustain the green revolution the government provided cheap electricity to pump groundwater and highly subsidized irrigational water. Today in large parts of the country the water table is falling 3-5 feet per year and now the government finds it impossible to deny farmers their "right" to electricity and to continue pumping groundwater. Similarly, subsidized irrigation continues to support inefficient and harmful practices (untimely and over watering by flooding the fields without means to deal with the salt buildup). There is talk of looming scarcity of ground water and increasing soil salinity due to irrigation that threaten many prime agricultural areas but there are no plans and little action to address the issues. For example, implementation of drip irrigation by farmers who cannot afford the capital cost requires funding (low risk loans) and training. In rural communities this requires mobilizing a very large fraction of the population which is best achieved if they understand what is at stake and how they can benefit from the investment. Thus, along with stabilization of population, it is very important to develop scientific tools that allow a comprehensive monitoring of watersheds and ecosystems so that stakeholders can be helped to understand, in time, the consequences and the cost of restitution, and based on this information are willing to put into place interventions before large-scale damage is done. Such systems monitoring and analyses is in its infancy and needs to be developed further to help poor communities manage their resources.

Compared to 50 years ago, the infrastructure for education has been very significantly enhanced but we have a long way to go both in numbers and quality. In the job market the premium today is on quality, versatility and novelty. Even a high school degree is insufficient to get a white collar job. Again, I will use India as an example of where we stand [15]. Of the roughly 25 million children born in India every year (this was the number in the 1990s; by 2004 it had decreased slightly to about 24 million) only about 30% study beyond eighth grade and only about 15% graduate from high school (grade XII). Of the roughly 4 million that graduate from high school yearly, less than 0.5 million receive education of sufficient quality that allows them to be competitive in various college entrance examinations. Of these, only about 25,000 students get university education in sciences, engineering and medicine that is of international standards whereas more than 10 times this number has the potential. With globalization, those who can afford to do so send their children to good universities in the industrialized world. This migration has many consequences, both positive and negative, that would be very illuminating to discuss in detail. For lack of time I will focus on one long-term benefit of this migration that I believe has played a very important role in seeding development in most of the success stories. It makes the case why American universities should continue to educate the world's best and why America should encourage migrant communities to invest in their countries of origin.

Successful, engaged, migrant communities that have large pools of entrepreneurs with skills, capital and networks have been key enablers in rapidly transforming their land of origin. Ethnic Chinese, dispersed over much of South-East Asia and North America, have been a major factor in the development of South-East Asia in general but Taiwan, Singapore, and now mainland China in particular. South Korea benefited from its special relationship with the US. More recently, migrants from India have invested in India and helped seed it's very rapid growth in information technology (IT). Ireland is another recent success story. Needless to say, in each case, in addition to such an empowering migrant community, a large well trained and educated labor force at home, usually under-employed, and a government willing to create an environment that encourages and protects investment and is willing to develop the infrastructure that facilitates trade and industry was necessary. Sub-Saharan Africa is a prime example of a region in which all these factors are missing and I contend that this lack of investment by the expatriates is one very important reason why there seems to be no end to the continued economic, social, political and environmental devastation. Instead of an educated but underemployed workforce that can be tapped, investors see large HIV/AIDS infected populations that need help to survive.

Even with the many success stories but under business-as-usual investment and intervention scenarios I can foresee only eight regions of about 400 million people each that will have access to 21st century opportunities in the next 50 years. These are North America, Western Europe, East Asia and Australia (including Japan, New Zealand, Taiwan, South Korea and Singapore), Eastern Europe and Russia, China, India, oil rich Middle East, and South America. This is a very encouraging and hopeful situation but it is not clear whether these regions will be able to lift the remaining five billion out of poverty (optimistically assuming that the global population stabilizes at roughly 8 billion). Not being an economist it is not clear to me whether there will be enough new technologies and services in place that will provide an ever growing market and meaningful employment to 8 billion people when most mechanization reduces the number of people needed to perform a task. Furthermore, as I have discussed briefly, it is also not clear whether under business as usual scenarios and without major technological breakthroughs there will be adequate resources (energy, water, biodiversity, healthy ecosystems, etc.) to sustain even 4 billion people at the standard of living of Western Europe in the next 2-3 decades.

Path Forward

With ever growing mechanization in all aspects of our lives and the very large investment needed to nurture and educate a person with technical skills valued in the 21st century, what niche or comparative advantage will the poor have to offer to investors. On the other hand it is hard to imagine that a world in which about 3 billion people live in industrialized societies geographically isolated from the remaining 5 billion people will be stable. Free societies will find it increasingly hard to stop large-scale migration, both illegal and legal, to fulfill needs in certain sectors of employment, and because people are willing to risk everything to move to where a better life is possible and, today, they know where to go to get it. Even if the skilled and unskilled were geographically meshed and integrated together, will there be enough natural resources so that everyone has a high enough standard of living such that no child with talent is denied the opportunity to rise to the top.

So what needs to be done? Should one concentrate on technological solutions or on social transformations and behavioral changes (conservation and sustainable management of resources, cooperation, reducing risky behaviors, etc.) or a judicious combination of the two? Technology is the much easier solution if it exists, but the poor are poor precisely because on their own they cannot access what technology exists. Also, as we are increasingly finding out, especially in environmental issues, technological solutions have been partial and often deployed before adequate understanding of their long-term consequences is in hand. Behavior change involves little or no cost but requires the thinking and maturity that comes with enlightened long-term planning and empowerment – again the poor don't often have the training or these skills. (A pressing example is HIV/AIDS – an effective vaccine would be a god send but its development may need decades of further research and is a big if for at least the next 10-20 years whereas behavior change leading to safe sexual practices would cost an individual nothing but has been hard to achieve within the poor and marginalized populations.) Thus the poor need help incorporating both – technology and behavior change. In this context I will briefly outline the four areas I am most passionate about.

(i) Reproductive health and Family Planning: Providing each person with education on safe sex and reproductive health and free and easy access to modern methods of birth control and disease prevention, i.e. condoms, microbicides and contraceptive pills. I cannot describe the joy shown by even the poorest of rural women in India once they begin to understand their reproductive system and realize the power to have planned families. Assuming that the delivery of a condom costs 5 cents and each sexually active male requires, on average, 100 condoms a year, the total resources required to supply 1.5 billion men would be $7.5 billion. Assuming 1.5 billion women need help buying contraceptive pills at $5.00 per year (price for generics), providing them free would require another $7.5 billion. Thus, including education and infrastructure costs, a $20 billion per year program in reproductive health would address issues of women's empowerment, planned families, population growth, and sexually transmitted diseases including HIV/AIDS. In this case adequate technology exists and what is needed is the political will in the industrialized countries to provide funding.

(ii) Confronting rogue economies: This requires well established and working laws and order systems within nations, education and awareness programs, and global cooperation between nations. Even though the industrialized nations are the end destinations of trafficking in narcotics and contraband and ill-gotten goods, they have not considered confronting rogue economies a priority. Post 9/11, more and more politicians and policymakers are understanding the nexus between terrorism, rogue economies and money laundering and it remains to be seen if this realization, coupled with an understanding of the widespread impact of these economies on development, leads to a global effort that addresses issues underlying the lure, the demand and the supply. In short, both the developed and developing world stand to benefit by bringing these economies under control. It is hard to estimate what additional resources are required to effectively confront rogue economies. Since their impact is global and controlling them is in everyone's direct interest I would not like to call money spent on them aid. So I will restrain from putting a dollar figure and simply say that we need to win this fight which requires education and social activism and functioning and responsible governments.

(iii) Enabling resources – energy, water, proper nutrition, vaccines and medicines: With dwindling or even constant supplies, access to oil and gas will be dominated by countries that hold the reserves and those that can afford to pay for them. Most of the poverty is concentrated within the tropics and this belt is rich in solar power but not in oil, gas or coal except for the Persian Gulf and North and Western African countries. An advantage of solar (photovoltaic) and wind turbine systems is they are easy to install and maintain. Another advantage is that they are local solutions that can be implemented without requiring a national energy grid. Significant investment in these systems has been lacking because fossil fuels have been cheap and energy from wind and solar is intermittent and therefore unacceptable as a source of base load in developed countries used to uninterrupted supply. Today, wind power is cost-competitive with fossil fuels while solar at $0.20-0.30 per kilowatt hour is still too expensive by roughly a factor of 3-4. The poor would, therefore, benefit tremendously if the industrialized world invested very heavily in reducing the cost of these systems and developing the remote monitoring and systems integration capability for making them less intermittent. Moreover, providing even 4-8 hours a day of electric power would change the lives of the poor, and they would not need any fancy system to tell them when to expect solar power.

Water is a more complicated problem with far larger regional variations. We should put priority on developing simple ways of improving water quality, better water management and preventing pollution. Teaching farmers better irrigation methods has to be coupled with guarantee of timely access to water and help with the initial capital cost of say drip irrigation systems. Such information and capability needs to be disseminated through a global education campaign and incorporated in school curricula so that there is public awareness and participation at all levels.

For the development of vaccines and medicines there should be a large global effort. I propose the creation of a Global Jackpot Fund. The organization or individual developing a successful and essential drug or vaccine should be adequately compensated from this fund which should then hold the patent and allow anyone to mass produce and market it as long as they can meet the standards.

Based on this simplistic analysis I estimate an investment of $80 billion a year ($40 billion for energy, $20 billion for water, and $20 billion for essential medicines and vaccines) would significantly accelerate the discovery of essential drugs and vaccines, bring down the cost of renewable energy and potable water systems and make them accessible to the poor.

(iv) Enabling services – basic health care, immunizations, proper nutrition and education: These services require an army of trained people, long-term financial commitment, and well developed institutions to deliver and maintain quality. Developing all these simultaneously poses the largest challenge and there does not seem to be any alternative to stable long-term commitment and support. The good news is that individuals can make tremendous contributions through local action – an individual can start a primary school or a health care clinic. In the long run a collaborative public-private partnership is necessary along with long-term national (and, in the case of poor regions, external) funding. I estimate that a $100 billion per year program targeting the 1 billion vulnerable children (at $100 per child per year) would help provide them with the required basic foundation and skills with which they can compete in the 21st century.

Thus, a $200 billion dollar a year program would make a very significant impact. This is a very large sum but roughly what the world is spending post 9/11 on confronting terrorism and also very close to the 0.7% of the GDP number for development aid that industrialized donor countries and the UN arrived at as the target. This emerging consensus in what is needed is what gives me hope that we today have the resources and the understanding to help the poor make the transition.

It is worth highlighting the similarities and differences between my priorities and, say, those stated in the Millennium Development Goals [11]. Both emphasize poverty reduction, nutrition, health care and education, especially for children. I prefer to put HIV/AIDS education and control within the context of reproductive health, family planning and STI control. I find it far easier to effectively engage men and women (rural and urban) in discussion on reproductive health that empowers them to plan when and how many children to have and how to protect themselves from STIs and other infections of the reproductive system compared to a HIV/AIDS spe-

specific discussion. People respond favorably to a frank discussion on reproductive health as they see it applicable to all, find it in everyone's interest, and do not feel bad or guilty as it does not invoke the stigma and social taboos associated with HIV/AIDS. Also, the benefits (population and STI control and women's empowerment) of providing the means and knowledge for improving reproductive health in general are much more comprehensive, bigger and longer-term. Free and easy access to HAART and medicines for opportunistic infections, TB and malaria control should be made part of the public health care systems.

I put very high priority on targeting rogue economies for they undermine law and order, impede development, undermine health, and help sustain communal violence and civil wars. I wish to focus on them in totality rather than follow a case by case approach, for example, dealing with the vulnerability of IDUs to HIV/AIDS. The consequences of the associated activities are severe for both developing and developed nations and the moral arguments against them are also, I hope, self-evident.

Next, I put a strong emphasis on energy and water as critical resources for which we need low cost solutions. Clean water and energy are essential for development and a healthy environment. They are increasingly becoming value added products whose prices are, without subsides and discounting taxes, roughly the same globally. To satisfy the growing demand for them by the haves, preventing environmental and climate related catastrophes, and simultaneously providing access to the poor requires major technological breakthroughs. Simple small-scale solutions, just like developing vaccines against malaria and tuberculosis, may not have been a priority in the developed world, but recognizing that long-term peace, prosperity and security depends on accepting the principle of shared fate demands that we now think globally. Development of the poor can only happen if the developed world makes very significant investment with the goal of providing low cost and environmentally sound solutions that satisfy the needs of both the haves and the have-nots.

In suggesting the above areas of priority and focus, it is important to restress that I believe money is necessary but it is not sufficient unless there is an unlimited supply. Money provides resources but we also need enough talented, caring and dedicated people to turn these resources into programs. And finally, even if we demonstrate success of a program we need stable institutions and civil societies to sustain these programs and to scale up the good ones to national levels. Money, people, and institutions are all important (developing and sustaining the latter two needs money, thus creating a chicken and egg problem), which is why the commitment has to be holistic and long-term.

In closing I would like to return briefly to the underlying premise of this article – that ordinary individuals can make a very significant difference. The easiest way is to donate money. There are a growing number of NGOs that today meet international standards of transparency and accountability and are carrying out excellent projects that make even small donations effective and worthwhile. (One reason for my optimism regarding India is the large number of good and innovative NGOs it has.) The problem can be identifying the good ones. This can be done through personal engagement or tapping into a growing network of involved people and organizations. The most effective contribution is through active involvement. Since change for the poor begins with a little better education, a little more health care, a little help getting a job and a little more civic involvement, each of us can contribute enormously. A doctor could donate one day a month to work in a village or slum; sports enthusiasts can help organize training sessions and tournaments for teams from neighboring villages once a week; an educationist could teach evening classes once a week; a caring person could monitor midday meals for children, etc.. More and more NGOs are willing to, and should be encouraged to, facilitate such part-time involvement. Furthermore, such engagement would lead to the creation of a civil society which could then partner with the government to create hope and opportunities for all.

Even the rosiest picture of development is predicated on the lack of any major wars or terrorism or international strife in the coming decades. Post 9/11 we have already seen that it takes hundreds of billions of dollars to deal with terrorism alone. Such large sums spent on development could transform the world but this will not happen as long as there is threat of war or terrorism. It is, therefore, very important that we, as scientists interested and active in development, work towards creating new technology, reducing strife and increasing global cooperation and global security. Physical and social science and scientists have a unique collaborative role to play in making this happen. There is an old saying that *the best time for planting the tree was twenty years ago and the second best time is today*. It is with this thought and intent that this conference was convoked and I thank the organizers for making this happen and for initiating this debate.

References

1. Population Reference Bureau, Washington D.C., http://www.prb.org/
2. A PDF version of the Microsoft Powerpoint presentation of the talk is available http://t8web.lanl.gov/people/rajan/HIV_Poor_rg_05.pdf.

3. See the United Nations Human Development Report 2005 at http://hdr.undp.org/reports/global/2005.
4. Address by Gro Harlem Brundtland to the 55th World Health Assembly. Available at http://www.who.int/directorgeneral/speeches/2002/english/20020513_addresstothe55WHA.html.
5. Commission on Macroeconomics and Health chaired by Prof. Jeffrey Sachs. World Health Organization 2001 report available at http://www.who.int/whosis/cmh/.
6. Jeffrey Sachs, "The End of Poverty: Economic Possibilities of Our Time", Penguin Press 2005.
7. National Intelligence Council report "The Next wave of HIV/AIDS: Nigeria, Ethiopia, Russia, India, and China", September 2002. Available at http://www.odci.gov/nic.
8. See reports on production, trafficking and control by the United Nations Office on Drugs and Crime at http://www.unodc.org/unodc/index.html.
9. Rajan Gupta, "Risky Sex, Addictions, and Communicable Diseases in India: Implications for Health, Development, and Security", Special Report 8 in the Health and Security Series by the Chemical and Biological Arms Control Institute (CBACI), Washington D.C., September 2004. Available at http://t8web.lanl.gov/people/rajan/AIDS-india/MYWORK/Gupta_HIV_India.pdf.
10. For a very insightful discussion on the connection between migration and development see Hernando de Soto, *The Mystery of Capital: Why Capitalism Triumphs in the West and Fails Everywhere Else*, Basic Books, 2000.
11. United Nations Millennium Development Goals at http://www.un.org/millenniumgoals/.
12. See http://www.eia.doe.gov and http://t8web.lanl.gov/people/rajan/energy_RG_06.pdf.
13. Sandra Postel, "Pillar *of Sand: Can the Irrigation Miracle Last?"*, W.W. Norton, 1999.
14. Jarad Diamond, "*Collapse: How Societies Choose to Fail or Succeed"*, Penguin Group, 2005.
15. Rajan Gupta, "Education, the Key to Development: Lessons from India". Available at http://t8web.lanl.gov/people/rajan/AIDS-india/MYWORK/education_India_Arab.pdf.

Ocean Settlements are a Step in the Future

Kenji Hotta

Nihon University College of Science & Technology, 8-14, Kanda-Surugadai 1-chome, Chiyoda-ku, Tokyo 101-8308, Japan, k-hotta@ocean.cst.nihon-u.ac.jp

Can man live in the sea? The history of the oceanic space use is longer than the history of the space use. Ocean space utilization is the one of our great challenge. In this talk, author will review history of ocean space utilization technology and introduce example of structure, energy and food supply from the ocean. Then, finally, we discuss the possibility of the ocean settlement in the future.

V. Burdyuzha (ed.), The Future of Life and the Future of Our Civilization, 401.

Futurology: Where is Future Going…?

Roman Retzbach Menachem

ZUKUNFT-INSTITUT / Future-Institute & Trend-Institute international, Franzosische Str. 8-12, 10117 Berlin, Germany, info@Future-Institute.com

> *"the future comes always differently, than one it in the spirit of the time thought in each case and also today thinks..."*

Ladies and gentlemen, may I introduce myself? My name is Roman Retzbach and I am a future and trend researcher. For the past 15 years I have been working for the internationally operating Future-Institute that was founded in 1920. I search both for short-term trends for the coming months with the help of trend scouts in America, Europe and Asia and determine Megatrends for the coming decades.

The "Future-Institute international" with the "Zukunft-Institut" and "Trend-Institute" was founded 1920 and is worldwide leading since 1988, organized as anon-profit donation; 2000 the "Trend-Academy & Future-University" began as an elite business school & university with education and training of trendscouts, today with 1250 coworkers, trendscouts and 12 directors in New York (American headquarter), Rio, London (European headquarter), Paris, Berlin, Moscow, Vienna, Zurich, Tokyo (Asian headquarter), Hong-Kong, Shanghai.

My associates and me deal with the same questions as everybody else: we are all looking for a meaning – where do we come from, what is our purpose in life, what direction are we heading and of course all this comes to nothing else than - the meaning of life.

In a small and modest way, future and trend research can start to answer some of these questions. Up to the present day dozens of instruments and means for this research have been developed and are in constant use.

Future scenarios are created in a x-multidimensional space model. One of the many axes in this model is for example the expert interviews ("Delphi")

V. Burdyuzha (ed.), The Future of Life and the Future of Our Civilization, 403–409.

phi") with people all over the world and the use of supercomputers as the ultimate opposite.

Of course there are an infinite number of future scenarios as well as very human and limiting future filters. Some of them show basic human needs such as security, convenience, supply, health, communication, wellness and leisure as well as luxury and Anthropology of the "secondary animal" - these supply both orientation and limitations. Future options are further influenced by mankind as such. Humans have been making the same mistakes over and over again throughout centuries of history – waging wars and searching for peace ever since they appeared on the cosmic stage of evolution.

Further axes are the very long term form of future trend research - looking ahead for thousands of years, and very short term consumer trend developments which lead to ever more big-, macro-, and shifting trends. Not to be underestimated is the impact of the entertainment industry and cyber worlds that continue to bring us a virtual reality future by trying to present and sell it in an overwhelming 3D fashion.

Another of the many future laws in this model of our future is the individual time frame from which these outlooks, visions, utopias and timelines are formulated. 70% of all future scenarios are extended from the present into the future coining the phrase "forwarding", 20% are contributed by experts and market research from every area of social, economical and scientific life, leading to forecasts and foresights, but not even 10% are made up of real futurings. This futuring is almost completely made up of science fiction, fantasy and fortune telling. Gaming and entertainment, conspiracy theories, matrix visions and deja vues are also a part and all of these are strongly influenced by certain genres and often have a military hierarchy in their future scenarios. What is interesting is that this kind of research is neutrally funded by the military in the United States, by the governments in Europe and the technology and entertainment industry in Asia.

One example for a "fabricated" future and a "wonderful new world" are statistics as well as scenarios that try to avoid any kind of panic by stating that the stagnation of world population at 12 to 15 billion within the next 70 years as well as a constantly growing economy is a comforting outlook. At the same time however different future scenarios are bought or created in an industrial countermovement concerning climate change, world hunger, change of magnetic poles, sun storms, dwindling nature, water shortage, aging societies, the widening gap between rich and poor, a destruction of supply networks, terrorism, China as the new super power as well as India, Brazil and Russia in their function as world BRIC supermarkets…

All these may be real but at the same time very limited current perspectives and mere reactions. Not one of them a possible solution for an improvement of our future. In short: the future is stopped by the present.

However, cosmological evolution and future research do not know any such limitations of any kind. A world population of 20 billion, problems that will only arise tomorrow but could be met today, these are all being made by humans themselves and are perceived as an inevitable future. And if we look at the world today, we find that anti trend is stronger than the official mainstream and pop trend. Also simultaneous as well as extreme polarization of the most diverse trend developments are becoming increasingly automated and today there no longer is one future ahead, but many different future realities next to each other – even in the same place, the same city, on the same – our planet earth.

There is one historical fact we know for sure. Namely the future has always been different than people could imagine in their time. How else could we have come so far but evolve so little. To put it short: There no longer is a difference between a glass that is half empty and a glass that is half full – because both states exist at the same time.

These are some of the cosmic laws that will help your understanding of future trend research of tomorrow: Let me present an excerpt of the 10 cosmic laws.

1. Law of the cosmos: All that mankind can think of, will become real. Good as well as bad, both real and virtual: for example, virus epidemics, catastrophes and new wars. But reality also has parallel universes and meta-dimensions in which futures can constantly change. Science fiction today is a warning as well as a mirror of times to come.

2. Law of the cosmos: everything is infinite. Only when man can conceive this fact, will he reach the first step of evolution. Until then we remain at the bottom.

3. Law of the cosmos: robots (droids, cyborgs, genoids and mutants) are the next generation in evolution. Man will someday look like the aliens he himself created in his future scenario. Just as everybody is related to every person just through three other people, we are all connected to every life-form in the galaxy.

4. Law of the cosmos: All life forms strive into the infinite space, back to where life began, where we came from. This future is our goal and this goal is our future.

Earth is only a temporary home for our kind. A nesting and birthing place – but we really are at the edge of space age.

Life began in space and it will also end there. Everything that humans can imagine about infinity can be realized and materialized. The universe can be a finite confining ball that reflects reality and multiplies like a mirror – making a planetary system of one star and millions of lives from just one.

But what lies behind all this? Even a vacuum or dark energy is in constant change and balancing with dark mass, form existences and states of reality with nothingness itself becoming an anti form of being.

Future trend research is the most complex and doubted science in the world, combining all human events and achievements. But behind all the empirical and precognitive aspects it follows high ideals. These are the laws of the cosmos. These are the rules by which future trend research plays. They state, that everything – be it good or bad, every abstract fantasies and possible apocalypses can come true one day, or are on their way to becoming true as we speak. Humans, in their rise and transcendental state in between nature, culture and spirituality – trend naming individuality and democratic ideals, do nothing else than follow their cosmic evolutionary code.

This states quite plainly that everything is possible. Everything you have ever heard, seen or thought of is true – either here on earth or somewhere in parallel universes on other planets and in other galaxies.

Future trend research seen from this perspective does nothing else but measure the probabilities and spirit of the present times in order to find out when any of these infinite number of possibilities might materialize, or else they are not desired by humanity as a whole and therefore suppressed, changed or nullified. In this sense the whole of humanity functions as one super intelligence (higher evolved as sponge or collective models) but at the same time on a level that cannot be perceived or remembered by any individual. On this highest level of intelligence, all humans strive for the same goals or dreams. They are on a search for peace, a way of living together and construction instead of destruction. At least that is the master plan. Seen from a cosmic point of view, humanity as a whole is a primitive race still in its first evolutionary stage long before space migration.

The mega trends in this case are the orientation and the actual path to this cosmological understanding and ways to measure or control what has been achieved and what still needs to be realized. The last century was a social and pre technological one, this century we have finally reached the technological age. By means of technology in the past, progress through machines, cars and ever changing energies as well as unlimited electricity and cultural as well as social improvement was made possible – but all this was only a necessary first step to approach higher technologies.

Technological jumps, meaning new macro megatrends trigger sociological changes or shifts and also wealth and new insights.

Super technologies as well as megatrends have multiple dimensions. On the one hand every technology serves as a means for advance and at the same time is a weapon of its own destruction. On the other hand it can never become flawless, not even in the highest order because its creator is not flawless and never will be. On top of that, new technologies mostly are an empty promise and only become useful at a later stage. This is why the achievements of bio-tech, namely immortality and anti-aging switching from youth fountain to being fit in old age, as well as nano-tech which will lead to the possibility of producing and recycling any given product – also artificial intelligence that does not really exist today and solar/quant-tech as cold fusion for free and unlimited power are praised too high and too loud, too soon.

Future trend research has the function and the responsibility to acknowledge that the future comes later and different than predicted, even from these researchers themselves.

But without a masterplan and without the necessary technologies, we will not be able to solve the Nostradamic catastrophes of the future. These are mainly the revival of nature, the vanquish of big brother like political systems and also to come: protection from meteorites, full scale war against global super viruses, huge volcanoes, nuclear war, cosmic super gamma neutron rays and the creation of black holes through super fusion energy…

The future trend research has something like a so-called "world formula" by which all future events can be predicted and estimated accordingly.

This world or age old formula with the future – and tech formula is similar to the universal formula of all energy forms. Of all the infinite possibilities, only a small number are active within a certain timeframe and are perceived and desired by humanity. This is where we come full circle.

The Zeitgeist of humanity already contains a certain future which is being followed and supported by future researchers and the entertainment industry. This scenario includes a 24-hour society, a united world, an autonomous robot civilization, immortality for humans and cloning of everybody and everything, including animals, food, organs and humans. Every civilization adapts its ethical and moral standards every 100 years or so. Even technologies are mostly only embraced in their 2nd or 3rd generations. And of course space travel and the ensuing contact and life with aliens of all colors, sizes, languages and living surroundings. The enormous number of evolutionary life forms on earth is just a miniature model of the possibilities in the whole universe.

Space travel and extraterrestrial contacts without doubt will be the next big step for mankind. Settling in the vast galaxies will begin, new colonies will be created, spaceships carrying whole generations will spread throughout space. And even now this cosmic call from the future is influencing the present day, forming a new goal and influencing what technologies are aiming at.

Modern future trend technology today and tomorrow looks further and further into the future and is therefore able to make increasingly accurate predictions, including the ups and downs of events. Wildcards – meaning unforeseen events – as well as coincidences are becoming very rare in the ever improving future trend research. In principle, the predictability of future events improves when one always takes into account that anything is possible at any time.

"The question is no longer what is going to happen in future… when it will happen is becoming the more interesting question…"

Human society is merely influencing the speed or deceleration of the future to come, but it cannot stop it in any way. Continuing retro movements and active delays can last from 10 to 15 years, but that is about the longest that future lets itself be held back.

How we are going to live, work and travel in future as well as the survival of our civilization can best be answered by the future trend research with its knowledge of cosmological evolution and technological megatrends.

"Everything that man has ever imagined has come true or is in the making."

Everything that man has imagined up to now – a horror in history, today can be seen as infotainment ranging from news to documentations, and tomorrow as visions and fiction.

Everything that man can imagine will become true – today as science fiction and tomorrow as implicitness.

Everything that man can imagine will become future reality – either in real life or at least in the shadow world of virtual reality in form of horror, thriller, drama or a fairy tale all in one.

Everything that we imagine is already in the past. Everything that we can dream of can become true. Everything we think of will become true, if we like it or not – sooner or later.

What today still is "new technology" such as teleportation, levitation, telepathy, stargates and beaming will be known as the "old traditional sciences" in 100 years time.

Looking at it this way, films like star wars are a nice introduction and popular entertainment, but sadly too militaristic – to think about the future in order to meet it accordingly.

Most people on earth today already believe in a future with space travel, alien contacts, "hitchhiking through the galaxy" and black holes leading to more galaxies…

Ladies and gentlemen – "the future is up to us and the future is in all of us…" Thank you very much.

Towards Sustainable Future by Transition to the Next Level Civilization

Andrei P. Kirilyuk

Solid State Theory Department, Institute of Metal Physics, 36 Vernadsky Avenue, 03142 Kiev-142, Ukraine, kiril@metfiz.freenet.kiev.ua

The universal and rigorously derived concept of dynamic complexity shows that any system of interacting components, including society and civilization, is a process of highly uneven development of its unreduced complexity. Modern civilization state corresponds to the end of unfolding of a big complexity level. Such exhausted, totally "replete" structure cannot be sustainable in principle and shows instead increased instability realising its replacement by a new kind of structure with either low or much higher complexity (degrading or progressive development branch, respectively). Unrestricted sustainability can emerge only after transition to the superior level of civilization complexity, which implies qualitative and unified changes in all aspects of life, including knowledge, production, social organization, and infrastructure. These changes are specified by a rigorous analysis of underlying interaction processes.

Keywords: Dynamic redundance, symmetry of complexity, sustainability transition, revolution of complexity, criterion of progress, noosphere

1. Future Quest in a High-Tech Epoch of Change

Although permanent change is inherent in a planet, life, and civilization existence, it has a highly uneven shape of "punctuated equilibrium", where larger periods of relatively smooth and slow evolution are interrupted by short periods of huge and abrupt, "revolutionary" change. Rapidly growing body of evidence shows that today the planetary life and civilization on Earth are approaching very closely the next "bifurcation point" of

V. Burdyuzha (ed.), The Future of Life and the Future of Our Civilization, 411–435.

development, or "generalized phase transition" [8], which is often referred to as "singularity" (though in terms of particular technological aspects) and marks a global change of unprecedented scale (see e.g. [24–27, 29]).

It is not surprising that the eternal humanity quest for its future gains today quickly growing importance and public interest [30] surpassed only by the global change dynamics itself. A part of this interest is driven by traditional "fear of the (unknown) future" amplified now because of the clearly felt huge scale of emerging change and related uncertainty [24, 26]. An important aspect of the present epoch of change and its "future shock" [27] is due to extraordinary growth of "high", but *empirically* based technologies that can now, *for the first time in history*, modify natural system complexity at its *full* depth, in quantum world (high-energy physics), biology (genetics), environment (industrial over-production) and human dimensions (psychology, media, information technologies), while remaining *effectively blind* with respect to underlying system dynamics [10, 11, 20, 21]. Even most serious attempts of future studies [30] fail to provide an *objectively* reliable, consistent and *unified* understanding of the emerging change, replacing it with empirical *interpolation* of *separate*, though important, aspects of current development, such as economic and technological tendencies, ecological system evolution, human behaviour, etc.

In this report we present the results of the *causally complete*, rigorous analysis of *unreduced* world dynamics based on the recently developed *universal concept of dynamic complexity* [8, 11–19, 21] and revealing the *unified*, many-sided picture, origin, dynamics, and purpose of the beginning *revolutionary* change [8, 9]. We start with an outline of the universal concept of complexity (Sect. 2) based upon *unreduced* solution of any *real* interaction problem (Sect. 2.1) and leading to the unified concept of system *development* as manifestation of the universal *symmetry (conservation)* of complexity (Sect. 2.2). In particular, the *sustainability transition* emerges today as inevitable and rapid "jump" to the next, *superior level of civilization complexity* (Sect. 3) prepared by its previous development and having only one alternative of irreversible destruction (Sect. 3.1). We then analyse various entangled aspects of life at the new complexity level and transition dynamics, including the qualitatively new kind of knowledge (Sect. 3.2), production (Sect. 3.3), social organization (Sect. 3.4), and infrastructure (Sect. 3.5). Finally, we pay homage to Carl Sagan and Joseph Shklovsky by showing that discovery of other forms of life and intelligence is related to the new future for our own civilization by the same universal concept of complexity (Sect. 4). We summarize the obtained results by concluding that the *causally complete* kind of knowledge of the universal science of complexity provides the *unique basis* for the truly scientific, *objectively reliable and intrinsically unified futurology* urgently needed at the modern critical point of development.

2. Universal Science of Complexity

2.1 Unreduced Interaction Dynamics

Any system dynamics and evolution are determined by the underlying interaction processes. The way of interaction analysis in usual science (including the scholar "science of complexity") involves rough simplification (reduction) of real interaction within a version of perturbation theory (or "model") that assumes effective *weakness* of interaction influence upon system configuration, which *kills* any possibility of essential *novelty emergence* from the beginning (with the evident fatal consequences for such approach ability to predict *any* nontrivial future). Further play with analytical or computer models of thus heavily reduced reality, empirically postulated (rather than derived) object properties and arbitrarily adjusted parameters cannot replace the *intrinsic creativity* of unreduced interaction processes. It is no wonder that the *qualitative* knowledge extension to the *causally complete understanding* of real phenomena provided by the universal science of complexity [8] is simply due to the proposed *non-simplified, truly "exact" analysis of unreduced, real interaction processes.* Its possibilities are confirmed by obtained *consistent solutions* to various stagnating, "insoluble" problems [8], from those of fundamental physics (causal and unified extensions of quantum mechanics, relativity, cosmology) [10, 15, 16] and unreduced many-body interaction (true quantum chaos, quantum measurement, many-body coherence) [7, 11], to reliable basis for nanobiotechnology [11, 17], genomics [21] and medicine [12], theory of genuine (including artificial) intelligence and consciousness [18], the new kind of communication and information systems [19], and realistic sustainability concept [9].

Real interaction is described by *existence equation*, generalising various models and simply fixing the initial system configuration in a "Hamiltonian" form (also confirmed below, see Sect. 2.2) [8, 11–19, 21]:

$$\left\{\sum_{k=0}^{N}\left[h_k(q_k)+\sum_{l>k}^{N}V_{kl}(q_k,q_l)\right]\right\}\Psi(Q)=E\Psi(Q), \quad (1)$$

where $h_k(q_k)$ is the "generalized Hamiltonian" for the k-th component, q_k is the degree(s) of freedom of the k-th component, $V_{kl}(q_k,q_l)$ is the (arbitrary) interaction potential between the k-th and l-th components, $\Psi(Q)$ is the system state-function, $Q \equiv \{q_0, q_1, \ldots, q_N\}$, E is the generalized Hamiltonian eigenvalue, and summations include all (N) system components. It is convenient to represent the same equation in another form by separating certain degree(s) of freedom, e.g. $q_0 \equiv \xi$, that correspond to a naturally selected, usually "system-wide" entity, such as "embedding" configuration (system of coordinates) or common "transmitting agent":

$$\left\{h_0(\xi) + \sum_{k=1}^{N}\left[h_k(q_k) + V_{0k}(\xi, q_k) + \sum_{l>k}^{N} V_{kl}(q_k, q_l)\right]\right\}\Psi(\xi, Q) = E\Psi(\xi, Q), \quad (2)$$

where now $Q \equiv \{q_1, \ldots, q_N\}$ and $k, l \geq 1$.

We pass now to a "natural" problem expression in terms of free-component solutions for the "functional" degrees of freedom ($k \geq 1$):

$$h_k(q_k)\varphi_{kn_k}(q_k) = \varepsilon_{n_k}\varphi_{kn_k}(q_k), \quad (3)$$

$$\Psi(\xi, Q) = \sum_{n \equiv (n_1, n_2, \ldots, n_N)} \psi_n(q_0)\varphi_{1n_1}(q_1)\varphi_{2n_2}(q_2)\ldots\varphi_{Nn_N}(q_N) \equiv \sum_n \psi_n(\xi)\Phi_n(Q), \quad (4)$$

where $\Phi_n(Q) \equiv \varphi_{1n_1}(q_1)\varphi_{2n_2}(q_2)\ldots\varphi_{Nn_N}(q_N)$, $\{\varepsilon_{n_k}\}$ are the eigenvalues and $\{\varphi_{kn_k}(q_k)\}$ eigenfunctions of the k-th component Hamiltonian $h_k(q_k)$, while $n \equiv (n_1, n_2, \ldots, n_N)$ runs through all eigenstate combinations. The system of equations for $\psi_n(\xi)$ equivalent to existence equation (1)–(2) is obtained in a standard way [8, 11–19, 21]:

$$\left[h_0(\xi) + V_{00}(\xi)\right]\psi_0(\xi) + \sum_n V_{0n}(\xi)\psi_n(\xi) = \eta\psi_0(\xi), \quad (5a)$$

$$\left[h_0(\xi) + V_{nn}(\xi)\right]\psi_n(\xi) + \sum_{n' \neq n} V_{nn'}(\xi)\psi_{n'}(\xi) = \eta_n\psi_n(\xi) - V_{n0}(\xi)\psi_0(\xi), \quad (5b)$$

where $n, n' \neq 0$ (also below), $\eta \equiv \eta_0 = E - \varepsilon_0$, $\eta_n \equiv E - \varepsilon_n$, $\varepsilon_n \equiv \sum_k \varepsilon_{n_k}$,

$$V_{nn'}(\xi) = \sum_k \left[V_{k0}^{nn'}(\xi) + \sum_{l>k} V_{kl}^{nn'}\right], \quad (6)$$

$$V_{k0}^{nn'}(\xi) = \int_{\Omega_Q} dQ \Phi_n^*(Q) V_{k0}(q_k, \xi) \Phi_{n'}(Q), \tag{7}$$

$$V_{kl}^{nn'}(\xi) = \int_{\Omega_Q} dQ \Phi_n^*(Q) V_{kl}(q_k, q_l) \Phi_{n'}(Q), \tag{8}$$

and we have separated the equation for $\psi_0(\xi)$ describing the generalized "ground state" of the system elements, i. e. the state with minimum complexity (defined below). The obtained system of equations (5) expresses the same problem as the starting Eq. (2), but now in terms of intrinsic variables. It can be obtained for various starting models, including time-dependent and formally "nonlinear" ones.

The usual, perturbative approach starts from simplification of "nonintegrable" system (5) down to a "mean-field" approximation:

$$\left[h_0(\xi) + V_{nn}(\xi) + \tilde{V}_n(\xi)\right]\psi_n(\xi) = \eta_n \psi_n(\xi), \tag{9}$$

where $|V_0(\xi)| < |\tilde{V}_n(\xi)| < |\sum_{n'} V_{nn'}(\xi)|$. General problem solution is then obtained as a linear or equivalent *superposition* of eigen-solutions of Eq. (9) similar to Eq. (4). If we want to avoid problem reduction, we can try to "solve" the unsolvable system (5) by expressing $\psi_n(\xi)$ through $\psi_0(\xi)$ from Eqs. (5b) with the help of standard Green function and then substituted the result into Eq. (5a) [1, 6]. We are left then with only one, formally "integrable" equation for $\psi_0(\xi)$:

$$h_0(\xi)\psi_0(\xi) + V_{\text{eff}}(\xi;\eta)\psi_0(\xi) = \eta\,\psi_0(\xi), \tag{10}$$

where the *effective potential (EP)*, $V_{\text{eff}}(\xi;\eta)$, is obtained as

$$V_{\text{eff}}(\xi;\eta) = V_{00}(\xi) + \hat{V}(\xi;\eta),\ \hat{V}(\xi;\eta)\psi_0(\xi) = \int_{\Omega_\xi} d\xi' V(\xi,\xi';\eta)\psi_0(\xi'), \tag{11}$$

$$V(\xi,\xi';\eta) = \sum_{n,i} \frac{V_{0n}(\xi)\psi_{ni}^0(\xi)V_{n0}(\xi')\psi_{ni}^{0*}(\xi')}{\eta - \eta_{ni}^0 - \varepsilon_{n0}},\quad \varepsilon_{n0} = \varepsilon_n - \varepsilon_0, \tag{12}$$

and $\{\psi_{ni}^0(\xi)\}$, $\{\eta_{ni}^0\}$ are complete sets of eigenfunctions and eigenvalues of a *truncated* system of equations:

$$\left[h_0(\xi) + V_{nn}(\xi)\right]\psi_n(\xi) + \sum_{n'\neq n} V_{nn'}(\xi)\psi_{n'}(\xi) = \eta_n \psi_n(\xi). \tag{13}$$

The *unreduced*, truly complete *general solution* to a problem emerges now as a *dynamically probabilistic* sum of *redundant* system *realizations*, *each* of them being equivalent to usual "general solution" [7–19, 21]:

$$\rho(\xi,Q) = \sum_{r=1}^{N_\Re}{}^{\oplus} \rho_r(\xi,Q), \tag{14}$$

where $\rho(\xi,Q)$ is the observed density, $\rho(\xi,Q) = |\Psi(\xi,Q)|^2$ for "wave-like" complexity levels and $\rho(\xi,Q) = \Psi(\xi,Q)$ for "particle-like" structures, index r enumerates system realizations, $N_\Re$ is realization number (its maximum value is equal to the number of components, $N_\Re = N$), and the sign $\oplus$ designates the special, dynamically probabilistic meaning of the sum (see below). The r-th realization statefunction, $\Psi_r(\xi,Q)$, entering the unreduced general solution, Eq. (14), is obtained as

$$\Psi_r(\xi,Q) = \sum_i c_i^r \left[\Phi_0(Q)\psi_{0i}^r(\xi) + \sum_{n,i'} \frac{\Phi_n(Q)\psi_{ni'}^0(\xi) \int_{\Omega_\xi} d\xi' \psi_{ni'}^{0*}(\xi') V_{n0}(\xi') \psi_{0i}^r(\xi')}{\eta_i^r - \eta_{ni'}^0 - \varepsilon_{n0}} \right], \tag{15}$$

where $\{\psi_{0i}^r(\xi), \eta_i^r\}$ are eigen-solutions of the unreduced EP equation (10), while the r-th EP realization takes the form:

$$V_{\text{eff}}\left(\xi;\eta_i^r\right)\psi_{0i}^r(\xi) = V_{00}(\xi)\psi_{0i}^r(\xi) + \sum_{n,i'} \frac{V_{0n}(\xi)\psi_{ni'}^0(\xi) \int_{\Omega_\xi} d\xi' \psi_{ni'}^{0*}(\xi') V_{n0}(\xi') \psi_{0i}^r(\xi')}{\eta_i^r - \eta_{ni'}^0 - \varepsilon_{n0}} . \tag{16}$$

Although the "effective" problem, Eqs. (10) – (16), is formally equivalent to its initial expression, Eqs. (1) – (5), it reveals *emerging interaction links*, in the form of EP dependence on the solutions to be found. It leads to a *new quality* of *unreduced* solution (as compared to usual reduction of Eq. (9)): it has *many* equally real, locally "complete" and therefore mutually *incompatible* solutions, or *realizations* [6–19, 21]. This quality of unreduced solution is designated as *dynamic multivaluedness (or redundance).* Standard theory tries to obtain problem solution in a "closed", "exact" form and therefore resorts to perturbative reduction of the original EP (see e.g. [1]), thus inevitably *killing* real system multivaluedness, complexity and related *creativity*.

Dynamic multivaluedness gives *dynamic, or causal, randomness*: multiple, but incompatible system realizations are forced, by the driving interaction itself, to *permanently replace each other* in a *truly random* order thus defined and giving the *unreduced* general solution, the *dynamically probabilistic* sum of Eq. (14). It implies that any quantity is *intrinsically unstable* and its value will *unpredictably* change (together with the system state) to another one, corresponding to the next, *randomly* chosen realization. We obtain thus a *consistently derived* and *universally valid* property of *novelty emergence*, or intrinsic *creativity* of *any* real system, absent in any its usual, dynamically single-valued model. We obtain also purely *dynamic* definition of realization emergence *event* and its *probability*:

$$\alpha_r = \frac{N_r}{N_\Re} \left(N_r = 1, \ldots, N_\Re ; \sum_r N_r = N_\Re \right), \quad \sum_r \alpha_r = 1 , \tag{17}$$

where α_r is the probability of emergence of r-th actually observed realization containing N_r elementary realizations ($N_r = 1$ for each of these).

The obtained picture of system dynamics is summarized by the *universal* definition of unreduced *dynamic complexity*, C, as a growing function of realization number $N_\Re$, or rate of change, equal to zero for the (unrealistic) case of only one realization: $C = C(N_\Re)$, $dC/dN_\Re > 0$, $C(1) = 0$. Major examples are provided by $C(N_\Re) = C_0 \ln N_\Re$, generalized energy/mass (temporal rate of realization change), and momentum (spatial rate of realization emergence) [8–19, 21]. Since dynamic redundance ($N_\Re > 1$) is at the origin of dynamic randomness, our dynamic complexity includes universally defined *chaoticity*. Whereas *all* real systems and processes are dynamically complex and (internally) chaotic ($N_\Re > 1, C > 0$), their "models" in usual science, including its versions of "complexity" and "chaoticity" (cf. [4, 5]), are invariably produced by artificial (and *biggest* possible) reduction of multivalued dynamics to the unrealistic case of *single* realization, *zero* complexity, *absence* of *genuine* chaos, *any* real, *intrinsic* change and related *time flow*. This *dynamically single-valued*, or *unitary*, science embracing the whole body of scholar knowledge is but a *zero-dimensional (point-like) projection* of multivalued world dynamics, which explains both relative (but never complete!) "success" of unitary science in *formal* description of *lower* complexity levels (≈ "fundamental physics") and its explicit failure to understand higher-level dynamics and unreduced complexity features (emergence, time, chaos, etc.) [8–19, 21].

Unreduced dynamic complexity thus defined includes other major features, such as essential (or dynamic) nonlinearity, dynamic entanglement, and probabilistic dynamic fractality. *Essential nonlinearity* designates *dynamically* emerging feedback links, described by EP dependence on the eigenvalues to be found (Eqs. (10)–(12), (16)). It is only incorrectly modelled by usual, mechanistic "nonlinearity" of the unitary theory and appears in interaction problems with a formally linear existence equation (1)–(2), such as quantum chaos [6–8, 11]. *Dynamic entanglement* is *physically real* mixing of interacting components reflected by the dynamically weighted products of functions depending on different degrees of freedom in Eq. (15). Essential nonlinearity and dynamic entanglement are amplified due to *multi-level* realization branching giving *probabilistic dynamical fractal*. It is obtained by application of the same EP method to solution of higher-level, (ever more) truncated systems of equations, starting from Eqs. (13) [8, 12, 18, 21]. Dynamical fractal differs from usual, dynamically single-valued fractals by its *permanently, chaotically changing realizations* at *each* level of fractal hierarchy, which leads to the key property of *dynamic (autonomous) adaptability* and includes *any* kind of structure.

Quantitative expression of dynamic adaptability includes *huge efficiency growth* of unreduced many-body interaction with respect to unitary models. The unreduced system efficiency P_{real} is determined by the link combination number in the multivalued fractal hierarchy [11, 17–19, 21]:

$$P_{\mathrm{real}} \propto N! \approx \sqrt{2\pi N}\left(\frac{N}{e}\right)^{N} \sim N^{N} \propto C \ , \tag{18}$$

where the number of links N is very large itself. Unitary (regular, sequential) dynamic efficiency grows only as $N^{\beta} \ll P_{\mathrm{real}}$ ($\beta \sim 1$). It is this huge efficiency advantage that explains such "magic" *qualities* in higher-complexity systems (very large N) as *life, intelligence, consciousness, and sustainability*. Obtained at the expense of *irreducible dynamic randomness*, these *causally derived* properties are *indispensable* for the *correct* analysis of planetary life and civilization dynamics.

Further development of the universal concept of complexity includes unified classification of all observed dynamic regimes and transitions between them [8, 11, 13, 19]. The limiting regime of *uniform, or global, chaos* is obtained for *comparable* interaction parameters (characteristic frequencies). If they differ essentially, one gets the opposite case of *dynamically multivalued self-organization, or self-organised criticality*

(SOC), where rigid, low-frequency components *confine* a fractal hierarchy of similar, but *chaotically changing* realizations of high frequency components. This case *unifies* the *essentially extended*, realistic and multivalued (internally chaotic) versions of usual, dynamically single-valued "self-organization" (that in reality does *not* describe any *new, explicit* structure *emergence*), SOC, fractality, "synchronisation", "chaos control", and "mode locking". We obtain also a rigorously derived and universal *criterion* of transition from SOC to the uniform chaos, occurring around the main frequency resonance, which reveals the true meaning of the "well-known" phenomenon of resonance [8, 11, 13, 19]. When the frequency ratio, or "chaoticity parameter", grows from small values for a quasi-regular SOC regime to unity in the global chaos case, system behaviour follows a gradual (though uneven) change towards ever less ordered patterns, reflecting the observed diversity of dynamical structures.

2.2 Universal Symmetry of Complexity and Evolution Law

Major features of *explicit structure creation* are emerging elements of *dynamically discrete, or quantized, space* (structure) and *irreversibly flowing time* (event, evolution). Space element, Δx, is given by realization eigenvalue separation, $\Delta_r \eta_i^r$, for the unreduced EP, Eq. (10): $\Delta x = \Delta_r \eta_i^r$. Time element, Δt, determines the duration of space element emergence, or realization change *event*, $\Delta t = \Delta x / v_0$, where v_0 is the signal propagation speed in component structure. A universal *integral* complexity measure is given by *action*, A, whose increment is *independently* proportional to Δx and Δt [8, 11, 14, 18]: $\Delta \mathsf{A} = -E\Delta t + p\Delta x$, where the coefficients E and p are identified as generalized system *energy (mass)* and *momentum*. They represent thus universal *differential* measures of complexity:

$$E = -\frac{\Delta \mathsf{A}}{\Delta t}\Big|_{x=\text{const}}, \quad p = \frac{\Delta \mathsf{A}}{\Delta x}\Big|_{t=\text{const}}. \tag{19}$$

Due to its irreversible (chaotic) character, any real interaction process can be described as *transformation and conservation (symmetry) of complexity*, where the potential (hidden) form of complexity, or *dynamic information I*, is transformed into the unfolded (explicit) form of *dynamic entropy S*, so that their sum, the *total system complexity* $C = I + S$, remains *unchanged*, $\Delta C = 0$, $\Delta I = -\Delta S$. Although both dynamic information and entropy are expressed in units of action, the latter corresponds rather to dynamic information, decreasing during system complexity development:

$$\Delta I = \Delta \mathsf{A} = -\Delta S < 0. \tag{20}$$

Dividing Eq. (20) by $\Delta t|_{x=\text{const}}$, we obtain differential expression of the symmetry (conservation) of complexity and *universal dynamic/evolution equation* in the form of generalized Hamilton-Jacobi equation:

$$\frac{\Delta \mathsf{A}}{\Delta t}\bigg|_{x=\text{const}} + H\left(x, \frac{\Delta \mathsf{A}}{\Delta x}\bigg|_{t=\text{const}}, t\right) = 0, \tag{21}$$

where the *Hamiltonian*, $H(x,p,t)$, expresses the differential complexity-entropy, $H = (\Delta S/\Delta t)|_{x=\text{const}}$. The *dynamic quantization* procedure relates complexity-action increment to that of the *generalized wavefunction (or distribution function)* Ψ, describing specific, "disentangled" system state *during* its chaotic jumps between realizations, and transforms Eq. (21) to the *universal Schrödinger equation* [8, 11, 14, 18]:

$$\mathsf{A}_0 \frac{\Delta \Psi}{\Delta t}\bigg|_{x=\text{const}} = \hat{H}\left(x, \frac{\Delta}{\Delta x}\bigg|_{t=\text{const}}, t\right)\Psi(x,t), \tag{22}$$

where A_0 is a characteristic action value by modulus (equal to Planck's constant at the lowest, quantum levels of complexity) and the Hamiltonian operator, $\hat{H}(x,p,t)$, is obtained from the Hamiltonian $H(x,p,t)$ by causal quantization. While the symmetry of complexity unifies and extends *all* (correct) laws and "principles" of unitary science, Hamilton-Schrödinger equations, Eqs. (21)–(22), connected by causal quantization, unify and extend all particular (model) dynamic equations [8, 11, 14, 18].

The key implication of the symmetry of complexity is that it provides the *universal meaning, dynamics, and measure of any system existence, evolution, and progress*, in the form of *complexity development* (internal transformation from dynamic information into dynamic entropy) as a result of its *conservation*, which gives a well-specified solution to such "difficult" and "ambiguous" problems as purpose of history, meaning of life, objective understanding of the future, etc. It is important that due to the internal chaoticity of *any* real (even externally "regular") system, every structure emergence process corresponds to *growth of complexity-entropy, i.e. chaoticity*, which resolves the longstanding contradiction between the (generalized) entropy growth law and visible order increase in structure creation processes. Another complexity development feature is that due to the *unreduced* interaction dynamics ("everything interacts with everything") it has a *dynamically discrete*, step-wise character [8]. The hierarchic structure creation and complexity development process is shown in Fig. 1. A sufficiently big step of complexity-entropy growth can be described as *generalized phase transition* to the superior *level of complexity* with a *qualitatively* different kind of structure and dynamics.

Let us consider complexity development stages in more detail, Fig. 2, in view of further application to (modern) *civilization development*. First of all, we can rigorously define periods of *progress* (accelerated complexity-entropy growth) and *decline* (relative stagnation of complexity development) constituting respectively the steep rise and plateau (saturation) of each discrete step of system complexity development. Whereas entropy S can only grow for both progress and decline, $H = \partial S/\partial t = -\partial \mathsf{A}/\partial t = E > 0$,

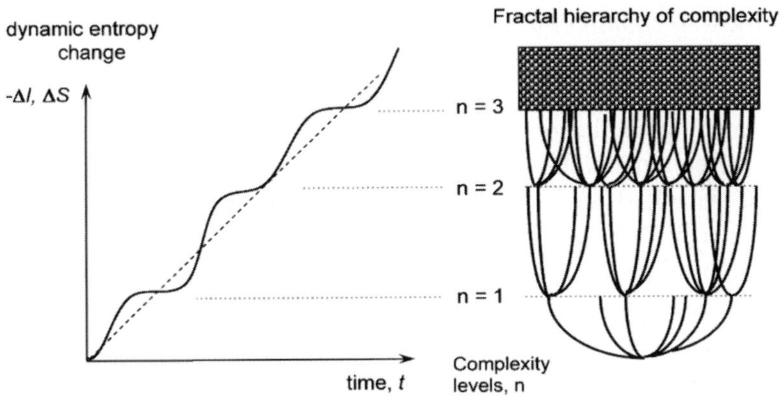

Fig. 1. Scheme of universal system development by transformation of its (decreasing) complexity-information (I) into (increasing) complexity-entropy (S).

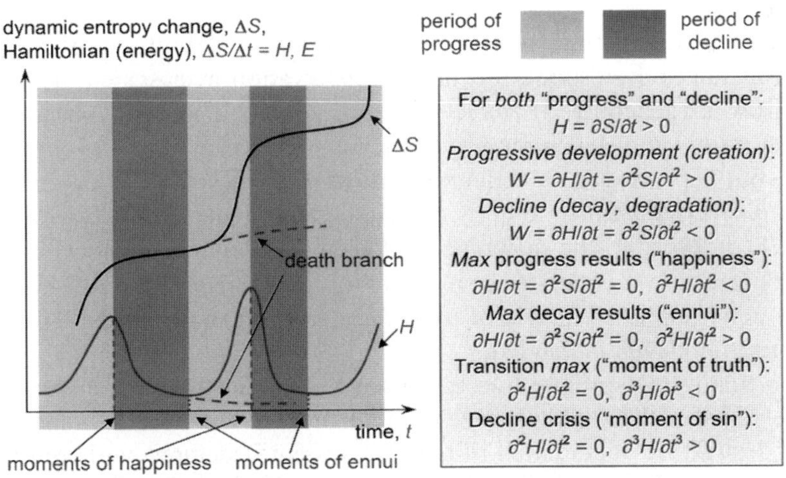

Fig. 2. Periods of system progress, decline, and transitions between them rigorously specified in terms of dynamic entropy change $\Delta S = -\Delta \mathsf{A}$, generalized Hamiltonian $H = \partial S/\partial t$, or energy $E = -\partial \mathsf{A}/\partial t = H$, and higher complexity-entropy/action derivatives.

acceleration of dynamic entropy growth, or the *power of development*, $W = \partial H/\partial t = \partial^2 S/\partial t^2$, is *positive for progress* (creative development), $W = \partial^2 S/\partial t^2 > 0$, and *negative for decline* (decay, degradation), $W = \partial^2 S/\partial t^2 < 0$. Points of *inflection* of the entropy growth curve, $\partial H/\partial t = \partial^2 S/\partial t^2 = 0$, separate adjacent periods of progress and decline and correspond, at the same time, to maximum (final) progress results ("point of *happiness*"), $\partial H/\partial t = 0,\ \partial^2 H/\partial t^2 < 0$, and maximum decay ("point of *sadness/ennui*"), $\partial H/\partial t = 0,\ \partial^2 H/\partial t^2 > 0$. One can also define the moment of *culmination* of a step-wise complexity increase, progressive transition climax, or the *moment of truth*, as the point of inflection of rising $H(t)$ curve, $\partial^2 H/\partial t^2 = 0,\ \partial^3 H/\partial t^3 < 0$, after which progressive complexity-entropy upgrade becomes eminent and irreversible. In a similar way, a critical inflection point within the period of decline, or the *moment of sin* $\partial^2 H/\partial t^2 = 0,\ \partial^3 H/\partial t^3 > 0$, marks the definite establishment of stagnation and decay. Whereas the points of happiness and sadness separate the periods of progress and decline as such, the moments of truth and sin, situated in the middle of progress and decline periods, separate intervals of maximum *subjective* perception of their *results* within the system. We can see that such "vague" and "inexact" notions as happiness, sorrow, and "psychological crises" between them are provided with unambiguous and *rigorous* definitions within the *unreduced* complexity concept (one should not forget, of course, the whole underlying *interaction analysis*, Sect. 2.1) [1].

Note that *partial* time derivatives in the above definitions of system evolution stages correspond to *external* observation over system development from a (generalized) *reference (rest) frame*. If now an observer is situated *within* the developing system, he will see similar development stages, but appearing on a different, "internal" time scale and determined by the respective *total* time derivatives, such as (generalized) *Lagrangian* $L = -dS/dt$ [8, 14]. The difference between those two time flows constitutes the *causal, complex-dynamic* basis of *generalized special relativity* effects emerging at *all* levels of dynamics, from quantum particle motion to civilization development [8].

Note finally that progressive transition to superior level of complexity can be replaced by another development branch, the "death branch" of purely destructive degradation of existing system structures, without qualitatively new, "progressive" structure emergence (Fig. 2). This scenario becomes real when the stock of complexity-information of the driving interaction process is exhausted or when further complexity development is seriously blocked in a deep impasse ("wrong way"). In the first case one deals with the generalized complex-dynamical system *death*, which is *rigorously* defined [8] and inevitable (for a closed system) because of a *finite* quantity of dynamic information, whereas in the second case one has a *bifurcation of development*, where *both* progressive transition to a higher complexity level *and* destructive degradation can happen with certain, dynamically determined *probabilities* (see Eq. (17)).

3 Sustainability Transition as Revolution of complexity

3.1 Modern Bifurcation of Civilization Development: Causal Apocalypse Now

We can apply now the unified development theory from the previous section to modern civilization development, including its recent past and forthcoming future. Observed features analysis shows that modern civilization, suitably represented by its advanced, "locomotive" parts, is situated in the vicinity of the last "point of sadness (ennui)" (Fig. 2) and maybe already slightly outside of it in the direction of a probable complexity-growth step (but well before its "moment of truth"). That modern world position at the beginning of emerging inflection of $H(t)$ curve after its deep minimum (development saturation) is supported by a variety of observed modern "ends", such as End of History, Science, Art, Religion, etc. (e.g. [2, 5, 24, 25]), appearing as stable absence of true novelty emergence (events) and pronounced degradation of existing structures [8]. In view of the close "death branch" beginning (Fig. 2), we get to the *great bifurcation of development* into the death branch of pure destruction and transition to a superior level of civilization complexity. Taking into account the huge, *ultimately complete* scale of all "ends" involved, we can say that we deal here with the rigorously specified version of Apocalyptic "End of the World", Doomsday, etc. appearing as development bifurcation into two branches of "(system) death" and "(new) life", where the latter emerges by transition to a *qualitatively* higher complexity level of the *whole* civilization dynamics [8, 9]. This change can also be designated as *Revolution of Complexity*, or *sustainability transition* (see Sects. 3.2–4).

Particular, *practically important* results of this *rigorously derived* development concept are specified below (Sects. 3.2–3.5). The *causally complete* nature of the underlying interaction analysis (Sect. 2) leaves practically no hope that the observed bifurcational, "Apocalyptic" state of modern civilization can be avoided by usual, "smooth" amelioration of life conditions, often subjectively defended by prosperous, "leading" civilization components (e.g. within standard, "protective" ecological actions, Sects. 3.3–3.5). Dynamic entropy growth *cannot* stop, but the failure to follow the strongly growing, *qualitative* development branch at the *current specific moment* will *inevitably* leave civilization on the death branch of irreversible destruction. We see that the *causally complete understanding* of *unreduced, unified* civilization dynamics within the universal science of complexity provides the unique and *vitally* important basis for *scientifically exact, rigorous futurology* (Sect. 4).

3.2 The Last Scientific Revolution

It is convenient to start our detailed analysis of sustainability transition and the resulting superior complexity level with the description of respective changes in the *system of knowledge*, the more so that the new level of complexity is characterized by a much greater, *decisive* role of a *new kind* of ordered, "scientific" knowledge in the whole civilization development.

Unitary, dynamically single-valued science approach dominating today (and including *zero-complexity imitations* of "complexity" and "chaoticity") is unable to provide consistent understanding of *any* real, *dynamically multivalued* system behaviour (Sect. 2.1), which becomes especially evident for higher-complexity cases (strong interaction, living organisms, intelligent behaviour, social, ecological systems, etc.). At the same time, the *purely empirical*, technological civilization power has attained today, *for the first time in history*, the critical threshold of the *full depth* of *any* real system complexity, from quantum world (elementary particles and fields) to the structure of life (genome and related cell processes, ecosystems, brain processes). This *effectively blind* but quantitatively powerful, "stupid" technology uses traditional trial-and-error empiricism to *strongly modify* systems whose real dynamic complexity exceeds by far the possibilities of zero-dimensional "models" of unitary science (they are still shamelessly promoted for "simulation" of ultimately complex behaviour of economic, social, and ecological systems!). The resulting contradiction creates *real and unprecedented dangers* at *all* complexity levels, from particle physics to genetics and ecology, which are *not* due to the "risk of science/ technology" in general (cf. [24]), but due to the *specific, artificial limitation* of the *unitary* science paradigm and results [8].

Transition to another, *causally complete* kind of knowledge is therefore *urgently needed* today and the failure to perform it will *inevitably* lead to destructive consequences, as the probability of successful empirical "guess" or unitary "simulation" of the huge power of real system complexity (see Eq. (18)) is very close to zero. It is clear that the new, *practically efficient* knowledge can only be based on the detailed understanding of the *unreduced* interaction process underlying any real system dynamics, which leads directly to the dynamic multivaluedness paradigm [6, 7] and universal science of complexity (Sect. 2) [8]. Being thus indispensable for real problem solution already at the existing level of development, the unreduced science of complexity becomes *unified and unique basis* for realisation of sustainability transition and resulting superior level of civilization complexity. It is that, ultimately complete and realistic kind of knowledge that can form a practical basis for the "*society based on knowledge*" at the superior complexity level. It is clear also that imitations of the unitary "science of complexity" can only be harmful because of their *biggest possible*, dynamically single-valued simplification of real system dynamics. Practical *organization of science* should follow the corresponding *qualitative* change towards a much more liberal, decentralized and adaptable system with *emergent* structure [8, 11, 20].

The essential extension of science content, role, and organization constitutes thus a major part of the forthcoming Revolution of Complexity. The latter can be considered, in this sense, as the *last* "scientific revolution" of the kind described by Thomas Kuhn [22], since the unreduced science of complexity realises the *intrinsically complete, permanently creative* kind of knowledge, devoid of *antagonistic* fight between "paradigms" and people (which originates, as it becomes clear now, from the *specific*, strongly *imitative* nature of the *unitary*, "positivistic" science, rather than scientific knowledge in general).

3.3 Complexity-Increasing Production: Growth Without Destruction and the Universal Criterion of Progress

Modern industrial production leads to evident and rapid degradation of environment and life quality, and therefore *cannot* provide *long-term progress*. As such progress is a *necessary* condition for planetary civilization *existence* one is brought to the idea of *sustainable development*. However, the self-protective approach of the current system tends to the tacit assumption that sustainability can be attained by gradual "purification" of production methods, without major, qualitative change of the dominating industrial mode as such [3, 23, 26, 28].

The unreduced interaction analysis of the universal science of complexity rigorously shows, first of all, that the latter hope is totally vain and sustainability *cannot* be attained within the current way of production, irrespective of details, simply because it is invariably reduced to complexity *destruction*, i.e. transformation of higher-complexity structures into lower-complexity ones [8, 9]. We also use here the rigorous and universal *definition and criterion of progress* as *optimal growth of complexity-entropy* according to the system development curve (Figs. 1, 2). At the modern moment of maximum/ending stagnation (Sect. 3.1) civilization progress can only proceed by self-amplifying complexity-entropy growth towards its superior level, without which the system will follow the death branch of catastrophic destruction.

This criterion of progress can be provided with exact formulation by recalling that transition to superior complexity level acquires a well-defined character after the "moment of truth", or Hamiltonian/ Lagrangian inflexion point, where the second time derivative of Hamiltonian/Lagrangian changes sign from plus to minus (Fig. 2). The ensuing criterion of progress (in its "internal" version expressed by total time derivatives) is

$$\frac{d^3S}{dt^3} < 0, \quad \text{or} \quad \frac{d^2L}{dt^2} > 0, \tag{23}$$

where: $L = -dS/dt$ is the system Lagrangian (Sect. 2.2) [8, 14]. Note that progressive *development* thus defined overlaps with both *periods* of progress and decline defined before (Sect. 2.2) and includes their "best" parts of *essential*, self-amplifying growth of complexity-entropy (even though its *rate*, dS/dt, decreases within the period of decline). A *narrow* meaning of "definite" progress includes only progressive development part *within* the period progress $d^2L/dt^2 > 0$, $dL/dt < 0$, while progress *in the whole* can be defined by the condition $dS/dt \gg (dS/dt)_{\text{death}}$, where $(dS/dt)_{\text{death}}$ is the maximum entropy growth rate for the death branch (or its minimum value for the decline period).

Impossibility of sustainable development at the current complexity level follows from the generalized entropy growth law: any, even "ecologically correct" production of modern, industrial way can at best only minimise the inherent complexity destruction (entropy growth), but can never reduce this high enough minimum to values around zero. But the *same* entropy growth law underlies *genuine sustainability* at the *superior* complexity level, after the key transition to *complexity-increasing production* methods and technologies. That's why it is called *sustainability transition* (Sect. 3.1). Indeed, in this case the inevitable complexity-entropy growth takes the form of intrinsically *progressive* creation of *ever more complex* structures ("period of progress" in Fig. 2), as opposed to a "period of decline" where entropy growth is dominated by destruction of previously created structures. It means that the criterion of progressive development, Eq. (23), remains practically *always valid* after the sustainability transition, and very short periods of formal "decline" are determined by decreasing, but *high* rate of entropy growth, $d^2S/dt^2 < 0$, $dS/dt \square (dS/dt)_{\text{death}}$, *within* progressive development, $d^3S/dt^3 < 0$.

Realistic basis for production sustainability is due to *complexity creation* and *complexity-based* kind of technology, where the unreduced complexity-entropy of *all* production results should be *essentially greater*, than that of the initial system configuration. An important example is provided by *irreducibly complex dynamics* of *realistic* sources of pure energy from *nuclear fusion reactions* (in its both "hot", less sustainable and "cold", more prospective versions), demonstrating the *unified, multi-level* structure of the Complexity Revolution.

Contrary to popular ideas about industrial production, its complexity-killing features are *not* due to *massive* use of man-made *machines* as such, but due to a certain, "unitary" way of using certain kind of machinery. Those particular, complexity-reducing tools and methods are closely related to *specific organization* of usual industrial production characterized by explicitly *reduced dynamic complexity* (tendencies of unification, regularity, etc.). Correspondingly, the new, intrinsically sustainable production at the superior complexity level should be organised in a qualitatively different way dominated by the permanently developing, hierarchic, distributed "ecosystem" of *dynamically* connected, generally small units of *individually* structured production (they certainly can form *loose, dynamic* associations at higher ecosystem levels that will replace modern inefficient and decadent corporate monsters). It becomes evident that such complexity-increasing production organization and content is inseparable from the accompanying *personal progress* of human complexity, i.e. the level of *consciousness* [18] (see also Sects. 3.4–4).

3.4 From Unitary to Harmonical Social Structure: Emerging Order Without Government

Due to *holistic* dynamics of unreduced interaction [8], sustainability transition involves a qualitative change of *social structure and dynamics*. In order to specify this change, we show first that "traditional" social structures, including *all* known (modern and ancient) social and political regimes, constitute a single kind of order called *Unitary System* [9, 20]. The term "unitary" (behaviour) has a *mathematically exact* interpretation in the universal science of complexity (Sect. 2.1) of "dynamically single-valued" and therefore *qualitatively uniform*, regular, zero-complexity, "effectively one-dimensional", sequential (dynamics, evolution, etc.). Although any social system cannot be strictly unitary in this rigorous sense, the Unitary System of social structure is close to it because it is a rigid, centralized system of *preferably* regular (controlled) dynamics that can change essentially (usually just to its another *version*) only by way of *destructive* "revolution". Such unitary social order includes *all* previously known social systems (usually considered to be very different), such as any totalitarian, democratic, or meritocratic political structure. Correspondingly, social structure resulting from sustainability transition should differ *qualitatively* from *any* of these, including allegedly the "best possible system" of modern democracy (as well as any meritocracy).

We call the social organization type of that qualitatively superior, higher-complexity level *Harmonical System* [9, 20]. Contrary to any version of Unitary System, the Harmonical System provides *emergent*, intrinsically *creative*, permanently developing kind of social order whose origin resembles that of the free market *economical* structure, but encompasses now the whole civilization structure. It is dominated by a system of interacting, independent units similar to those of the complexity-increasing production structure (Sect. 3.3), but including *all* spheres of activity. *Global* system dynamics is monitored by the *same* kind of *independent*, interactive units, very different from any unitary "government" (or even "non-government organizations", NGO) in that they are *forced* to produce *explicitly useful* services, *compete* with each other and bear *individual, well-specified responsibility* for their results (similar to *small* enterprises within market economy). Loose associations of such units, as well as "high councils", may exist only where needed and without any formal power *exceeding* that of *emergent* actions of independent enterprises (including various "forces of order").

Note that some seeds of such emergent social order may exist within modern "developed" version of unitary democracy, but any its most "liberal" version or component (like NGO) is severely limited by the *imposed* rigid, formal ("obligatory"), centralized power ensuring the status of unrealistic dream to any true liberty (=unreduced, natural, progressive development). The Harmonical System of *emergent order* realises what is considered as *impossible* by the conventional, unitary democracy, a *qualitatively higher* kind of liberty and "democratic" order obtained *without* any "majority vote" (always manipulated by "minority games"). This "miracle" becomes possible only at the described superior level of civilization complexity realized by unreduced interaction of independent units pervading *all* spheres of activity (Sects. 3.2–3.5).

The harmonical social order has *intrinsically progressive* or *sustainable* structure due to *permanent*, non-antagonistic and *essential* complexity development in the sense of our rigorously defined progress (Sect. 3.3). The very character of civilization development changes *forever* after sustainability transition, from painful alternation of "stagnation" and "revolution" periods to the *permanent unreduced creativity* (that could also be described as "distributed complexity revolution"). By contrast, the modern "developed" unitary democracy, apparently repeating respective periods of ancient civilization development, represents not the "best possible" social system (according to its own praise, thoroughly maintained by self-privileged "powers that be"), but rather the *definite end*, generalized complex-dynamical *death-equilibrium* [8] of the Unitary System as such, in *any* its version, followed inevitably by either sustainability transition to a superior complexity level of Harmonical System, or irreversibly destructive death branch (Sect. 3.1, Fig. 2). In fact, this "final", decadent, *equilibrium* character of unitary democracy, clearly seen today, does result from its *highest possible* development of the *unitary* kind of social structure that does not need to be, however, its *only* possible kind and actually represents the *simplest*, basically "tribal" (imposed, compulsion-based) kind of social order. The latter becomes insufficient today just *due to* the ultimately high development of the industrial Unitary System creating self-amplifying, and therefore *insurmountable*, dynamic barriers to its own progress.

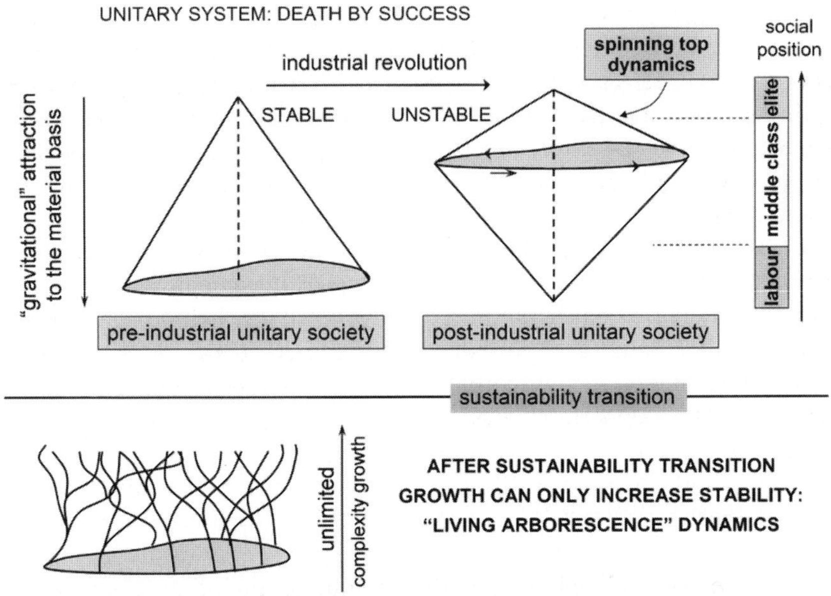

Fig. 3. Schematic representation of the critical instability of developed (modern) Unitary System followed by the globally stable structure of Harmonical System.

The origin of modern, *inevitably* emerging *critical instability* of the *developed*, industrial Unitary System can be conveniently demonstrated with the help of schematic presentation of its social structure dynamics, Fig. 3. Pre-industrial, "traditional" Unitary System can be presented by a pyramidal structure stably resting on its large base of labour classes due to the "gravitational attraction" towards material production/consumption. In the post-industrial society, the same unitary pyramid acquires a strongly deformed, "inverse" (upside-down) configuration due to huge productivity growth as a result of technological revolutions. But since the material "gravity" force preserves the same downward orientation, that monstrous construction with now *quantitatively* dominating layers of non-productive "elite" becomes *critically unstable* and can preserve its normal, "vertical" position only due to the high-speed spinning motion of production-consumption cycles (similar to spinning top stability). However, this *artificially* maintained, *relative* stability has its limits, especially due to basically *dissipative*, chaotic dynamics of any social system, which means that the unitary "whipping top" *will* fall in a destructive manner within a reasonably small time period (like few tens of years from now). By contrast, the harmonical social structure, shown at the bottom of Fig. 3 as a distributed arborescence, does not create destructive instability: instead, its *local, creative* instability provides sustainable progress.

Harmonical System represents thus the *unique* way of *any* further progress, and in order to realise it one should have a *realistic* sustainability transition. Such realistic transformation takes the form of generalized "phase transition" of *higher order*, where the *qualitatively big* structure change occurs not in the whole system volume simultaneously (as in "first-order transitions"), but starts with *small, growing* "seeds" of the "new phase", which strongly facilitates the transition process. The dynamics of both sustainability transition and resulting Harmonical System can be properly understood and monitored only with the help of the *causally complete* understanding of unreduced complexity (Sect. 3.2), which emphasizes once more the high role of this *new kind of knowledge* at the forthcoming development stages.

3.5 New Settlement and Infrastructure

It is not surprising that civilization infrastructure at the unitary level of development, including dynamical structure of settlements, production and communications, reflects major features of the Unitary System, such as centralization, rigidity, development rather by destruction, pronounced tendency towards mechanistic simplification, and the resulting urban decadence in the phase of "developed" unitarity. Indeed, there is the evident degradation to over-simplified, "squared" and "smooth" configurations and operation modes in modern infrastructures, despite *much greater* practical possibilities for their diversity in the developed industrial technology. Whereas this effective complexity destruction is a part of the emerging "death" tendency of the ending level of development (Sect. 3.1), it is also evident that the forthcoming harmonical level of complexity should be based on a *qualitatively different* type of settlement with a distributed, decentralized, and *progressively developing* structure (see Sect. 3.3 for the universal progress definition).

That another kind of settlement can only be realized as a manmade structure intrinsically and strongly *submerged* into the "natural environment" and *constructively* interacting with it, so as to *increase* complexity-entropy of the whole system. Such sustainable civilization structure can be described as omnipresent, man-controlled, progressively evolving *forest*, or "natural park", with submerged, *distributed* settlement, production and transport infrastructure, which excludes anything much resembling modern cities, towns, and villages, with their *centralized* structure tendency. Transport networks in such "living" infrastructure will be well hidden among other, more "natural" and complexity-bearing elements, contrary to their domination in the unitary infrastructure. The omnipresent and intense *creation* of "natural", i.e. *complex-dynamic* environment, rather than its unitary "protection" (inevitably failing), constitutes the essence of complexity-increasing settlement and infrastructure dynamics. The latter correlates directly with the complexity-increasing production mode (Sect. 3.3) because it can be considered as a specific sphere of production with strong involvement of "human dimensions".

Progressively growing dynamic complexity of this new kind of "natural" but totally man-controlled environment and infrastructure has a positive reverse influence upon dynamic complexity of man's consciousness and life style. This positive feedback loop in the man-environment system leads to a dynamic complexity boost that can be described as *Supernature* at the level of "environment" structure and as (*realistically* specified) *Noosphere* at the level of human consciousness (including its individual and "social" aspects). Supernature can have the same or even *much greater* dynamic complexity than the "wild" nature (contrary to any "protected" environment of the unitary ecology), while Noosphere emerges as inseparable, fractally structured and progressively evolving *dynamic entanglement* (Sect. 2.1) of superior consciousness and Supernature. In this sense one can say that nature, in its new form of Supernature, should become again man's home, at that superior, harmonical (complexity-increasing) level of their interaction.

4 Cosmic Intelligence, Future, and Complexity: Concluding remarks

Summarising the universal science of complexity [6–21] (Sect. 2) and its application to modern development problems (Sect. 3), one should emphasize *intrinsic unification* of *causally specified* meaning and purpose of life, future, progress, nature, cosmos, and our destiny within the *universal symmetry of complexity* (Sect. 2.2) forming thus the *practical guiding principle for civilization development.*

Application of the unreduced science of complexity to the problem of cosmic life and extraterrestrial civilizations shows that life realizations in cosmos should be *multiple and diverse*, while unique civilization existence is highly improbable: it follows already from the basic property of *dynamic multivaluedness* (Sect. 2.1). The *complexity correspondence principle* resulting directly from the universal symmetry of complexity [8,11] provides a *rigorous* basis for the statement that real, constructive contact between different civilizations is possible if they have similar levels of unreduced complexity (consciousness) that should certainly be high enough for the contact at a cosmic scale. Therefore the complexity/consciousness upgrade of particular civilization of the planet Earth, which is necessary for its own development (Sect. 3), can be a much more efficient way of establishing contact with extraterrestrial intelligence than usually applied technical means ("find an alien within yourself").

There is no other way to a sustainable, non-destructive future than essential growth of civilization complexity taking the form of *Revolution of Complexity* in all fields of human activity (Sect. 3.1). But since the latter is determined by the level of consciousness that can be causally understood itself as a high enough level of complex interaction dynamics [8, 18], it becomes evident that modern *bifurcation of development* is centred around that *critical consciousness upgrade*, which constitutes today the main factor of *civilization survival*: real Future comes as a superior level of individual consciousness. It shows the emerging dominant role of *individually* specified results of *global interaction* processes, as opposed to "mass consciousness" effects of the unitary society at previous development stages. In fact, only consciousness complexity development provides the basically *unlimited progress perspective* after the objective end of the unitary history of "hot" events (cf. [2]).

As *every* future becomes *uncertain* at a qualitative transition point of modern Apocalyptic scale (Sect. 3.1), one should understand now *all* possible futures within a *unified* vision, in contrast to innumerable "scenarios" and one-dimensional unitary interpolation "threads" for *separate* aspects of development that become totally inefficient and misleading just at such critical point of "generalized phase transition" [1] (cf. [24–27, 29, 30]). Providing a unique possibility of such unified, causally complete vision of multiple interaction processes determining civilization development, the *universal science of complexity* constitutes the *truly scientific* basis for *consistent, provably reliable futurology* and its critically important applications to modern development problems [8, 9].

References

1. Dederichs P.H. (1972) Dynamical diffraction theory by optical potential methods. In: Ehrenreich H., Seitz F., Turnbull D. (eds) Solid state physics: Advances in research and applications, vol. 27. Academic Press, New York, pp. 136–237.
2. Fukuyama F. (1992) The end of history and the last man. The Free Press, New York.
3. Jäger J. (2002) Summary: Towards global sustainability. In: [26], p. 201.
4. Horgan J. (1995) From complexity to perplexity. Scientific American, June: 74–79.
5. Horgan J (1996) The end of science. Facing the limits of knowledge in the twilight of the scientific age. Addison-Wesley, Helix.
6. Kirilyuk A.P. (1992) Theory of charged particle scattering in crystals by the generalized optical potential method. Nucl. Instr. and Meth. B69: 200–231.
7. Kirilyuk A.P. (1996) Quantum chaos and fundamental multivaluedness of dynamical functions. Annales de la Fondation Louis de Broglie 21: 455–480. ArXiv:Quant-ph/9511034–38.
8. Kirilyuk A.P. (1997) Universal concept of complexity by the dynamic redundance paradigm: Causal randomness, complete wave mechanics, and the ultimate unification of knowledge. Naukova Dumka, Kyiv, 550 p, in English. For a non-technical review see also: ArXiv:Physics/9806002.
9. Kirilyuk A.P. (1999) Unreduced dynamic complexity, causally complete ecology, and realistic transition to the superior level of life. Report at the conference "Nature, Society and History", Vienna, 30 Sep – 2 Oct. 1999. See http://hal.ccsd.cnrs.fr/ccsd-00004214.
10. Kirilyuk A.P. (2000) 100 years of quanta: Complex-dynamical origin of Planck's constant and causally complete extension of quantum mechanics. ArXiv:Quant-ph/0012069.
11. Kirilyuk A.P. (2002) Dynamically multivalued, not unitary or stochastic, operation of real quantum, classical and hybrid micro-machines. ArXiv: Physics/0211071.
12. Kirilyuk A.P. (2003) The universal dynamic complexity as extended dynamic fractality: Causally complete understanding of living systems emergence and operation. In: Losa G.A., Merlini D, Nonnenmacher T.F., Weibel E.R. (eds)

Fractals in Biology and Medicine, vol. III. Birkhäuser, Basel, pp. 271–284. ArXiv:Physics/0305119.
13. Kirilyuk A.P. (2004) Dynamically multivalued self-organization and probabilistic structure formation processes. Solid State Phenomena 97–98: 21–26. ArXiv:Physics/0405063.
14. Kirilyuk A.P. (2004) Universal symmetry of complexity and its manifestationsm at different levels of world dynamics. Proceedings of Institute of Mathematics of NAS of Ukraine 50: 821–828. ArXiv:Physics/0404006.
15. Kirilyuk A.P. (2004) Quantum field mechanics: Complex-dynamical completion of fundamental physics and its experimental implications. ArXiv:Physics/0401164.
16. Kirilyuk A.P. (2004) Complex-dynamic cosmology and emergent world structure. Report at the International Workshop on Frontiers of Particle Astrophysics, Kiev, 21–24 June 2004. ArXiv:Physics/0408027.
17. Kirilyuk A.P. (2004) Complex dynamics of real nanosystems: Fundamental paradigm for nanoscience and nanotechnology. Nanosystems, Nanomaterials, Nanotechnologies 2: 1085–1090. ArXiv:Physics/0412097.
18. Kirilyuk A.P. (2004) Emerging consciousness as a result of complex-dynamical interaction process. Report at the EXYSTENCE workshop “Machine Consciousness: Complexity Aspects”, Turin, 29 Sep. – 1 Oct. 2003. ArXiv:Physics/0409140.
19. Kirilyuk A.P. (2004) Complex dynamics of autonomous communication networks and the intelligent communication paradigm. Report at the International Workshop on Autonomic Communication, Berlin, 18–19 October 2004. ArXiv:Physics/0412058.
20. Kirilyuk A.P. (2004) Creativity and the new structure of science. ArXiv:Physics/0403084.
21. Kirilyuk A.P. (2005) Complex-dynamical extension of the fractal paradigm and its applications in life sciences. In: Losa G.A., Merlini D., Nonnenmacher T.F., Weibel ER (eds) Fractals in Biology and Medicine, vol. IV. Birkhäuser, Basel, pp 233–244. ArXiv:Physics/0502133.
22. Kuhn T. (1962) The structure of scientific revolutions. University of Chicago Press, Chicago.
23. Lillo J.C. (2002) Challenges and road blocks for local and global sustainability. In: [26], pp. 193–195.
24. Rees M (2003) Our final hour: A scientist’s warning: How terror, error, and environmental disaster threaten humankind’s future in this century—on earth and beyond. Basic Books, New York.
25. Soros G. (2000) Open society: Reforming global capitalism. Public Affairs Press, New York.
26. Steffen W., Jäger J., Carson D.J., Bradshaw C. (eds) (2002) Challenges of a changing earth. Proceedings of the Global Change Open Science Conference, Amsterdam, 10–13 July 2001. Springer, Berlin Heidelberg New York.
27. Toffler A. (1984) Future shock. Bantam, New York.
28. Vellinga P. (2002) Industrial transformation: Exploring system change in production and consumption. In: [26], pp. 183–188.

29. Vinge V. (1993) Vernon Vinge on the singularity. http://www.ugcs.caltech.edu/~phoenix/vinge/vinge-sing.html.
30. World Future Society, http://www.wfs.org/; The Arlington Institute, http://www.arlingtoninstitute.org/; Spiral Dynamics Integral, http://www.spiraldynamics.net/; The Global Future Forum, http://www.thegff.com/; Future-Institute & University, http://www.futureinstitute. com/; Finland Futures Academy, http://www.tukkk.fi/tutu/tva/; World Future Council, http://www.worldfuturecouncil.org/; Infinite Futures, http://www.infinitefutures.com/; Futuribles, http://www.futuribles.com/; Potsdam Institute for Climate Impact Research, http://www.pik-potsdam.de/.

The Future of Solar System and Earth from Religious Point of View

Kamel Ben Salem

Department of Computer Science - Faculty of Sciences of TunisUniversity, Tunisia, kamel.bensalem@fst.rnu.tn

In this paper, we analyse the Qur'an description of phenomena linked to the evolution of the solar system in the light of our present knowledge. The Qur'an specifies that towards the end of its life, the Sun will start to swell. The consequences of this swelling are the following: the sky will turn red and will look like a scarlet rose; light intensity will increase and it will not be possible to observe the stars from the surface of the Earth. The Qur'an also specifies that the Sun will get closer to the Moon (that is to the Earth). The Moon will be destabilized by the Sun; it will be rent and will ultimately fall onto the Earth. Moreover, the orbits of the solar system planets will be modified. As for our planet Earth, it will witness an overheating and boiling of both seas and oceans and an expansion of waters, a disintegration of mountains, expansion of Earth crust... The Qur'an also indicates that the Earth will survive in spite of these cataclysms.

Foreword: Qur'an & Astronomy

Speaking of the future of the solar system and Earth from a religious point of view may seem a bit peculiar. However, it may be very instructive to study how religions spoke of the end of life. If we consider the Muslim faith, the Qur'an is bursting of interesting statements that deserve to hold our attention for they may enlighten our knowledge on the above astronomy topic. Indeed, the Qur'an affirmates that ***"The creation of the heavens and of the Earth is verily more grandiose than the creation of Man"*** (40:57).

V. Burdyuzha (ed.), The Future of Life and the Future of Our Civilization, 437–453.

First of all, let us briefly remind some facts on Islam and Qur'an. We'll then examine and comment a sample of Qur'anic verses concerning the future of the solar system by focusing on both Moon and Earth. Our comments will be based on an exegesis of the words used in the studied texts that respects their right meanings in Arabic. It has to be underlined that our study is only but a tentative one whose aim is to give rise to further works.

The remainder of this paper is organized as follows. In section 1, we present some historical facts dealing with the birth of Islam and the Qur'an revelation. In section 2, we detail general facts on universe as described in the Qur'an. Finally, the Qur'anic view on the solar system and its future is deeply discussed in section 3.

1. Some Historical Facts

The Muslim religion was born in the seventh century C.E and the Qur'an revelation lasted about 22 years (610-632 C.E). At the pre-islamic epoch, the majority of Arab people in the Arabian peninsula was living a humble and primitive life. However, their particular specificity was their outstanding mastering of their language. Poets and eloquent people were very respected and poetry competitions were very frequent. Best poems were even hanged at Mecca on the walls of the Kaabah shrine. In this context, the Qur'an was revealed "in a clear Arabic language". It even challenged the addressees to produce a verse or a surah (chapter) similar to it. History tells us that they failed to do so. Thanks to the Qur'an, the life of the then Arab people knew a dramatic metamorphosis that involved all aspects of life. Indeed the first revealed verse says. ***"Read! In the Name of your Lord Who has created (all that exists)"*** (96:1). Addressees are even invited to focus first on what exists in the heavens and then on what exists on Earth ***"Say: "Behold all that is in the heavens and the Earth...".*** (10:101). It is after this invitation for reflection that Moslems began to develop Astronomy. There is also an invitation to conquer space and explore the depths of the Earth.

> ***"O assembly of Jinns and Humans, if you can pass beyond the zones of the heavens and the Earth, then pass beyond (them)! Not without a power shall ye be able to pass!"*** (55:33).

Hence, verses related to scientific facts, particularly astronomy, are frequent. In our study, we extracted an exhaustive sample of verses speaking of the subject we are interested in. It is useful to add that we consulted known Qur'an exegesis books, widespread Qur'an translations (in both English and French) and most used Arabic dictionaries, in order to be as close as possible to the original sense of the words and not force them to meanings they cannot support.

2. General Facts on Universe

2.1 Birth of Universe

In the following verses, the Qur'an says that the universe was created from a gaseous mass with fine particles (first verse) the elements of which were initially soldered to form one unit that will later be divided (second verse).

> ***"Moreover He turned to heaven when it was smoke...*** (41:11)
> ***"Do not the unbelievers see that the heavens and the Earth were joined together, then We clove them asunder and We got every living thing out of the water. Will they not then believe?"*** (21:30)

It is clear that the Qur'an thesis concerning the creation evidently contradicts the classical Big Bang theory according to which the universe would have been first concentrated at a very hot and dense but infinitely small point.

It is known that stars and planets were formed from fine particles of a gas cloud and from dusts of celestial bodies according to an agglomeration and gathering process. It is to be noted that this very process is at the origin of the formation of galaxies and consequently of the Universe.

As to the Big Bang, it generated the formation of both several Earth-like planets and galactic matters. Concerning the first, it is important to mention that the Qur'an explicitly refers to the existence of numerous planets like ours in the universe as it is mentioned in the following verse:

> ***"God is the One Who created seven heavens and of the Earth a similar number. The command descends among them so that you know that God has power over all things and comprehends all things in His knowledge."*** (65:12)

The number of heavens and planets is designated by the digit 7. The choice of this digit as a symbol originates from its particular status since it is the smallest prime number after 1, 3 and 5 which are more ordinary ones. All commentators agree that the digit 7 and its multiples designate plurality. This is explained in the Qur'an itself as indicated in the verses below:

> ***"Whether thou ask for their forgiveness, or not, (their sin is unforgivable): if you ask seventy times for their forgiveness, God will not forgive them: because they have rejected God and His Apostle: and God guideth not those who are perversely rebellious."*** (9:80)
>
> ***"The parable of those who spend their substance in the way of God is that of a grain of corn : it groweth seven ears, and each ear hath a hundred grains. God giveth manifold increase to whom He pleaseth: And God careth for all and He knoweth all things."*** (2:261)

As to the galactic matter, the Qur'an announces, in the following verse, that the universe, including the galaxies matter, was created in six periods (yawn) and yet God did not experience any weariness.

> ***"We created the heavens, the Earth and what is between them in six periods, and no weariness touched Us."*** (50:38)

2.2 Starting/Continuity/Expansion

Expansion started after the build up operation and it continues until now (the present is used for the verb expand)

> ***"The heaven, We have built it with power, Verily We are expanding it."*** (51:47)

Remark that the word "heaven" designates all that is exterior to Earth.

2.3 Creation Process Repetition

The Qur'an states that the creation phenomenon in the universe is a continuous process

> ***"It is God Who begins (the process of) creation; Then repeats it."*** (30:11)

This notion of repetition in the creation process is found in several verses and may be interpreted in two ways: the creation process is a continuous operation in our universe, because the latter is permanently transformed (stars die and others are born, etc..), and whatever is produced in a point of the space would also be produced in another point, in the same way. The second interpretation is the cyclical birth of our universe or perhaps the creation of other universes.

3. The Solar System in the Qur'anic Text

3.1 General Statements

The three bodies belonging to the solar system are Earth, Sun and Moon. It has to be noted here that the Qur'an has emphasized the harmony in the movements of the Sun and the Moon (but not in those of the other planets of the solar system). The explanation of this point resides in the two following facts: on the one hand because the Moon and the Sun are the most visible and evident celestial bodies and on the other hand because both these stars will undergo a grandiose astrophysical phenomenon. In fact, at the final stage of the Sun's life, the transformation of hydrogen atoms into helium will be over and as a result of that the outer layers of the Sun will expand. As we will deduce it later on, this will lead to the swelling of the Sun that will destabilize the Moon.

Before going further, let us mention as a digression before going deeper in the Sun-Earth-Moon point, that a little time after its formation, the still soft and hot Earth probably underwent an intensive and continuous bombarding by metallic meteorites. The following verse specifies that **the presence of iron on the Earth is the result of a "descending" process**.

> ***"…And We sent down iron, a Source for great might, as well as many benefits for mankind..."*** (57:25)

It is to be noted that the verb **to "descend" has been used only for iron** (and not for all the other metals referred to in the Qur'an such as gold, silver, copper etc…). This suggests that the iron origin is extra-terrestrial.

Now, concerning the Sun and the Moon, the Qur'an makes a clear distinction between Sun (heat-light source) and Moon/other planets (light reflectors):

"God is the One Who made the Sun a shine and the Moon a light…" (10:5)
"God is the light of the heavens and the Earth. The similitude of His light is as if there were a niche and within it a luminary. The luminary is in a glass. The glass is as if it were a planet glittering like a pearl." (24:35)

It is furthermore precise that, firstly, the Sun was evolved to a steady state:

"The Sun runs its course towards a proper state (or site) of stability. This is the decree of the Almighty, the Full of Knowledge." (26:38)

Secondly, the Sun moving is submitted to computing:

"…The Sun must not catch up the Moon… Each one is travelling in an orbit with its own motion." (36:39-40)
"The Sun and Moon (are subjected) to calculations." (55:5)

In fact, these human calculations are but symbolic representations of these fundamental physical subjacent laws.

3.2 Future of the Solar System

The Qur'an precise that the Sun becomes like a swelling ball:

"When the Sun becomes like a swelling ball (kuwwirat)." (81.1)

Here, the use and the selection of the Arabic word "kuwwirat" (for the phenomenon of the swelling Sun, deriving from the verb "kawwara" leading to the word "kura(t)", meaning "ball"), is intentional. This verb is used especially to designate the ball-shaped winding (for example of a silk string or a cloth band). It is known that the further the winding the greater the volume of the ball. This means a swelling of the ball. In fact, towards the end of its life, the Sun will exactly undergo the same phenomenon. Its volume will increase with time until constituting a red giant. Specialists indicate that the expansion of the Sun will result in the coming closer of the surface of the Sun and the Earth's orbit. This coming closer of the Sun's surface and the Earth's orbit is not explicitly mentioned in the Qur'an, but it is quoted in a "Hadith" (tradition of the prophet Muhammad).

"When the apocalypse gets closer, the surface of the Sun and the Earth ('s orbit) will get closer" [Hadith: Prophet Muhammad's saying].

Many consequences of the Sun's expansion and of its movement closer to the Earth's surface are mentioned in the Qur'an. However, the successive events are not referred to in sequence. They can be classified into three categories: impacts on (i) the Sky in general (Heaven), (ii) the Moon and (iii) the Earth. These points are detailed in the following.

3.2.1 General Impacts on the Sky

a) The Sky Looking like a Scarlet-Rosy

When the Sun turns into a red giant, the solar disc will occupy a great part of the heavens. The sudden appearance of this red giant with a remarkable volume will give the illusion of an invasion of the heavens. At this stage, the Qur'an indicates that the sky will look like a rose (the Sun representing the heart of the rose) and the rest of the sky will become like red hide (the rose's petals). This is mentioned in the following verse:

"Then, when the heaven is rent asunder, becoming scarlet-rosy like red hide." (55:37)

b) The Sky Shining like to a Molten Metal

"The Day that The sky will be like Molten metal" (70:8)

c) Light Intensity Increase

"So, when the sight is dazzled." (75:7)

It is to be noted, however, that the previous verse, like those relating to the final episode of the Sun's life, **does not express what will be "observed" or "experienced"** by eventual inhabitants of the Earth, because the Earth will have become inhabitable a long time before this event. Sight dazzling is therefore only a way to explain the increase of light intensity and does not imply that there will still exist human beings to witness this phenomenon. In fact, the Qur'an specifies that the occurrence of this event is not isolated but it is associated with the appearance of other phenomena, such as the boiling of the seas and oceans due to temperature increase. This will lead to the Earth's surface water evaporation, thus eliminating all forms of life.

d) Stars are no Longer Visible

The following verses indicate that it will be difficult to see the stars from the Earth, probably because of the light intensity (of the Sun) which is very strong, as it has already been mentioned.

> ***"Then when the stars become dim."*** (77:8)
> ***"And when the stars, lose their lustre"*** (81:2)

e) Modification of the Solar System Planets Orbits

The Qur'an specifies that there occurs a dispersion in the movement of the solar system planets (modification of orbits), including the Earth.

> ***"...And when the planets will disperse"(intatharat)."*** (82:2)

This is confirmed by the use of the Arabic term "intatharat" which means an organized dispersion, by opposition to the (not used) "tanatharat" which means a random dispersion.

3.2.2 Impacts on the Moon

a) Moon is no Longer Visible

The difficulty of observation caused by the increase of the light intensity does not in fact concern only the stars; it applies to the totality of the other celestial bodies. In fact, when the Earth is covered by the extreme intensity of the Sun's light, it will be impossible to observe even the Moon, as it is explained in the following verse:

> ***"And the Moon will be eclipsed."*** (75:8)

The fact that this verse immediately follows verse 7 of sura 75 mentioned above (cf. §3.2.1-c) and speaking about sight dazzle confirms once again that the Moon's eclipse is certainly due to the increase in light intensity. This is, in fact, a final and irreversible eclipse.

b) Sun and Moon are Joined

After the start of the Sun's expansion process, the Moon will be totally immersed in the Sun's atmosphere. In the celestial dome, only the red giant will be present. The two celestial bodies will be joined as it is specified in the following verse:

> ***"... And the Sun and Moon are joined together."*** (75:9)

The enormous increase of the Sun's volume will in fact result in its "coming closer" to the Moon (joining of the two celestial bodies), therefore closer to the Earth.

Let us remark here that this latter verse is also explained by a "Hadith" (tradition of the prophet Muhammad) which specifies that the Sun and the Moon will both be bathing in fire. This refers in fact to the immersion of the two celestial bodies in the very high temperature of the Sun's atmosphere. The same "Hadith" refers to the increase of the volume of both celestial bodies. As regards the Sun, the Qur'an has already dealt with this phenomenon; concerning the Moon, this volume increase is due to the Moon's expansion as a result of the Sun's intense heat.

c) Moon Fragmentation

In the presence of the star's (Sun and Moon joined) tenuous atmosphere, the Moon will be submitted to viscous forces and will gradually lose its energy and also its angular moment. This will result in the destabilization of its orbit (orbital braking) the Moon's orbit will shrink little by little resulting in its progressive fall onto the Earth's surface after having been fragmented.

Long before its fall onto the Earth, the effects of the Earth's tide on the Moon will ultimately lead to its fragmentation below a certain distance, known as Roche limit. This Moon's "break up" would occur suddenly and brutally once the limit is reached, as it is specified in the following verse:

> ***"The Hour has drawn near and the Moon has been cleft asunder."*** (54:1)

An alternative explanation of this fact consists in assuming that the Moon's fragmentation is a result of its expansion. But if we take into account the verses that suggest that the Moon's fragments will fall onto the Earth's surface (cf. §3.2.3-e), the thesis of fragmentation resulting from the effect of orbital braking would be more plausible.

The Moon's fragmentation will constitute its final stage of disappearance from the solar system. It is obvious that the Sun will destabilize the Moon only after having swallowed up both Mercury and Venus, which are closer to it than the Moon.

3.2.3 Impacts on Earth

a) Exodus Movement and Gathering of Wild Beasts

Once a certain degree of heating up of the Earth's surface is reached, forest fires will become more and more frequent. Feeling the danger, wild beasts will start general and continuous escapes, which will correspond to great gatherings as it is specified in the following verse.

> ***"...And when the wild beasts are herded together."*** (81:5)

b) Boiling and Bursting Forth of the Seas and Oceans

The Qur'an indicates that the Sun's expansion will result in the warming up of the seas and oceans as it is specified in the verse below:

> ***"And when the seas become as blazing fire."*** (81:6)

The warming up of the seas and oceans will result in their boiling up as it is explained in the following verse:

> ***"(I swear)... by the canopy raised high, and by the boiling sea."*** (52: 5-6)

After the boiling up of the seas and oceans, water expansion will follow as it is said in the verse below:

> ***"... And when the seas are burst forth."*** (82:3)

The warming up and expansion of the seas and oceans' waters are certainly indicators of a first stage before their evaporation. According to us, these phenomena will take place at the beginning of the Sun's swelling process and in all cases before the destruction of the Moon by the Sun, because at that moment the Earth's surface would already be totally turned into lava.

We have to add here that it is obvious that the seas and oceans' evaporation will result in a considerable loss of the Earth's mass. But, we should also note that when the Sun turns into a red giant and reaches its maximum volume, it would already have lost a part of its mass. According to Newton's law of gravitation a fraction of mass loss (in relation with the initial mass) greater for the Sun than for the Earth will result in the orbit of the Earth moving off from the Sun's and vice-versa. Because the solar atmosphere will consequently be lost in thc intergalactic space, the previous conclusion remains a fortiori valid for any (surviving) planet. Thus, the term "will disperse" (cf. §3.2.1-e) should be interpreted as "will move farther away from the Sun". The changing of orbit for the surviving planets of the solar system, and especially for the Earth, allows us to suggest that modifications in its revolution and rotation will occur leading to a change of the length of the day, the night and the year.

c) Expansion of the Earth's crust

In the same way, the Earth's overheating will result in its expansion, hence a repulsion and rejection of its content (probably the most volatile elements i.e. atmosphere, water and some trapped elements presently below the crust).

> ***"...And when the Earth expands. And has cast forth what is within it."*** (84:3-4)

If the Earth's expansion is not explicitly referred to in the Qur'an, it is explicitly mentioned in a "Hadith".

d) Pulverization of Mountains

The fate of mountains, known for their rigidity is explained in the following verse:

> ***"... And the mountains will be carded wool."*** (101:5)

The choice of the expression **"carded wool"** denotes in fact the change of the mountains-forming rocks' nature; these initially highly compact and tough rocks will ultimately become, because of the heat, **viscous** and more **voluminous** (like carded wool). Hence, it is clearly shown that the change in the rocks' state is due to the excessively high temperature prevailing over the Earth after the Sun's expansion.

Another fact corroborating precedent one is referred to in another verse indicating that the mountains will melt like **"crumbling sand dunes"** and will gradually lose height and rigidity:

> ***"...And the mountains will be like crumbling sand dunes."*** (73:14)

The comparison with **crumbling sand dunes** aims to show the progressive loss of height and rigidity of mountains. These two features certainly indicate that the rocks constituting these mountains will be in the process of fusing. This fusion will go on until these mountains are totally flattened as it is indicated in the following verse:

> ***"They ask thee concerning The mountains, say: "My Lord will demolish them and He will leave them as plains smooth and level"."*** (20:105-106)

In other words, mountains will be totally disintegrated under the effect of heat. The same reference to "disintegration" is found in the following verse:

> ***"And the mountains turning into crumbles to end up as sparse particles."*** (56:6)

Thus, the tough and high mountains will ultimately be frittered away and will look like a mirage as is explained in the following verse:

> ***"And the mountains shall vanish, as if they were a mirage."*** (78:20)

From the preceding verses, it can be deduced that the Earth's crust will undergo the same fate, that is, it will fuse.

e) Fall of Lunar Fragments on Earth

In another verse, it is said that the Earth (the land's surface) and the mountains will suddenly and fatally be hammered (the Qur'an uses the verb "dakka" to describe what will happen); this verb means to bombard, to hammer.

> ***"And the soil and the mountains are lifted, then violently hammered at one stroke. On that day, shall the (great) event come to pass."*** (69:13-14)

The lifting cited above is to be related to the expansion of the Earth's crust. The hammering of the land's surface and of the mountains, as a result **of one stroke**, which, according to the Qur'an is **fatal**, would probably refer to the result of the fall of the biggest (giant) lunar fragment after the Moon's disintegration onto the Earth's surface; this event would have the same destructive effect as the explosion of several atomic bombs. This will certainly constitute a remarkable and unique event in Earth's life cycle: **"On that day, shall the great event come to pass"**.

The Qur'an even specifies the occurrence of this violent shaking. In fact, according to verse 73:14, the Earth (the soil) and the mountains will quake only after the mountains have become comparable **to crumbling sand dunes** that is after having lost much height. In other words, the occurrence of the **"great event"** will follow the crucial phase of mountain disintegration.

> ***"The day when the earth (the soil) and the mountains are shaken up, once the mountains have become like crumbling sand dunes."*** (73:14)

It is to be noted that the Qur'an stipulates that, after the "great event" (violent shaking of the land's surface), the land surface will undergo successive poundings, which are in fact but the results of the fall of lunar fragment remains.

> ***"Nay! When the earth (the soil) undergoes hammering after hammering."*** (89:21)

f) Earth Covered by very Dense Smoke

Another consequence of Earth's hammering by the enormous lunar meteorites is the formation of "swirling clouds" in the sky as it explained in the following verse:

> ***"The day when the sky is agitated by a swirling cloud."*** (52:9)

The swirling clouds in the sky will be in fact accompanied by the formation of a real ashy and dusty screen of clouds, the thickness of which would probably reach tens of kilometers and which would cover the whole Earth. This covering of the Earth by a great mass of smoke ("clouds") is referred to in the following verse:

> ***"And the day when the sky shall be rent asunder with clouds..."*** (25:25)

The next verse specifies that the smoke in question is not a common one similar to that known by the prophet's contemporaries. This cloud of smoke will certainly occur after very violent impacts resulting from enormous collisions of great rock bodies (meteorites) falling on the Earth. This cloud of smoke will then be very dense and unique in its kind (a very particular kind of smoke).

> ***"Then wait you for the day when the sky will bring forth a visible smoke."*** (44:10)

g) Slowdown of Earth's Rotation Speed

We know that the Moon insures the stability of the Earth's movement. The disturbance exerted by the other planets: Venus, Mars, Jupiter and the others, might, in the absence of the Moon, enter in resonance with the movement of the Earth's axis of rotation. The Earth would not leave its present orbit, but its movement might become chaotic and totally unpredictable.

It is specified in a "Hadith" that, when the apocalypse approaches:

> ***"A day will become like a year, then like a month, then like a week, and finally like a present day."*** [Hadith]

According to us, the friction and the violent impact of giant lunar meteorites with a diameter of some hundreds of kilometers, added to the forces exerted by the tide of the nearby Sun, which once reaching its maximum size, will cause a slowdown in the speed of the Earth's rotation round its axis. This slowing down would result in a longer day. A literal interpretation of the "Hadith" mentioned above means that a day becoming like a year implies a slowdown and a reduction of the initial rotation speed with a ratio of about 1/365. This slowdown will gradually continue until the initial rotation speed is re-established.

h) Changing of the Earth's Rotation Direction

A second "Hadith" specifies that:

> ***"At the approach of the apocalypse, the Sun will rise in the West."*** [Hadith]

It may be deduced that the huge collisions of lunar fragments with our planet's surface, will first end the rotation of our planet round its axis and will ultimately lead to the inversion of the rotation direction. It is obvious that the resumption of rotation in another direction will be progressive, which explains the gradual return to a stability speed as mentioned in the first "Hadith". It is clear that without taking into account the Moon, the inversion in the Earth's rotation direction cannot be explained. The Sun's expansion has but morphological transformations on the Earth, and a changing (enlargement) of its orbit.

Let us remark here that as already noticed, the Qur'an verses referring to the consequences of the turning of the Sun into a red giant on the evolution of the Earth's crust and on the other solar system planets, describe a multitude of facts without precise chronology. For example, when the Qur'an specifies that there will occur an expansion of the Earth's crust, and that the mountains will be pulverized, it should be understood that these phenomena will take place only after the boiling and evaporation of the seas and oceans. They are in fact separated by thousands (or millions) of years.

3.3 The Aftermath of the Solar System Collapse

The Qur'an only mentions that the solar system planets will be affected by the swelling Sun. No mention of other impacts on stars, except the fact that they lose their lustre (no more visible).

The Qur'an has clearly dissociated the apocalypse that is a phenomenon proper to the solar system and the universe collapse phenomenon In fact, the verses that we have just exposed mention only the Sun, the Moon, the Earth and the bodies that turn around the Sun. The Qur'an has clearly specified in Sura 82, verse 2, the planets of the solar system will disperse. **Then only the solar system planets will be affected by the expansion phenomenon of the Sun**. The Qur'an does not mention the stars that are distant from our solar system. The only references to the stars are found in Sura 77, verse 8 and Sura 81, verse 2: they specify that during the Sun swelling stage, the stars will lose their luster and will become difficult to observe because of light intensity increase on Earth.

It is very interesting to note that the Qur'an has specified that, after the apocalypse, the Earth will be transformed into another Earth. Its general aspect will be different from the present one:

> *"The Day when the Earth turns out into another Earth and so will be the heavens."* (14:48)

The Earth will undoubtedly undergo great transformations as a result of Sun's expansion and its closer position from the Earth's surface. But these transformations concern certainly the Earth's crust only. This may be consolidated by the fact that the Qur'an says in Sura 82, verse 2 *"...that the planets will disperse"*. In other words, the Earth and its sisters Mars, Jupiter and the others will move farther from the Sun (change of orbit) and will continue to turn round the Sun which will progressively use up its energy.

It is also interesting to note that the use of the expression "transformed into another Earth" in the previous verse does not give further details, which might shock the prophet's contemporaries: some of them accused him of madness and magic because the Qur'anic revelation ran counter to their beliefs and modes of life.

4. Conclusion

As a conclusion to this paper, one wonders whether the Qur'an was written by a human being who had lived a period of obscurantism (cf. §1) or it is really a divine revelation. If the Qur'an was a pure invention of a highly intelligent person, why would this person take the risk to put forward scientific theses which would prove false afterwards?

What we can say is that the study of the Qur'an verses related to scientific notions has not, so far, revealed any discordance with our present knowledge, contrary to the ideas that prevailed 15 centuries ago. But it should be noted that such questions that cross the mind of any human being had already been asked in the Qur'an that gives an answer in the following two verses:

> ***"We will show them Our signs in the universe, and in their own selves (souls), until it becomes manifest to them that this (the Qur'an) is the truth. Is it not sufficient in regard to your Lord that He is a Witness over all things?"*** (41:53).
>
> ***"Do they not then consider the Qur'an carefully? Had it been from other than God, they would surely have found therein many a contradiction."*** (4:82).

Our study is far from being exhaustive. Our aim was especially to show that the Qur'an includes verses related especially to science, which no objective reader can deny. We can quote here the points dealing with Astronomy, e.g. the apocalypse scenario and the final fate of the Earth and the Moon. It is interesting to note that one of the interesting aspects of this study is the fact that there exists a difference between the Qur'an theses and some present scientific knowledge concerning points that have not definitely been settled.

«Verily! In the creation of the heavens and the Earth, and in the alternation of night and day, there are indeed signs for men of understanding. Those who remember (celebrate) the praises of God standing, sitting, and lying down on their sides, and think deeply about the creation of the heavens and the Earth, (saying): Our Lord! You have not created (all) this without purpose, glory to You!» (3:190-191)

5. References

1. M. Bucaille, (1990) The Bible, the Qur'an and Science, Seghers.
2. F. Gabtni, (2000) Le soleil se lève à l'Occident - Science pour l'Heure. (http://www.cirs.tm.fr/index-fr.htm) Presse scientifique, 1er sept. 2000.
3. Al Bukhari, Sahih, translation by M.M. Khan (9 Vol), ed. Dar-us-Salam, Saudi Arabia. (http://www.witness-pioneer.org/vil/hadeeth/bukhari/index.htm).
4. A. Rifaï, Le Miracle, Tel-Chiheb, Deraa, Syria (in Arabic).
5. T.X. Thuân, (1991) La mélodie secrète, Gallimard.
6. http://www.akamaiuniversity.us/PJST6_1_37.pdf.

Remark: When translating the Qur'anic text, we mainly adopted the works of M. Hamidullah (in French), Youssef Ali (in English) and Maurice Bucaille (in his book "The Bible, the Qur'an, and Science"). In some translations, the word God is replaced by ***Allâh*** which means, in Arabic, ***the Unique God*** of Abraham, …of Moses and Jesus.

VI

ARE WE ALONE?

The Life-Time of Technological Civilizations

A Description of the Terrestrial Technological Adolescence Case

Guillermo A. Lemarchand

Centro de Estudios Avanzados and FCEN, Universidad de Buenos Aires; C.C. 8 Sucursal 25; C1425FFL Buenos Aires, Argentina, lemar@correo.uba.ar

We present an empirical study of the long-term evolution of several social indicators (e.g. human population growth, statistics of deadly quarrels, diffusion of democratic systems, etc.). We assume that the human species emerge, develop and become extinct with similar evolutionary patterns that other terrestrial species. We propose that the long-term indicators are showing some sort of macro-transition in their long-term behavior that we defined as "Technological Adolescent Age." We present an estimation of this period. Assuming the "Principle of Mediocrity" and using the Drake Equation we calculate a lower threshold for the number of technological civilizations in the galaxy.

Key words: SETI, Technological Civilization Lifetime *(L)*, Human Population, Deadly Quarrels, Evolution of Democracies, Drake Equation.

1. Introduction

Human beings have broken the ecological 'law' that says that big, predatory animals are rare. Two crucial innovations in particular have enabled us to alter the planet to suit ourselves and thus permit unparalleled expansion: speech (which implies instant transmission of an open-ended range of conscious thoughts) and agriculture (which causes the world to produce more human food than unaided nature would do). However, natural selection has not equipped us with long-term sense of self-preservation. Based

V. Burdyuzha (ed.), The Future of Life and the Future of Our Civilization, 457–467.

in our previous works (Lemarchand, 2000, 2004), here, we will show that the human species is facing a new type of macro-transition: the technological one.

Sagan (1980) defined the *Technological Adolescent Age* (TAA) as the stage in which an intelligent species has the capability to become extinct due to technological misuse (e.g. global war), environmental degradation of the home planet (e.g. global warming, overpopulation, etc.), or simply by the misdistribution of physical, educational and economical resources (difference between the degree of development among developed and developing societies) that may cause the collapse of the civilization due to the tensions generated by the inequities among different fractions of the global society.

Here we present a semi-empirical approach to estimate the period of time that may last with this new macro-transition. Our humankind needs to pass from the TAA into a *Technological Mature Age* (TMA), in which we would learn how to live in harmony with the members of our species and the environment, and learn how to manage efficiently the increase of our power over nature at their different dimensions. The comprehension of the evolution of these long-term social patterns may help us to design different strategies to avoid the self-annihilation. The last is an imperative requirement needed to extend the lifetime of the present civilization.

The so-called "Principle of Mediocrity" proposes that our planetary system, life on Earth and our technological civilization are about average in the universe, and that life and intelligence will develop by the same rules of natural selection wherever the proper surroundings and the needed time are given (Hoerner, 1961). In other words, anything particular to us is probably average in comparison to others. From a Lakatosian epistemological point of view[19], this hypothesis is within the "hard core" of the research programs which main purpose is the search for life in the universe (e.g. exobiology, bioastronomy, astrobiology, SETI). If we considered "average" all the steps that let the appearance of our technological civilization in the universe, we may also assume that this TAA transition may be a typical evolutionary stage for all the hypothetical galactic civilizations.

---

[19] According to Imre Lakatos (1922-1974), within the "hard-core" of any "Research Program" are those hypotheses, that the community of experts in the field, assume – consciously or unconsciously – as valid, without the explicit requirement of an experimental or observational test (see I. Lakatos, "Falsification and the Methodology of Scientific Research Programs" in I. Lakatos and A. Musgrave (eds.), *Criticism and the Growth of Knowledge,* Cambridge University Press, Cambridge, 1974).

In 1961 Frank Drake (Pearman 1963) proposed an equation to estimate the number of technological civilizations in our galaxy. Several estimations assigning different values to each factor of the *Drake Equation* showed that the number of technological galactic civilizations (N) has a strong dependence with the last factor of the equation, L, or the lifetime of a technological civilization in years (Kreifeldt 1973; Oliver 1975). A technological civilization is the one that has the technological capability to communicate, in any way, among interstellar distances. Less than seventy years ago, our species has reached the last stage, only when our first strong radio transmissions left the terrestrial ionosphere into outer space. Most of the authors, using the *Drake Equation*, would agree that the possible number of technological civilizations in the galaxy, would be close to $N \sim (\pi_f \times L)$, where π_f is the product of: the rate of galactic star formation, the fraction of stars forming planets, the number of planets per star with environments suitable for life, the fraction of suitable planets on which life develops, the fraction of life-bearing planets on which intelligence appears, and the fraction of intelligent cultures which are communicative in an interstellar sense. The best present estimation of this product is $0.1 < \pi_f < 10$.

A systematic study of the relevant indicators of the long term behavior of our technological civilization will be useful, not only to make an estimation of the possible value of N, but most importantly to identify which are the most relevant variables that we should encourage to improve and change, in order to optimize the life time of our present global civilization.

In the following sections, we present some preliminary results that show the long-term evolution of several social indicators and we applied those results to estimate the value of L and N.

2. The Determination of the Technological Adolescent Age Transition

Several very long-term biological and ecological studies have shown that different species on Earth, emerge, develop and become extinct with similar evolutionary patterns (Charnov 1993; Gurney & Nisbet 1998). Human species may be not an exception. Our hypothesis is that we are living in a very special moment of our species history, the transition from the TAA to the TMA (Lemarchand 2000, 2004).

In order to estimate the period of time of this TAA transition we have used three different societal indicators: (1) the human demographic transition, (2) the distribution of deadly quarrels during the last five centuries and (3) the diffusion of democracies within nations during the period

1800-2004 (considering the "democracy" as a disembodied technology of government).

The available temporal series come from a few hundred years (e.g. world democracies) to several thousand years (e.g. the evolution of human population). We have found that the temporal series analyzed for the first two social indicators follow a Self-Organized Criticality (SOC) system with a scale free behavior, a characteristic presented in most of complex systems (Bak 1996; Jensen 1998). On the other hand, the third indicator shows a logistic growth pattern, similar to those found in the diffusion of new technologies in a close market (Fisher & Pry 1971).

The data corresponding to the three indicators shows similar phase-transition patterns starting after the World War II (WWII) and ending by the year 2100.

Within this particular period of our human species history, we have developed sophisticated war-technologies, so efficient, that we may become extinct at a rate of 500 million people per hour (nuclear war). We may destroy the natural balance of our biosphere by the industrial pollution, the greenhouse effect and other environmental degradation activities generated by the human behavior. We may cause a world population explosion that may rapidly exhaust the natural resources or we may increase the gap between developed and underdeveloped societies in such way to generate the collapse of different regions of our home planet. All these examples may have a long-term description using several mathematical approaches that show the emergence of patterns very similar to those that appear in the ecological dynamics of other species.

3. Empirical Results and Data Analyses

3.1 The Demographic Transition

We have used a mathematical model of the world population growth that shows a blow-up and self-similar regime that was developed by Kapitza (1996). We found that this model and the available historical data of the global human population exhibit a SOC behavior (Lemarchand, 2004). Another intrinsic property of the model is the so-called *Demographic Transition* or the well established change in the pattern of growth of all populations, when they reach a certain stage and rate of development. This transition has been experienced by all developed countries; it began there at the end of 18th Century with the Industrial Revolution. At the present time, we are at the height of the transition on a global scale. Following the mathematical model, the absolute rate of global population growth is expected to peak in the year 2007, but the relative growth rate reached its maximum value of 1.7 per year in 1989. The equations show that – for a planetary behavior – the human population started with its demographic transition by 1960 and would end by 2050. After that a new reproductive regime is expected to appear, different to the previous one that dominated the dynamics of the last 12,000 years (since the emergence of agriculture and the cities).

3.2 The Distribution of Deadly Quarrels (1500-2000)

Richardson (1945) discovered that the distributions of wars over time follow a power-law (using data from 1800 to 1930). This is a characteristic of all SOC systems. In one way, this shows that the dynamics on inter-human violence is governed by the same type of processes present in the human population growth dynamics and in a great variety of other complex systems examples (Bak, 1996).

Although there is a pattern in the dynamics of inter-human violence, following a SOC system, we also have to take into account the evolution of the technologies of war against time. We have worked with a coefficient of lethality in order to normalize the evolution of weapon technologies from the sword in 400 B.C. to the atomic bombs at the end of the twentieth Century. We have also analyzed the distribution of wars for the period 1495-2000, extending the original Richardson work to 500 years of data.

We introduced a definition of intensity of a war, I, as the ratio of battle deaths to the population at the time of the war (Levy 1983). We have represented the distribution of the number of battles N_c against the intensity I (Lemarchand, 2004). However, when considering distributions that may exhibit SOC it is preferable to use non-cumulative data. An equivalent approach is to take the mathematical derivative of the cumulative distribution with respect to intensity dN_c/dI. Here the fractal dimension resulted in 1.28. An alternative approach to the analysis of this data is in terms of return periods T, the time we should wait to have an event of intensity I. When we consider these distributions of deadly quarrels using normalized values of technologies, we found a correlation between the coefficient of lethality and the slope of the SOC distribution. A rough extrapolation of this analysis shows that – using 500 years of deadly quarrels data – we would need to wait between 30 to 500 years in order to generate a violent event at which the whole human population will disappear. These results are in agreement with a couple of auxiliary indicators of the war processes:

(a) the distribution of the destructive power available in nuclear arsenals per capita or the number of tons of TNT per person in the world during 1954-2000 and (b) the evolution of world military expenditures during 1950-2000. Both indicators show an extraordinary coincidence in their distributions with the demographic transition period with a highest peak around 1980-1990 (Lemarchand, 2000).

3.3 Democratic Diffusion within Nations (1800-2000)

We have analyzed the evolution of this societal indicator in order to understand the time constants at which societies organized themselves in order to produce changes in a macro-behavior level. Democracy is a mechanism of collective choice and a form of social organization that can be considered a superior substitute for other such mechanisms or forms of organization. In this way, democracy may be considered as a *disembodied technology* of government and social organization. As such, as well as any other embodied or disembodied technology, democracy may be expected to grow, or diffuse, over time, amongst the world's population, and the hypothesis posed in the present study is that the growth follow a regular pattern, according with the Fischer & Pry (1971) substitution model of technological change.

To test this hypothesis we have used, for the period 1800-2004, the list of national democracies (www.bsos.umd.edu/cidcm/polity) provided in the POLITY IV dataset. The POLITY IV survey covers all independent members of the international system, those that have attained independence by 1975 and whose population exceeded one million by the mid-1990s. It gives, for each such polity, an annual score of institutional democracy, on a scale ranging from zero to ten.

Using this data set, we calculate the fraction of the world population at each year of those polities with scores over 4; 5; 6; 7; and 8 in the POLITY IV classification. Applying the logistic growth model, the best curve fit was obtained only with those countries that have a score over 7 (e.g. a group that includes all the countries with scores of 7, 8, 9 and 10). The obtained results are shown in Figure 1. Clearly, the dynamics of the diffusion of democracies among the world population follows a logistic growth curve. The correlation factor of the data fitting is $r = 0.96$. The calculation of the mathematical derivative of the logistic curve also shows a bell-shape curve centered at the year 1989. This, again, is similar to the distribution found for the demographic transition data, the deadly quarrels analysis, the

1945-2005 distribution of nuclear stockpiles (megatons per capita), the 1950-2004 world military expenditures, etc.

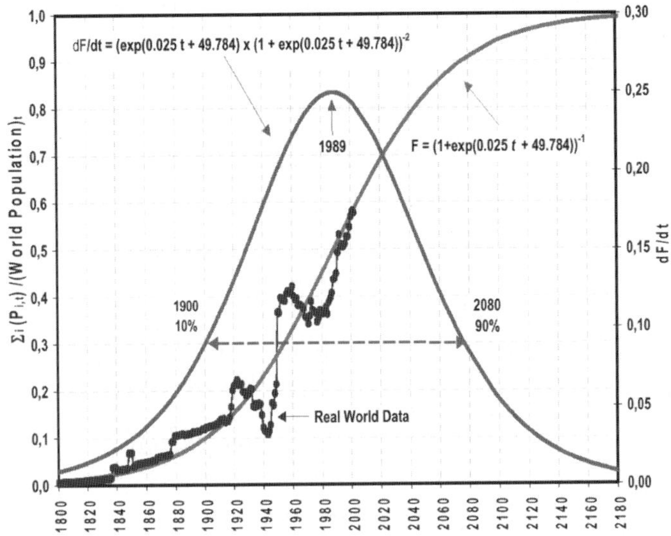

Figure 1. Distribution of democracies represented against time (1800-2000). Here, we calculated the fraction of the world population under democratic governments with a score over 7 points (from a scale that goes from 0 to 10), according to POLITY IV data-base. If we consider the democracy as a disembodied technology of government we see that diffuses against time as any other technology in a closed market, following a logistic-type growth. Here F is the fraction of the world population that is under a democratic government at time t and may be represented by a logistic equation that describes the evolution and diffusion of democracies against time. The bell-shape curve is the density function *(dF/dt)*. The takeover time *(Δt)* is defined as the time required for the technology to increase from $F=0.10$ to $F=0.90$. In our case $\Delta t=176$ years.

4. From Kantian Ethics to *Lex Galactica*

If most of the galactic societies have a similar destructive evolutionary trajectory as the patterns that we are facing in our human race, the number of communicative civilizations may be very small. If the metaphor of *technological adolescence* is correct, we may consider that we are living in a very particular moment of our civilization's history where we can start our self-annihilation. In order to avoid it, our species must make a deep transformation of the human individual behavior towards, at least, three main aspects[20]: *Intra-individual* or *Somatic, Inter-individual* or *Social* and *social* or *Habital*[21].

In order to improve the lifetime of a technological civilization it is impossible to have superior science and technology and inferior morals. This combination is dynamically unstable and we can guarantee a self-destruction within the lifetimes of advanced societies (10^5 to 10^6 years?). At some point, in order to avoid their self-destruction, all the intelligent species in the universe must have to produce this ethical breakthrough among the members of their societies in order to live in harmony between them and their planetary environment. Otherwise, the probability of global extinction would be very high and consequently their social life expectancy very short. What kind of ethical principles should guide this transformation or *social mutation?* We consider that Kantian Ethics provides some good elements to start the discussion.

Kant's outstanding contribution to moral philosophy was to develop with great complexity the thesis that moral judgments are expressions of practical as distinct from a theoretical reason. For Kant practical reason or the rational will does not derive its principles of actions by examples from the senses or from theoretical reason; it somehow finds its principles within its own rational nature. Kant argues that willing is truly autonomous if but only if the principles which we will are capable of being made universal laws. Such principles give rise to *categorical imperatives*[22], or duties binding unconditionally, as distinct from hypothetical imperatives, or commands of reason binding in certain conditions, such as that we desires for certain ends. Kant seems to hold that universalizability is both necessary and sufficient for moral rightness. Kant arrives at the ideal of the kingdom of ends in themselves or of people respecting each other's universalizing wills. This has been an enormously influential idea, and its most distinguished recent exponent has been John Rawls (1980).

Some useful ideas in the direction of the evolution of social ethical stages – applied to the study of several terrestrial cultures- were developed originally by Piaget (1971) and extended by Kohlberg (1973). In his pio-

---

[20] For a complete description of these aspects see C.A. Mallmann, "On Human Development, Life Stages and Needs Systems" in F. Mayor (Edi.), *Human Development in its Social Context*, UNESCO, Paris, 1986.

[21] Here we use the word *"Habital"* in reference to the concept of *Habitat.*

[22] In its most famous formulation, it states that "the *maxim* implied by a proposed action must be such that one can will that it become an universal law of nature."

neer works, Kohlberg established a correspondence between Piaget's cognitive evolutionary stages and his moral judgment stages. According to his view, the final ethical evolutionary stage is based on "universal principles."

Our thesis is that all the civilizations should evolve ethically at the same time they evolve technologically. When these civilizations reach their TAA, they must perform the *social mutation* or become extinct. After learning how to reach a synergetic harmony among the individual members, theirs groups and their habitat, they would extend this *praxis* to the rest of living beings, including their hypothetical galactic neighbors. Their own evolutionary history will teach them the Kantian principle of respecting each other's universalizing wills. Being aware that each planetary evolutionary path is unique, these advanced civilizations will have a non-interference policy with the evolutionary process of underdeveloped societies. This galactic quarantine hypothesis – based in the Kantian Ethics – was defined as *Lex Galactica* (Lemarchand, 2000).

If some of these ideas are correct, they will produce some observational consequences within the SETI (Search for Extra Terrestrial Intelligence) radioastronomical programs. If the *Lex Galactica* principle is applied by galactic civilizations after the macro-transition between TAA and TMA, we might expect that very limited amounts of practical technical information would be distributed among the galaxy by them. This behavior will be reflected within the contents of any intended electromagnetic galactic message, sent to us, that may be detected by any of the SETI projects that are carrying out from different observatories worldwide. An access to technologies thousands of years more advanced that the present ones, could cause our self-destruction if those technologies become available to terrorists or some other crazy leaders. These hypothetical advanced civilizations would not want to place potentially destructive knowledge at the disposal of any 'ethically underdeveloped' society. Such knowledge could be a threat to the emerging societies' survival. Any civilization needs time to work out adequate moral restraints on their own behavior.

Earthlings' electromagnetic transmissions have only revealed their existence to the universe within a sphere of approximately 70 light-years around the Sun. Distant advanced societies will be unable to recognize that some primitive intelligent life is around our Sun, and consequently they will be unable "to calibrate" our technological and ethical evolutionary level to start sending their "advance knowledge" to us.

5. Conclusions

With the invention of the technologies of mass destruction after WWII we have – for the first time in the human history – the technological capability to annihilate our species and most of the life forms on Planet Earth. The long-term evolution of the three different societal indicators analyzed, showed a transition that started after the WWII and that may end after the middle of XXI Century. If we want to avoid self-extinction we must change the rules of the inter-human interaction within the limits of this particular period of time, defined as TAA.

After the WWII our terrestrial civilization reached the technological capability for interstellar communication via electromagnetic waves (radio and laser signals). In a broad sense, the bottleneck for the evolution of any technological civilization in the galaxy would be the TAA.

If we assume the so-called "Principle of Mediocrity" (Hoerner, 1961) that proposes that our planetary system and our civilization are about average and that life and intelligence will develop by the same rules of natural selection wherever the proper surroundings and the needed time are given. Then we may also assume that the average lower boundary for a technological civilization lifetime with interstellar communication capabilities would be close to $L \sim 150$ to *200* yrs. If this is so, we may use the *Drake Equation* to determine the lower limit of the number of technological civilizations in our galaxy as $N \sim (\pi_f \times L) \sim (2 \text{ to } 2000)$.

Acknowledgments

This research is supported by a Foundation for the Future (Seattle, USA) research grant and CONICET (Argentina). My participation in this meeting was possible thanks to the generosity of the organizers and The Planetary Society (Pasadena, USA).

References

Bak, P. 1996, *How Nature Works*, Springer-Verlag, New York.

Charnov, E. L. 1993, *Life History Invariants*, Oxford University Press, Oxford.

Fisher, J. C., & Pry, R. H. 1971, A simple substitution model of technological change, *Tech. Forecast. & Social Change,* 3: 75-88.

Gurney, W. S. C., & Nisbet, R. M. 1998, *Ecological Dynamics,* Oxford University Press, Oxford.

Hoerner, S. von 1961, The search for signals from other civilizations *Science*, 134: 1839-43.

Jensen, H. J. 1998, *Self-Organized Criticality*, Cambridge. Univ. Press, Cambridge.

Kapitza, S. P. 1996, The phenomenological theory of world population growth, *Physics-Uspekhi*, 29: 57-71.

Kohlberg, L. 1973, The claim to moral adequacy of the highest stage of moral judgment, *Journal of Philosophy*, 70: 630-645.

Kreifeldt, J. G. 1973, A formulation for the number of communicative civilizations, *Icarus*, 14: 419-430.

Lemarchand, G. A. 2000, Speculations on the first contact, in *When SETI Succeeds: The Impact of High-Information Contact*, ed. A. Tough, The Foundation for the Future, Bellevue, pp. 153-163.

Lemarchand, G. A. 2004, The technological adolescent age transition in *International Astronomical Union Symposium 213: Bioastronomy 2002, Life Among the Stars*, eds. R. P. Norris and F. H. Stootman Astronomical Society of the Pacific, San Francisco, pp. 460-466.

Levy, J. S. 1983, *War in the Modern Great Powers*, Univ. of Kentucky Press, Lexington.

Oliver, B. 1975, Proximity of galactic civilizations, *Icarus,* 25, 360-367.

Pearman, J. P. T. 1963, Extraterrestrial intelligent life and interstellar communication: an informal discussion, in *Interstellar Communication: The Search for Extraterrestrial Life,* ed. A. G. W. Cameron, W. A. Benjamin, New York, pp. 287-293.

Piaget, J. 1971, *El Criterio Moral en el Niño*, Fontanella, Barcelona.

Rawls, J.1980, Kantian constructivism in moral theory, *Journal of Philosophy*, **77:** 515-572.

Richardson, L. F. 1945, Distributions of wars in time, *Nature*, 155**:** 610.

Roberts, D. C., & Turcolotte, D. L. 1998, Fractality and self-organized criticality of wars, *Fractals*, 6, 351-357.

Sagan, C. 1980, *Cosmos,* Random House, New York.

Calculating the Number of Habitable Planets in the Milky Way

Siegfried Franck, Werner von Bloh, Christine Bounama and Hans-Joachim Schellnhuber

Potsdam Institute for Climate Impact Research (PIK), P.O. Box 60 12 03, 14412 Potsdam, Germany, franck@pik-potsdam.de

It is traditional to discuss about extraterrestrial intelligence in terms of the celebrated Drake equation, with its estimates of habitable planetary systems, origins of life, evolution of technology, and so on. Although several factors are highly speculative, a subset of them, describing the number of habitable planets can be now stated more precisely with the help of new results from investigations in extra-solar planetary systems. Probabilistic estimations of the first factors provide about 50 millions habitable planets in the present Milky Way. A second approach is based on an integrated Earth system analysis. Combining the formation rate of Earth-like planets with estimations of extra-solar habitable zones (HZ) gives the number of habitable planets in the Milky Way over cosmological time scales. There was a maximum number of habitable planets around the time of Earth's origin. If at all, interstellar panspermia was most probable at that time.

Keywords: Drake equation, habitable zone, extrasolar planets, panspermia.

Introduction

The extraterrestrial life debate spans from the ancient Greek world of Democritus over the 18th century European world of Immanuel Kant to the recent discoveries of more than 150 extra-solar planets. Already now, it can be stated that the search for extraterrestrial life, both in our solar system and in extra-solar planetary systems, will be one of the predominant themes of science in the 21st century.

V. Burdyuzha (ed.), The Future of Life and the Future of Our Civilization, 469–482.

The definition of habitability is closely related to the very definition of life. Up to now, we only know terrestrial life, and therefore the search for extra-terrestrial life is the search for life, as we know it from our home planet. Life can be defined as a self-sustained system of organic molecules in liquid water immersed in a source of free energy. It is well known that organic molecules are rather common in the solar system and even in interstellar clouds. There is also no problem to find any source of free energy for extra-terrestrial life. Therefore, the existence of liquid water is the central point in the search for extra-terrestrial life and in the definition of habitability. Nevertheless, it is evident that liquid water and basic nutrients are essential but not sufficient requirements for life.

In our calculation of the HZ we are following an integrated system approach. On Earth, the carbonate-silicate cycle is the crucial element for a long-term homeostasis under increasing solar luminosity. In most studies (see, e.g. Caldeira and Kasting, 1992), the cycling of carbon is related to the present tectonic activities and to the present continental area as a snapshot of the Earth's evolution. On the other hand, on geological time-scales the deeper parts of the Earth are considerable sinks and sources for carbon. In addition, the tectonic activity and the continental area change noticeably. Therefore, we favor the so-called geodynamical models that take into account both the growth of continental area and the decline in the spreading rate (Franck et al., 2000a). Our numerical model couples the stellar luminosity, the silicate-rock weathering rate and the global energy balance to allow estimates of the partial pressure of atmospheric and soil carbon dioxide, P_{atm} and P_{soil}, respectively, the mean global surface temperature, T_{surf}, the biological productivity, Π, as a function of time, t, and the distance, R, from the central star (Fig. 1).

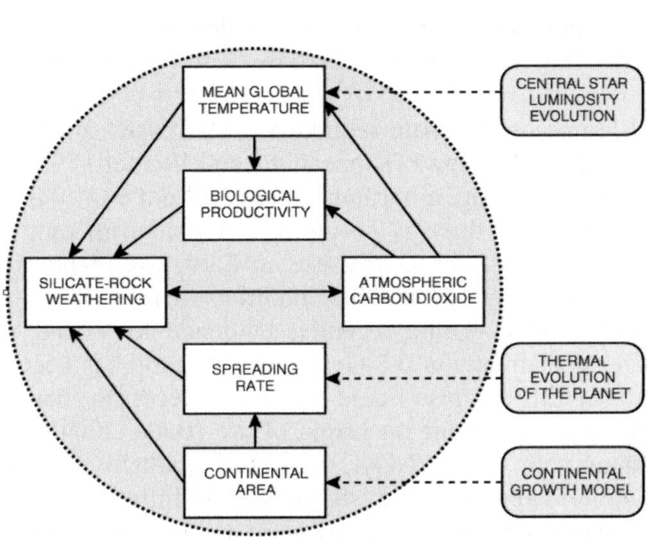

Fig. 1. Box model of the integrated system approach (Franck et al., 2000a). The arrows indicate the different forcings (dotted lines) and feedback mechanisms (solid lines).

The HZ around an extra-solar planetary system is defined as the spatial domain where the planetary surface temperature stays between 0°C and 100°C and where the atmospheric CO_2 partial pressure is higher than 10^{-5} bar to allow photosynthesis. This is equivalent to a non-vanishing biological productivity, $\Pi > 0$, i.e.

$$HZ := \{R | \Pi(P_{atm}(R,t), T_{surf}(R,t)) > 0\}. \tag{1}$$

According to the definition in Eq. 1 the boundaries of the HZ are determined by the surface temperature extremal, $T_{\text{surf}} = 0°\text{C} \vee T_{\text{surf}} = 100°\text{C}$, or by the minimum CO_2 partial pressure, $P_{\text{atm}} = 10^{-5}$ bar. Therefore, the specific parameterization of the biological productivity plays a minor role in the calculation of the HZ. In the approach by Kasting et al. (1993) the HZ is limited only by climatic constraints invoked by the luminosity of the central star, while our method relies on additional constraints. First, habitability is linked to the photosynthetic activity of the planet (Eq. 1) and second, habitability is strongly affected by the planetary geodynamics. In principle, this leads to additional spatial and temporal limitations of habitability. To present the results of our modeling approach we have delineated the HZ for an Earth-like extra-solar planet at a given but arbitrary distance R in the stellar mass-time plane (Fig. 2). A detailed discussion about the influence of continental growth models can be found in Franck et al.

(2000a). The qualitative behavior does not depend on the choice of a specific scenario. In general, HZ is limited by the following effects:

- Stellar lifetime on the main sequence τ_H decreases strongly with mass. Using simple scaling laws (Kippenhahn and Weigert, 1990), we estimated the central hydrogen burning period and got $\tau_H < 0.8$ Gyr for $M >$ 2.2 $M_\odot$. Therefore there is no point in considering central stars with masses larger than 2.2 $M_\odot$ because an Earth-like planet may need approximately 0.8 Gyr of habitable conditions for the development of life (Hart 1978; 1979). Quite recently, evidence for liquid water at the Earth's surface already at 0.2 Gyr after formation has been found (Wilde, et al., 2002). Because liquid water is a necessary condition for life this is the lower limit for the origin of life (Bada, 2004). If we perform calculations with $\tau_H < 0.2$ Gyr, we obtain qualitatively similar results, but the upper bound of central star masses is shifted to 3.4 $M_\odot$.
- When a star leaves the main sequence to turn into a red giant, there clearly remains no HZ for an Earth-like planet. This limitation is relevant for stellar masses in the range between 1.1 and 2.2 $M_\odot$.
- In the stellar mass range between 0.6 and 1.1 $M_\odot$ the maximum life span of the biosphere is determined exclusively by planetary geodynamics, which is independent (in a first approximation, but see next limiting effect) of R. So we obtain the limitation $t < t_{max}$.
- There have been discussions about the habitability of tidally locked planets. We take this complication into account and indicate the domain where an Earth-like planet on a circular orbit experiences tidal locking. That domain consists of the set of (M, t) couples, which generate an outer HZ boundary below the tidal-locking radius. This limitation is relevant for $M < 0.6$ $M_\odot$. As an illustration we depict the HZ for $R = 2$ AU in Fig. 2.

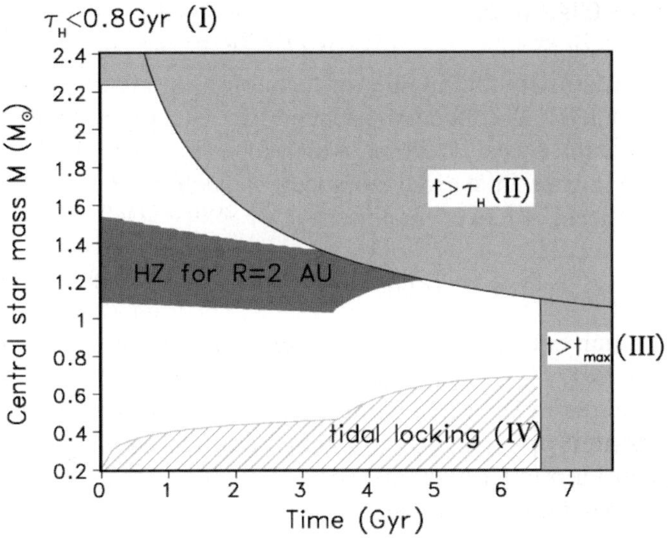

Fig. 2. Shape of the HZ (gray shaded) in the mass-time plane for an Earth-like planet at distance R=2 AU from the central star. The potential overall domain for accommodating the HZ for planets at some arbitrary distance is limited by a number of factors that are independent of R: (I) Minimum time for biosphere development (τH < 0.8 Gyr excluded); (II) Central-star life time on the main sequence (t > τH excluded); (III) Geodynamics of the Earth-like planet ($t > t_{max}$ excluded); (IV) Tidal locking of the planet (non-trivial sub-domain excluded). The excluded realms are marked by light gray shading in case of the first three factors, and by gray hatching for the tidal-locking effect. The picture is taken from Franck et al. (2000b).

Based on Eq. 1, the continuously habitable zone (CHZ) is defined as the band of orbital distances where the planet is within the HZ for a given time interval τ.

In the next chapter we will use our results presented in Fig. 2 to calculate the average number of planets per a planetary system, which are suitable for the development of life. This number will be first used to calculate the number of habitable planets in the Milky Way with the help of a subset of the Drake equation. In a second approach we calculate the Number of habitable planets over cosmological time scales with the help of a convolution integral over the planet formation rate of Earth-like planets and the probability that a stellar system hosts a habitable Earth-like planet at a certain time after its formation.

The Drake Equation

How can we estimate the number of technological civilizations that might exist "out there"? A convenient scheme for extra-terrestrial intelligence prospecting is the Drake equation, which identifies specific factors considered to play a role in the development of such civilizations and allows estimating at least orders of magnitude:

$$N_{CIV} = N_{MW} \cdot f_P \cdot n_{CHZ} \cdot f_L \cdot f_{CIV} \cdot \delta . \tag{2}$$

This equation counts the number of contemporary technical civilizations in the Milky Way, N_{CIV}, whose radio emissions may be detectable, but note that some factors in Eq. 3 are highly speculative: Depending on more pessimistic or more optimistic assumptions, one can end up with either no candidates at all or a surprisingly large number of possible entities. Let us discuss the specific factors in detail:

- N_{MW} is the total number of stars in the Milky Way.
- f_P is the fraction of stars with Earth-like planets.
- n_{CHZ} is the average number of planets per a planetary system, which are suitable for the development of life.
- f_L is the fraction of habitable planets where life emerges and a full biosphere develops, i.e. a biosphere interacting with its environment on a global scale (Gaia's sisters).
- f_{CIV} denotes the fraction of sisters of Gaia developing technical civilizations. Life on Earth began over 3.85 billion years ago. Intelligence took an extremely long time to develop. On other life-bearing planets, it may happen faster, it may take longer, or it may not develop at all.
- δ describes the average ratio of civilization lifetime to Gaia lifetime (Lemarchand, 2005).

As already mentioned above, f_{CIV} and δ are highly speculative items: There is just no information available about the *typical* evolutionary path of life, or the characteristic "life span" of communicating civilizations. Regarding the fate of ancient advanced civilizations on Earth, the typical lifetime was limited by increasing environmental degradation or over-exploitation of natural resources. One can also speculate that the development and utilization of certain techniques, which facilitate the emergence of a higher civilization, may be accompanied by new vulnerabilities or hazard potentials that jeopardize the very subsistence of advanced cultures. As a consequence, the lifetime of any communicating civilization may be limited to the range of few hundreds of years, yet this is not even an educated guess.

On the other hand, the first three factors seem to be assessable by geophysiological theory and observation. Therefore, from the view of Earth system analysis, we will focus on estimation for the number of habitable planets in the Milky Way:

$$N_{hab} := N_{MW} \cdot f_P \cdot n_{CHZ} \,. \tag{3}$$

The key factor in Eq. 3 is n_{CHZ}. For the assessment of this factor it is necessary to investigate the habitability of an extra-solar planetary system.

Now we can start to calculate N_{hab} by discussing the three factors in Eq. 3:

1. The total number of stars in the Milky Way is rather well known (see, e.g. Dick, 1998). It can be derived from the star formation rate (Zinnecker et al., 1993). We pick the value $N_{MW} \approx 4{\cdot}10^{11}$. According to Gonzalez et al. (2001) there exists also a so-called "Galactic Habitable Zone" (GHZ), i.e. that region in the Milky Way, where an Earth-like planet can be habitable at all. The inner limit of the GHZ is defined by exogenous perturbations destroying life (e.g. supernovae, gamma ray bursts, comet impacts). The outer limit is set by the chemical evolution of the galaxy, in particular the radial disk metallicity gradient. Up to now, Gonzalez et al. (2001) have investigated only the outer limit quantitatively. Therefore, a quantitative estimation of the GHZ is still not possible and we cannot reduce the value for N_{MW} given above.
2. Current extra-solar planet detection methods are sensitive only to giant planets. According to Marcy and Butler (2000) and Marcy et al. (2000) approximately 5% of the Sun-like stars surveyed possess giant planets. These discoveries show that our solar system is not typical. Several stars are orbited by giant planets very close and highly eccentric. Up to now, the fraction of stars with Earth-like planets can be estimated only by theoretical considerations. Lineweaver (2001) combines star and Earth formation rates based on the metallicity of the host star. Using his results we can find a rough approximation for the fraction of stars with Earth-like planets from the ratio of Earth formation rate to star formation rate. Since the Sun has formed this ratio is always between 0.01 and 0.014. In the framework of a conservative approximation we pick the value $f_P \approx 0.01$.
3. The average number of Earth-like planets per a planetary system which are suitable for the development of life, i.e. residing in the CHZ, can be calculated in the following way (Franck et al., 2001): First one computes the probable number of planets, $P_{hab}(M,t)$, which are within the CHZ, $[R_{inner}(\tau), R_{outer}(\tau)]$, of a central star with mass M at a certain time t. For this calculation it is assumed that the planets

are distributed uniformly on a logarithmic scale (Kasting, 1996). This distribution is a good approximation for the solar system and not in contradiction to our knowledge of already discovered planetary systems. Knowing $P_{hab}(M,t)$, one has to integrate over all stellar ages t on the main sequence, and after that over all central star masses that are relevant for HZs, i.e. between 0.4 M_s and 2.2 M_s. The stellar masses M are distributed according to a power law $M^{2.5}$ (Scheffler and Elsässer, 1988). Introducing certain normalization constants and using τ = 500 Myr as the necessary time interval for the development of life (Jakosky, 1998), Franck et al. (2001) find, for the geodynamic model, n_{CHZ} = 0.012. This means that only about 1% of all the extra-solar planets are habitable. This calculated number is actually one order of magnitude smaller than the number 1/4 implied by the situation in our solar system.

With the help of the three numbers just discussed we finally arrive at

$$N_{hab} \approx 4.8 \cdot 10^7, \tag{4}$$

which is indeed a rather large number (Fig. 3).

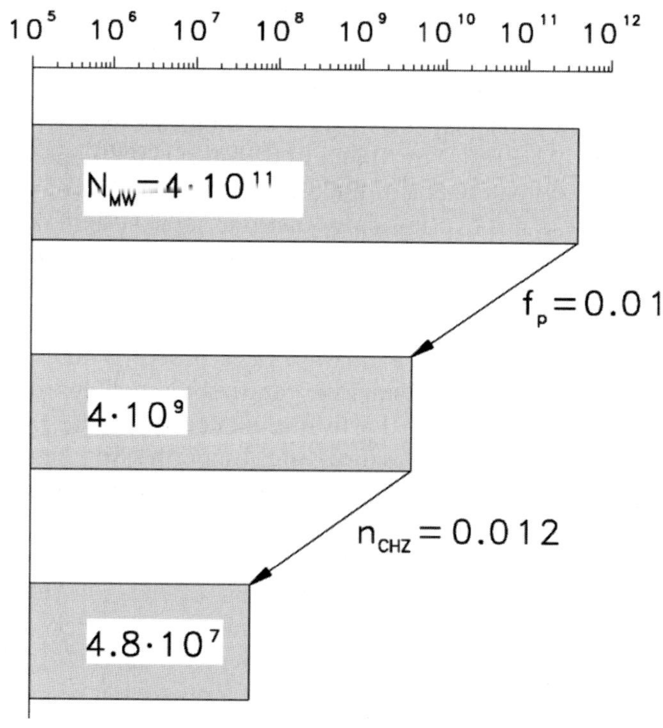

Fig. 3. Determination of the number of habitable planets N_{hab}, bases on the number of stars in the Milky Way by application of the factors in Eq. 3.

Star Formation, Planet Formation Rates and Panspermia

The question of the star formation history (SFH) has been the subject of a number of studies. While some authors favor a smooth and constant SFH, others favor a bursty SFH fluctuating around a constant mean (Twarog, 1980; Scalo, 1987; Barry, 1988; Rocha-Pinto et al., 2000). However, cosmological simulations result in an exponentially decaying star formation rate (SFR) with intermittent spikes (Nagamine et al., 2001). Based on the most recent observational data, Lineweaver (2001) fits the SFR for the universe to an exponentially increasing function for the first 2.6 Gyr after Big Bang followed by an exponential decline. He uses this fit to quantify star metalicity as an ingredient for the formation of Earth-like planets. The metalicity μ is built up during cosmological evolution through stars, i.e.,

$$\mu \sim \int_0^t SFR(t')\, dt' . \tag{5}$$

Then the planet formation rate (PFR) can be parameterized in the following way:

$$PFR = 0.05 \cdot SFR \cdot p_E(\mu) \cdot [1 - p_J(\mu)], \tag{6}$$

where p_E is the probability that Earth-like planets are formed and p_J the probability for hot Jupiter formation with orbits at which they would destroy Earth-like planets. The pre-factor 0.05 reflects the assumption that 5% of the stars are in the range of 0.8…1.2 solar masses (M_s). The relation between metalicity and the probability p_E $(1-p_J)$ is a so-called Goldilocks problem: if the metalicity is too low, there is not enough material to build Earth-like planets; if the metalicity is too high, there is a high probability of forming hot Jupiters. Taking all these effects into account, one can derive the time-dependent PFR.

The number of stellar systems containing habitable planets in the Milky Way, *P(t)*, can be calculated with the help of a convolution integral:

$$P(t) = \int_o^t PFR(t') \times p_{hab}(t - t')\, dt', \tag{7}$$

where p_{hab} is the probability that a stellar system hosts a habitable Earth-like planet at time Δt after its formation. The probability that an Earth-like planet is in the HZ, p_{HZ}, is (Whitmire and Reynolds, 1996):

$$p_{HZ}(M,\Delta t)=\frac{1}{C_1}\int_{R_{inner}(M,\Delta t)}^{R_{outer}(M,\Delta t)} R^{-1}\,dR\,, \tag{8}$$

where C_1 is a normalization factor. p_{hab} can be expressed as follows:

$$p_{hab}(\Delta t)=\frac{1}{C_2}N_P\int_{0.8M_s}^{1.2M_s} M^{-2.5}\left(1-\left(1-p_{HZ}(M,\Delta t)\right)^{N_P}\right)dM\,, \tag{9}$$

where C_2 is again a normalization factor, and N_p is the average number of Earth-like planets per stellar system. Under the condition of small p_{HZ} Eq. 9 can be simplified to:

$$p_{hab}(\Delta t)=\frac{1}{C_2}N_P\int_{0.8M_s}^{1.2M_s} M^{-2.5}\,p_{HZ}(M,\Delta t)\,dM\,. \tag{10}$$

The product of the normalization factors $C_1C_2=1.57M_s^{-1.5}$ results from solving Eq. 10 between the central-star-mass-dependent minimum and maximum HZ boundaries $0.1\cdot M/M_s$ AU and $4\cdot M/M_s$ AU, respectively. With this definition of p_{hab} Eq. 7 yields also the number of habitable planets in the Milky Way. In order to estimate p_{hab} , the following assumptions are made:

- The stellar masses M are distributed according to a power law (Scheffler and Elsässer, 1988) $\sim M^{2.5}$;
- the distribution of planets can be parameterized by $p(R)\sim R^{-1}$, i.e. their distribution is uniform on a logarithmic scale in the distance R from the central star (Whitmire and Reynolds, 1996; Kasting, 1996);
- following Lineweaver (2001) we restrict our attention to the set of Sun-like stars in the mass range from 0.8 to $1.2M_s$;
- R_{inner} and R_{outer} are the inner and outer boundaries of the HZ, respectively. They are explicit functions of the central star mass and the age of the corresponding planetary system (Franck et al., 2000b); and
- the average number of Earth-like planets per stellar system, N_p, is set to 4 according to the number of Earth-like planets in our solar system.

In Fig. 4a we show the PFR recalculated from Lineweaver (2001) and rescaled to the present star formation rate in the Milky Way of about one solar mass per year. The results for the calculation of the number of stellar systems containing habitable planets in the Milky Way, *P(t)*, are presented in Fig. 4b.

The value $P(t=13.4$ Gyr) of about $4\cdot 10^7$ is of the same order of magnitude as given in Eq. 4. In principle, *P(t)* depends on the chosen function for the star formation history. The assumption of a constant SFH would lead to qualitatively different results. In contrast, the addition of "spikes" to our applied SFH has only minor effects. This results from the convolution integral, which damps out fluctuations in the PFR in the sense of a low-pass filter.

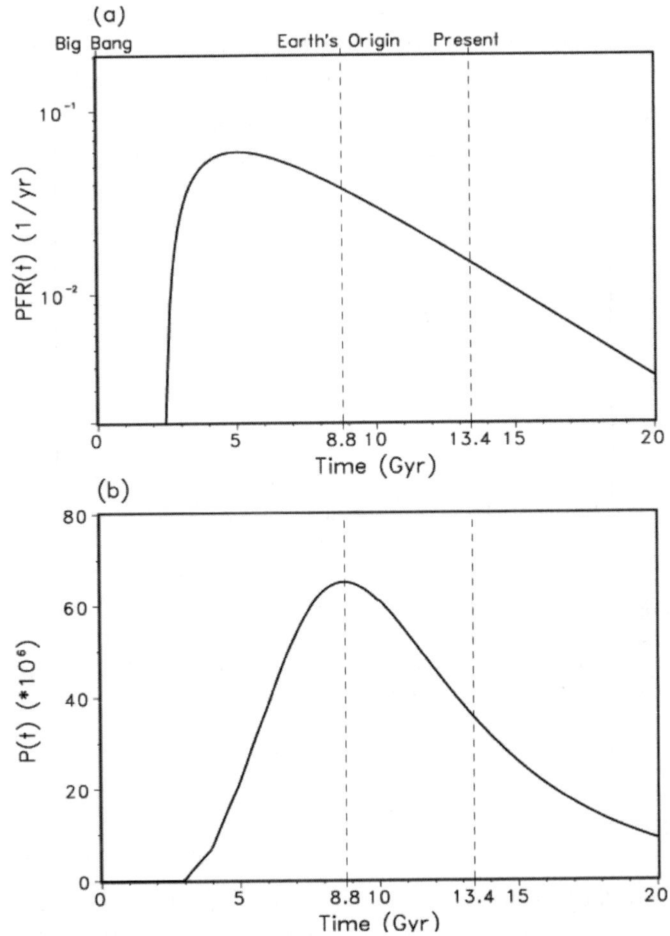

Fig. 4. (a) Earth-like planet formation rate (PFR) (Lineweaver, 2001), and (b) number of stellar systems containing habitable planets, *P(t)*, as a function of cosmological time for the Milky Way. The vertical dotted lines denote the time of Earth's origin and the present time, respectively.

The probability of interstellar panspermia depends on several factors. First of all, the emitting planet must be habitable because it must be a source of viable microorganisms. Secondly, the microorganisms must survive the interstellar journey. This depends strongly on their survival rate. At least, the target planet must be also habitable to allow the seeding by a single organism. In a Gaian perspective a habitable planet is related to the instability of a terrestrial planet in a dead state, i.e. a small perturbation by a seed forces the system to a state with a globally acting biosphere. Therefore, interstellar panspermia events are related to the average density of stellar systems containing habitable planets (Fig. 4b).

On the basis of our results (von Bloh et al., 2003) and the results of Melosh (2001) we can calculate the average number of interstellar panspermia events in the Milky Way, $N(t)$. The result is shown in Fig. 5.

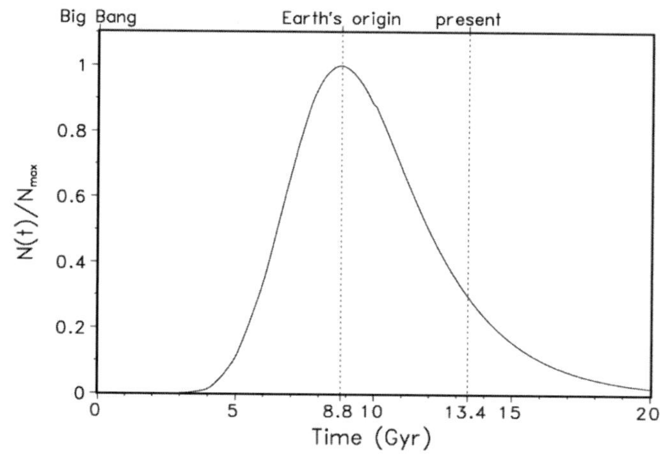

Fig. 5. The number of interstellar panspermia events in the Milky Way, $N(t)$, rescaled to N_{max} as a function of cosmological time. Dotted vertical lines denote the time of Earth's origin and the present time, respectively. If at all, panspermia was most probable at the time around Earth's origin.

Conclusions

The present number of habitable planets in the Milky Way is in the order of magnitude of 10^7, while there was a distinct maximum at the time of Earth's origin. In this way the number of interstellar panspermia events has also a maximum at the time of Earth's origin. Therefore, if at all, panspermia was most probable at this time. This supports the idea that panspermia might have caused a kick start to the processes by which life originated on Earth: there is palaeogeochemical evidence of a very early appearance of life on Earth leaving not more than approximately 1 Gyr for the evolution of life from the simple precursor molecules to the level of the prokaryotic photoautotrophic cells (Schidlowski, 1990; Brasier et al., 2002).

References

Bada, J. L., (2004) How life began on Earth: a status report, *Earth Planet. Sci. Lett.* 226: 1-15.

Barry, D. C., (1988) The chromospheric age dependence of the birthrate, composition, motions, and rotation of late F-dwarf and G-dwarfs within 25 Parsecs of the Sun, *Astrophys. J.* 334: 436-448.

Brasier, M. D., Green, O. R., Jephcoat, A. P., Kleppe, A. K., Van Kranendonk, M. J., Lindsay, J. F., Steele, A. and Grassineau, N. V., (2002) Questioning the evidence for Earth's oldest fossils, *Nature* 416: 76-81.

Caldeira, K. and Kasting, J. F., (1992) The life span of the biosphere revisited, *Nature* 360: 721-723.

Dick, S. J., (1998) *Life on other worlds*, Cambridge University Press, Cambridge.

Franck, S., Block, A., von Bloh, W., Bounama, C., Schellnhuber, H.-J., and Svirezhev, Y., (2000a) Reduction of biosphere life span as a consequence of geodynamics, *Tellus* 52B: 94-107.

Franck, S., Block, A., von Bloh, W., Bounama, C., Steffen, M., Schönberner, D., and Schellnhuber, H.-J., (2000b) Determination of habitable zones in extrasolar planetary systems: where are Gaia's sisters?, *JGR-Planets* 105 (No. E1): 1651-1658.

Franck, S., Block, A., von Bloh, W., Bounama, C., Garrido, I., and Schellnhuber, H.-J., (2001) Planetary habitability: is Earth commonplace in the Milky Way?, *Naturwissenschaften* 88: 416-426.

Gonzalez, G., Brownlee, D., and Ward, P., (2001) The Galactic Habitable Zone: Galactic Chemical Evolution, *Icarus* 152: 185-200.

Hart, M. H., (1978) The evolution of the atmosphere of the Earth, *Icarus* 33: 23-39.

Hart, M. H., (1979) Habitable zones about main sequence stars, *Icarus* 37: 351-357.

Jakosky, B., (1998) *The Search for Life on Other Planets*, Cambridge University Press, Cambridge.

Kasting, J. F., (1996) Habitable zones around stars: an update, in: Circumstellar Habitable Zones, L. R. Doyle, ed., Travis House Publications, Menlo Park, pp. 17-18.

Kasting, J. F., Whitmire, D. P., and Reynolds, R. T., (1993) Habitable zones around main sequence stars, *Icarus* 101: 108-128.

Kippenhahn, R., and Weigert, A., (1990) *Stellar Structure and Evolution*, Springer, Berlin Heidelberg.

Lemarchand, G., (2005) The technological adolescent age transition: a boundary to estimate the last factor of the Drake equation, this volume.

Lineweaver, C. H., (2001) An estimate of the age distribution of terrestrial planets in the universe: quantifying metallicity as a selection effect, *Icarus* 151: 307-313.

Marcy, G. W., and Butler, R. P., (2000) Millennium essay: planets orbiting other suns, *Publ. Astron. Soc. Pac.* 112: 137-140.

Marcy, G. W., Cochran, W. D., and Mayor, M., (2000) Extrasolar planets around main-sequence stars, in: *Protostars and Planets IV*, V. Mannings, A. Boss, S. Russel, eds., University of Arizona Press, Tucson, pp. 1285-1311.

Melosh, H. J., (2001) Exchange of meteoritic material between stellar systems, 32nd Lunar and Planetary Science Conference, March 12-16, 2001, Houston, Texas, Abstract No. 2022.

Nagamine, K., Fukugita, M., Cen, R., and Ostriker, J. P., (2001) Star formation history and stellar metallicity distribution in a Λ cold dark matter universe, *Astrophys. J.* 558: 497-504.

Rocha-Pinto, H. J., Scalo, J., Maciel, W. J., and Flynn, C., (2000) Chemical enrichment and star formation in the Milky Way disk. II. Star formation history, *Astron. Astrophys.* 358: 869-885.

Scalo, J. M., (1987) The initial mass function, starbursts, and the Milky Way, in: *Starbursts and Galaxy Evolution,* T. X.. Thuan, T.. Montmerle, and Tran Thanh Van, eds., Editions Frontières, Gif-sur-Yvette., pp. 445-465.

Scheffler, H., Elsässer, H., (1988) *Physics of the Galaxy and Interstellar Matter*, Springer Verlag, Berlin.

Schidlowski, M., (1990) Life on early Earth: bridgehead from cosmos or autochthonous phenomenon?, in: From Mantle to Meteorites, K. Gopalani, V. R. Gaur, B. L. K. Somayajulu, and J. D. MacDougall, eds., Indian Acad. Sci., Bangalore, 189-199.

Twarog, B. A., (1980) The chemical evolution of the solar neighbourhood. II - The age-metallicity relation and the history of star formation in the galactic disk, *Astrophys. J.* 242: 242-259.

Von Bloh, W., Franck, S., Bounama, C., and Schellnhuber, H.-J., (2003) Maximum number of habitable planets at the time of Earth's origin: new hints for panspermia?, *Origins Life Evol. Biosph.* 33: 219-231.

Whitmire, D. P., and Reynolds, R. T., (1996) Circumstellar habitable zones: astronomical considerations, in: *Circumstellar Habitable Zones*, Doyle, L. R., ed., Travis House Publications, Menlo Park., 117-143.

Wilde, S. A., Valley, J. W., Peck, W. H., and Graham, C. M., (2002) Evidence from detrital zircons for the existence of continental crust and oceans on the Earth 4.4 Gyr ago, *Nature* 409: 175-178.

Zinnecker, H., Mc Caughrean, M. J., and Wilking, B. A., (1993) The initial stellar population, in: *Protostars and planets III*, E. H. Levy, J. I. Lunine, and M. S. Mathews, eds., University of Arizona Press, Tucson, pp. 429-495.

Participants

Yuri Babich
Center of Biomedical Engineering, Kurska 12A/38, Kiev 03049, Ukraine, babich@ua.fm

David Begun
Department of Anthropology, University of Toronto, Toronto, Ontario, M5S 3G3, Canada, bedun@chass.utoronto.ca

Vladimir Burdyuzha
Astro-Space Center of Lebedev Physical Institute, Russian Academy of Sciences, Profsoyuznaya 84/32, Moscow, Russia, burdyuzh@asc.rssi.ru

Siegfried Franck
Potsdam Institute for Climate Impact Research (PIK), P.O. Box 60 12 03, 14412 Potsdam, Germany, frank@pikpotsdam.de

Claudius Gros
Department of Physics, Frankfurt University, 60438 Frankfurt am Main, Germany, gros05@itp.uni-frankfurt.de

Don-Edward Beck
Box 797, Denton, Texas 76202, USA, drbeck@attglobal.net

Jarle Breivik
Interactive Life Science Laboratory and Section for Immunotherapy,
University of Oslo at the Norwegian Radium Hospital, 0310 Oslo, Norway.
jbreivik@medisin.uio.no

Christian de Duve
Christian de Duve Institute of Cellular Pathology, ICP 75.50, Avenue
Hippocrate 75, B-1200 Brussels, Belgium, deduve@icp.ucl.ac.be

Menachen Goren
Department of Zoology, The George S. Wise Faculty of Life Sciences,
Tel Aviv University, 69978 Tel Aviv, Israel,
gorenm@post.tau.ac.il

Anatoly Gupal
V.M. Glushkov Institute of Cybernetics, Kyiv, Ac.Glushkov str. 40, Ukraine,
gupal_anatol@yahoo.com

Rajan Gupta
Theoretical Division, Los Alamos National Laboratory, Los Alamos, NM 87545, USA, rajan@lanl.gov

Kay Hamacher
Center for Theoretical Biological Physics, University of California, San Diego, La Jolla, CA 92093, USA, hamacher@matisse.ucsd.edu

Andrei Kirilyuk
Solid State Theory Department, Institute of Metal Physics, 36 Vernadsky Avenue, 03142 Kiev-142, Ukraine; kiril@metfiz.freenet.kiev.ua

Mojib Latif
Leibniz Institute of Marine Sciences at Kiel University (IFM-GEOMAR), Dusternbrooker Weg 20, 24105 Kiel, Germany,
mlatif@ifm-geomar.de

Guillermo Lemarchand
Centro de Estudios Avanzados and FCEN, Universidad de Buenos Aires, C.C. 8 Sucursal 25; C1425FFL Buenos-Aires, Argentina,
lemar@correo.uba.ar

Volodymir Magas
Departament de Fisica Teorica, Universitat de Valencia, C. Dr. Moliner 50, E-46100, Burjassot (Valencia), Spain.
Volodymyr.Magas@ific.uv.es

Mohammad Hafeznia
Department of political geography, Tarbiat Modarres University (TMU), Tehran, Islamic Republic Iran,
hafezn_m@modares.ac.ir

Kenji Hotta
Nihon University College of Science & Technology, 8-14, Kanda-Surugadai 1-chome, Chiyoda-ku, Tokyo 101-8308, Japan,
k-hotta@ocean.cst.nihon-u.ac.jp

Josip Kleczek
Astronomical Institute Czech Ac. Sci., 25165 Ondřejov, Czech Republic, kleczek@asu.cas.cz

Jacek Leliwa-Kopystinskiy
Warsaw University, Institute of Geophysics, ul. Pasteura 7, 02093 Warszawa, Poland, and Space Research Center of Polish Academy of Sciences, ul. Barty-Bartycka 18A, 00-716 Warszawa, Poland, jkopyst@mimuw.edu.pl

Kim Losev
Geological Department of Moscow State University, Leninskie gori,
119992 Moscow, Russia,
kim@viniti.ru

Brian Marcotte
Eco-Ethics International Union, Strategic Analysis, Inc., 401 Cumberland Avenue, Suite 1102, Portland, ME 04101-2875, USA,
brianmarcotte@thecollegium.org

Mauro Messerotti
INAF-Trieste Astronomical Observatory, Loc. Basovizza n. 302, 34012 Trieste, Italy and Department of Physics, University of Trieste, Via A.Valerio.
N. 1, 34133 Trieste, Italy,
messerotti@ts.astro.it

Nami Mowlavi
INTEGRAL Science Data Center, Ecogia, 1290 Versoix, Switzerland, Observatoire de Gen`eve, Sauverny, 1290 Versoix, Switzerland, Nami.Mowlavi@obs.unige.ch

Roman Retzbach
ZUKUNFT-INSTITUT / Future-Institute & Trend-Institute international, Franzosische Str. 8-12, 10117 Berlin, Germany, info@Future-Institute.com

Ting-Kueh Soon
Malaysian Scientific Association, Room 2, 2nd Floor, Bangunan Sultan Salahuddin Abdul Aziz Shah, 16, Jalan Utara, 46200 Petaling Jaya, Malaysia, malsci@tm.net.my

Bernard Swynghedauw
Hôpital Lariboisière. 41 Bd de la Chapelle. 75475, Paris Cedex 10, France, Bernard.Swynghedauw@larib.inserm.fr

Chandra Wickramasinghe
Cardiff Center for Astrobiology, Cardiff University, 2 North Road, Cardiff, CF10 3DY, UK, wickramasinghe@Cardiff.ac.uk

Boris N. Zakhariev
Nils-Axel Morner
Paleogeophysics & Geodynamics, Stockholm, Sweden, morner@pog.su.se

Olexandr Potashko
Marine Geology and Sedimetory Ore Formation Section of National
Academy of Sciences of Ukraine, Kiev, Ukraine, potashko@mail.ru

John Skinner
University of Pretoria, Faculty of Veterinary Science, Private Bag X04, ONDERSTEPOORT, 0110, South Africa Republic, john.skinner@up.ac.za

Bidare Subbarayappa
Science and Spiritual Research in India, 30, M. N. Krishna Rao Road, Basavangudi, Bangalore-560004, India, phispc@bgl.vsnl.net.in

Rudiger Vaas
Center for Philosophy and Foundations of Science, University of Gießen, Germany, Ruediger.Vaas@t-online.de

Hoi-Lai Yu
Institute of Physics, Academia Sinica, Taipei, Taiwan,
hlyu@phys.sinica.edu.tw

Maciej Zalewski
Bogolyubov's Laboratory of Theoretical Physics, Joint Institute for Nuclear Research, 6 Joliot Curie, 141980 Dubna, Russia,
zakharev@thsun1.jinr.ru

International Centre for Ecology Polish Academy of Sciences, Tylna str. 3, 90-364 Lodz, Poland; Department of Applied Ecology, University of Lodz, Banacha str. 12/16, 90-237 Lodz, Poland, mzal@biol.uni.lodz.pl

Index